# REVERSING THE ARMS RACE

# SCIENCE · & · GLOBAL · SECURITY MONOGRAPH · SERIES

edited by Harold A. Feiveson
Princeton University, New Jersey, USA

volume 1
**REVERSING THE ARMS RACE**
How to Achieve and Verify Deep Reductions in the Nuclear Arsenals
*edited by Frank von Hippel and Roald Z. Sagdeev*

This book is part of a series. The publisher will accept continuation orders that may be canceled at any time and that provide for automatic billing and shipping of each title in the series upon publication. Please write for details.

# REVERSING THE ARMS RACE

## How to Achieve and Verify Deep Reductions in the Nuclear Arsenals

*A book reporting results from the Cooperative Research Project on Arms Reductions of the Federation of American Scientists and the Committee of Soviet Scientists for Peace and Against the Nuclear Threat*

*edited by*

**Frank von Hippel**
Princeton University
New Jersey, USA

**Roald Z. Sagdeev**
Space Research Institute
Moscow, USSR

Gordon and Breach Science Publishers
New York Philadelphia London Paris Montreux Tokyo Melbourne

## Gordon and Breach Science Publishers

Post Office Box 786
Cooper Station
New York, New York 10276
United States of America

58, rue Lhomond
75005 Paris
France

5301 Tacony Street, Drawer 330
Philadelphia, Pennsylvania 19137
United States of America

3-14-9, Okubo
Shinjuku-ku, Tokyo 169
Japan

Post Office Box 197
London WC2E 9PX
United Kingdom

Private Bag 8
Camberwell, Victoria 3124
Australia

Some of the articles appearing in this book, as indicated in the table of contents, were previously published in volume 1 of the journal *Science & Global Security*.

Set in 10/15 pt New Century Schoolbook

Library of Congress Cataloging-in-Publication Data

    Reversing the arms race : how to achieve and verify deep reductions in the nuclear arsenals / edited by Frank von Hippel, Roald Z. Sagdeev.
          p.      cm. — (Science & global security monograph series, ISSN 1048-7042 ; v. 1)
         "A book reporting results from the Cooperative Research Project on Arms Reductions of the Federation of American Scientists and the Committee of Soviet Scientists for Peace and Against the Nuclear Threat."
         "Some of the articles ... were previously published in volume 1 of the journal Science & global security"—Verso of t.p.
         Includes bibliographic references.
         ISBN 2-88124-436-X. — ISBN 2-88124-390-8 (pbk.)
         1. Nuclear arms control.  2. Nuclear arms control—Verification.
I. Von Hippel, Frank.  II. Sagdeev, R. Z.  III. Federation of American Scientists.  IV. Komitet sovetskikh uchenykh v zashchitu mira, protiv iadernoĭ ugrozy.  V. Science & global security.  VI. Series.
JX1974.7.R486  1990
327.1'74—dc20
                                                90-3047
                                                 CIP

*A nuclear war cannot be won and must never be fought*

Mikhail S. Gorbachev
Ronald Reagan

Joint summit statements
Geneva, 21 November 1985
Washington DC, 10 December 1987
Moscow, 1 June 1988

# CONTENTS

\*   *Earlier versions of these chapters appeared in* Science & Global Security, *1*
*(1989–90)*

# Foreword

This book is a result of a joint research project of the Federation of American Scientists (FAS) and the Committee of Soviet Scientists for Peace and Against the Nuclear Threat (CSS).

The Federation of American Scientists was established in 1946 by some of the original Manhattan Project scientists to work for nuclear arms control. It has a full-time professional staff based in Washington DC and a voting membership of about 4,000 US natural and social scientists, including more than one half of all living US Nobel Prize winners.

The Committee of Soviet Scientists is a nongovernmental group of senior Soviet scientists established in 1983 for the purpose of studying the technical feasibility of disarmament agreements and discussing these questions with Western groups.

In February 1987, the FAS and CSS signed an "Agreement to Carry Out a Joint Scientific Study of the Feasibility of Implementing and Maintaining Disarmament." Since that time, the two groups have held seven joint workshops.[1] Between these joint workshops, each group has held its own separate meetings, and the individual scientists have worked on their research projects.

The joint research reported in this book concerns the technical basis for agreements to drastically reduce the US and Soviet nuclear arsenals. This has involved studies on both the stability and verifiability of such reductions. Another joint research project has analysed the feasibility and verifiability of a ban on space reactors in earth orbit as a means of restraining the weaponization of space and protecting the environment below.[2]

Much of the research reported in this book is focused on laying the basis of arms-control agreements that would go beyond those currently being negotiated. However, in two cases (verifying limitations on nuclear-armed sea-launched cruise missiles and on the numbers of warheads on deployed ballistic missiles) we have explored the technical possibilities for verifying agreements that have already been the subject of negotiations between our two governments.

The book as a whole represents a unified analysis of the technical requirements for deep reductions in the nuclear arsenals. Most of the chapters were written by solely US or solely Soviet authors; individual chapters reveal some differences in perspective and opinion. On the purely technical issues there were no differences of opinion, however; accordingly the chapter "Detecting Nuclear Warheads" represents truly joint work between the two groups.

It is our hope that our work will make further nuclear arms-control agreements more credible to both the governments and the citizenry of the US and USSR. We are gratified to note that our work has already encouraged our governments to expand their own research efforts on many of these subjects.

Some of the results of our joint research have already been used in the design of a warhead-detection experiment jointly sponsored by the Soviet Academy of Sciences and US Natural Resources Defense Council (NRDC) on the Soviet cruiser *Slava* off Yalta in July 1989, and efforts to mount further demonstration projects are currently under way. Such efforts raise the visibility of proposals for cooperative verification arrangements and provide relevant monitoring data and experience. The nongovernmental research whose results are reported here is also helping to lay the foundations for research and cooperative experiments by our governments.

All the work that we have done is based on basic physical principles and publicly available information. However, we do not feel that we have been greatly disadvantaged by our lack of access to classified governmental information. While the governments know a great deal about their own weapon systems, most of what they know about the interior workings of the weapon systems of the other side is based on the same general physical principles that we have used. Even where more is known, there are no limits on future changes in weapon designs other than those provided by natural laws.

Earlier versions of a number of the chapters in this book have appeared in *Science & Global Security*, a new journal with an initial editorial board of Soviet and US scientists that has been established to examine the technical basis for arms-control and environmental policy initiatives.[3] Both the journal and this book are to be published in Russian by Nauka ("Science") Publishers, the publishing house of the Soviet Academy of Sciences.

Alex De Volpi and William Higinbothom contributed to this project through numerous comments and suggestions. During most of the duration of this collaboration, administrative arrangements were made with unfailing skill and good cheer by Cely Arndt (FAS) and Elena Loshchenkova (CSS). Design and production editing was by John Shimwell.

Financial support for the FAS involvement in this project has been provided by the Carnegie Corporation, the W. Alton Jones Foundation, and an anonymous philanthropist.

## NOTES AND REFERENCES

1. During 1988: 3–9 February in Key West, Florida; 11–12 May in Washington DC; June 20–21 in Moscow; September 8–13 in Moscow; and November 20–22 in Washington; during 1989: 6–8 May in Washington; during 1990: 20–21 April in Washington.

2. See the six articles on the subject in *Science & Global Security*, 1, 1–2, 1989.

3. *Science & Global Security* is published in English by Gordon and Breach Science Publishers (PO Box 786, Cooper Station, New York, NY 10276, USA or PO Box 197, London WC2E 9PX, UK).

# Chapter 1

# Overview

The nuclear arsenals of the US and Soviet Union today (1990) contain at least tens of times the destructive power required to destroy either country as a modern state. During the past 30 years, we have become accustomed to these huge levels of "overkill." Indeed, many nuclear-weapon–policy makers worry that deep reductions might somehow destabilize the nuclear balance.

Stability is important. Therefore this book begins with two chapters investigating the stability of the nuclear balance after deep reductions, in which the implications for stability of reductions in the Soviet and US nuclear arsenals from their current levels of over 10,000 nuclear warheads on each side to 2,000 or fewer nuclear warheads each. It is found that the current level of stability could be maintained and probably increased if the arsenals were restructured to be less threatening to each other.

Today's strategic arsenals are so huge in part because most launch platforms carry many independently targetable warheads: modern ICBMs each carry up to 10 warheads, modern ballistic-missile submarines each carry up to 200 warheads on 16–24 missiles, and modern bombers can each carry up to 28 nuclear-armed air-launched cruise missiles, bombs, and short-range attack missiles. This large number of warheads makes it possible to consider such nuclear-warfighting tactics as "barraging" a large area in an effort to destroy a bomber or mobile missile launcher whose location is only approximately known. If the number of strategic warheads per launch platform were reduced, such tactics would become less feasible and stability would be enhanced.

Thus, after stabilizing reductions, both sides would have strategic forces containing primarily single-warhead ICBMs and ballistic-missile submarines and bombers each carrying fewer nuclear warheads than today.

But would such reductions be verifiable? For example how could we verify that bombers or missiles would be equipped with fewer than the maximum number of nuclear weapons that they could physically carry or that ordinary transport aircraft, ships—or even trucks—might not be converted to launchers for long-range cruise missiles? And what about the tactical nuclear weapons that account for approximately half the current nuclear arsenals but have thus far been virtually untouched by arms control? If the number of strategic warheads were dramatically reduced, tactical warheads could come to dominate the nuclear arsenals, and concerns would develop aboutsuch issues as the potential strategic capabilities of forward-based long-range fighter-bombers.

All these problems suggest that if deep cuts in the nuclear arsenals are to be achieved, verification will have to extend beyond launchers, missiles, and bombers to limitations on nuclear warheads themselves.

Nuclear warheads are comparatively small and relatively easy to hide. But, as has been shown by the verification arrangements for the Inter-mediate-range Nuclear Forces (INF) Treaty and the emerging Strategic Arms Reductions (START) Treaty, cooperative on-site verification arrangements can now be included in US–Soviet arms-control agreements. Many of the articles in this book explore the extent to which such approaches would make it feasible to extend arms control to nuclear warheads.

## VERIFYING REDUCTIONS OF NUCLEAR WARHEADS

We therefore address the general aspects of the problem of establishing a verifiable system of limits on nuclear warheads. We analyze the verifiability of:

♦ A cutoff of the production of new fissile materials for warheads

♦ The dismantlement of warheads and safeguarded disposal of their fissile materials

♦ Declarations of the sizes of the warhead stockpiles and the total quantities of fissile material available for their manufacture.

*A cutoff in the production of new fissile material for weapons* would not by itself cause any shrinkage in the nuclear arsenals because fissile material can be (and is) recycled from old weapons to new ones almost indefinitely. However, the flow of new fissile material into the production complex must be halted if the dismantlement of warheads and the placement of their fissile material under safeguards are to ensure reductions.

Verification of a fissile cutoff would require on-site safeguards of the type developed for the International Atomic Energy Agency to verify the compliance of non–nuclear-weapon states with the Non-Proliferation Treaty. National intelligence systems, perhaps supplemented with challenge inspections, would make it very difficult to hide a clandestine production facility of significant size.

*Verifiable elimination of warheads* is not included in the INF treaty of 1987 or the START treaty, which was being completed as this book went to press. The warheads or the fissile materials that they contain will therefore be available to be made into warheads for systems that are not limited by treaty, such as shorter-range air- or sea-launched missiles.

Dismantlement of nuclear warheads could be done by technicians of the owning country, with the other side's inspectors controlling only the perimeter of the dismantlement facility, checking later that the warheads had indeed been eliminated and all their fissile material placed under safeguards.

The only potentially intrusive part of the verification arrangements would occur during confirmation that the objects taken into the dismantlement facility were indeed intact warheads of the agreed type. A number of possible solutions to this problem are discussed in chapter 5.

A potential obstacle to very deep reductions would be uncertainties as to the size of the remaining stockpiles. It is therefore necessary to explore the verifiability of *declarations of total stockpiles of warheads and fissile materials.*

As is explained in chapter 4, uncertainties concerning declarations of the total amount of fissile material available for nuclear weapons on each side could be reduced by a cooperative approach to the reconstruction of past production and disposition of such materials. There would be many possible checks if the records of these activities were made available, including cross-checks between records from separate facilities, historical intelligence information and physical measurements at production facilities ("nuclear archeology").

The numbers and locations of the warheads in each side's stockpile would be declared and a random sample of warheads of each type would be "fingerprinted" so that any counterfeits turned in later during the dismantlement process could be identified. The total amount of fissile material in declared warheads could be estimated by having a weighted sample of the stockpiles dismantled and assaying the total amount of fissile material recovered.

The task of detecting any clandestine stockpile would be left to national intelligence methods, perhaps supplemented by challenge inspections.

## VERIFICATION OF NUMBERS OF WARHEADS ON DEPLOYED BALLISTIC MISSILES

Verification of the numbers of warheads on deployed ballistic missiles will already be required in the START treaty because that treaty is to limit the number of warheads carried by some missiles to levels below the numbers with which they have been tested.

The simplest way to check the number of warheads on a deployed ballistic missile would be to remove the nose cone and allow the other side to see the number of re-entry vehicles—perhaps covering individual re-entry vehicles with shrouds to protect any secret details of their design. If, for any reason, this is considered unacceptable, however, chapter 6 examines other possibilities, including the use of beams of particles that could penetrate the shroud and reveal the number of masses of fissile material that it contains. A declaration of the number of warheads deployed on each missile could be verified with an agreed number of short-notice checks on randomly selected deployed missiles.

In the longer term, however, deep cuts should not be carried out through the reduction of warheads on missiles capable of carrying large numbers of warheads. That process is too easily reversible. Eliminating "heavy" ballistic missiles in favor of "light" ones and reducing the number of ballistic-missile launchers per submarine makes "breakout" less likely.

## VERIFYING LIMITS ON NUCLEAR CRUISE MISSILES

Verification of limits on deployed nuclear-armed cruise missiles has been and will continue to be a contentious issue in nuclear arms control because modern long-range cruise missiles are small and can be launched from almost any type of ship or aircraft. Furthermore, the deployment of large numbers of non-nuclear cruise missiles is adding the verification requirement that the nuclear-armed cruise missiles must be distinguishable from the conventionally armed cruise missiles.

As is discussed in the chapters on the verification of limits on nuclear-armed sea-launched cruise missiles (SLCMs) and air-launched cruise missiles (ALCMs), declarations of all cruise-missile production and storage sites and controls at the choke points in the cruise-missile life-cycle would contribute in a relatively nonintrusive way to the verification of limits on nondeployed as well as deployed nuclear-armed SLCMs. Such a system could involve portal and perimeter controls at assembly and maintenance facilities in combination with a tagging system. New or refurbished nuclear-armed cruise missiles could be tagged when they left these facilities. Only as many tags would be available as the agreed number of permitted nuclear-armed SLCMs or ALCMs. Storage depots might be subjected to on-site challenge inspections to verify that all the nuclear-armed cruise missiles they contained were indeed tagged.

The system might be supplemented by a certain additional allowance of challenge inspections to allow inspection of suspected clandestine production or storage facilities. It could also be supplemented by a quota of random challenge inspections of launchers on ships, submarines, and bombers to verify that they did not contain untagged nuclear cruise missiles.

## THE TECHNICAL BASIS FOR NUCLEAR-WARHEAD DETECTION

Most of the verification systems discussed above require the detection of nuclear warheads or fissile material—at least at short range for portal controls around production facilities. Since the ingredients unique to nuclear warheads are multikilogram quantities of fissile material, the same techniques can be used for the detection of warheads as are used in the detection of fissile materials. In chapter 10 we describe applications of these techniques to the problem of detecting fissile material being transported through a portal. Techniques that could detect the chemical explosives in a nuclear warhead are also described. In chapter 11 and its associated appendixes a comprehensive discussion of the technical basis of the ways in which fissile material in nuclear warheads can be detected.

"Passive detection" techniques are based on the fact that all fissile materials emit some penetrating radiation. For unshielded warheads containing quantities of weapon-grade plutonium, this radiation would be detectable at a distance of up to several tens of meters. Such techniques are already used to distinguish the Soviet three-warhead intermediate-range SS-20 missile, banned under the INF treaty, from the permitted long-range single-warhead SS-25 missile.

Such techniques were also tested in the July 1989 "Black Sea Experiment," in which the gamma and neutron radiation from a nuclear warhead in a Soviet cruise-missile launcher were measured by US and Soviet teams. The results of the gamma measurements are presented in chapters 13 and 14.

Warheads can be designed that emit very little penetrating radiation, however, and the radiation from any warhead could be shielded with a few hundred kilograms of material or less. "Active detection" techniques are therefore also considered. Such methods divide into two classes: transmission radiography and induced fission. In transmission radiography, high-energy x-rays, neutrons, or other penetrating particles would be beamed at an object suspected of containing a warhead, and the pattern of transmitted radiation examined to reveal the presence of lumps of heavy material—much as an ordinary dental x-ray reveals the fillings in one's teeth. Such radiographs would also reveal the presence of thick radiation shields. The induced-fission approach is somewhat more flexible because it does not require access to both sides of the object being scanned. Detectors would be set up to detect the penetrating neutrons or gamma rays emitted when the particle bombardment caused some of the atoms of any fissile material present to fission.

# Chapter 2

*Roald Z. Sagdeev*
*Andrei A. Kokoshin*

# Stability of the Nuclear Balance after Deep Reductions I

Strengthening strategic stability has become one of the central objectives of nuclear arms control. It must therefore be a primary consideration in the design of arrangements for radical reduction of nuclear arms right up to their complete elimination.

In this chapter, we discuss a number of scenarios for the radical reduction of nuclear weapons that would not only maintain the stability of the nuclear balance, but also enhance it.

However, there are several factors that could make stabilizing reductions impossible. Most prominent is the unilateral or bilateral deployment of antiballistic-missile (ABM) systems—especially space-based systems. Radical reductions can only be realized if such destabilizing factors are excluded.

En route to the goal of the universal elimination of nuclear weapons, it would be necessary to pass through several stages of reductions to the penultimate step where each side would have only sufficient nuclear potential for a retaliatory strike that could cause unacceptable damage to an attacker. In order to ensure strategic stability at these intermediate stages, the nuclear forces of both sides would have to be sufficiently invulnerable and under reliable control. The reduction agreements and their associated verification procedures would have to guarantee that neither side could obtain a significant advantage from a surprise violation.

## GENERAL CONSIDERATIONS

One of the main determinants of strategic stability is a rough quantitative balance. Even if a quantitative advantage did not enable one side to threaten the other's retaliatory capabilities, a major quantitative asymmetry would still encourage the side with the smaller forces to build up, thereby fueling the arms race and hampering arms-control agreements.

Strategic stability must also be considered in the context of the political relationships between the USSR and the US, and between the Warsaw Pact Organization and NATO. The stability of the military-strategic equilibrium would be reduced in a period of escalating international tensions.

The main route towards increased strategic stability is through the limitation and significant reduction of nuclear arms while simultaneously preventing the spread of the arms race to other spheres. However, improperly designed reductions, which, for example, increased the fraction of forces that were vulnerable to a first strike, could actually result in *decreased* stability.[1]

Increased stability can best be achieved by coordinated, mutually acceptable, and mutually beneficial actions based on a shared understanding of the character of the contemporary military-strategic balance and of the principles for ensuring its stability.

## FACTORS INFLUENCING STABILITY

Many political and military figures in Western countries reduce the problem of strategic stability to the issue of the vulnerability of silo-based intercontinental ballistic missiles (ICBMs) to strikes by the other side's MIRVed ICBMs. However, while the vulnerability of silo-based ICBMs is important, it is not the only determinant of strategic stability. On both sides, ICBMs comprise only one of the three components of the strategic forces (the other two being submarine- and bomber-based forces). In the US, 16 percent of nuclear warheads are on ICBMs, while in the USSR 60 percent of such warheads are on ICBMs.

Even if it were physically possible to destroy all the other side's ICBMs in their silos, ballistic missiles on submarines at sea, which are almost invulnerable to destruction today, and heavy bombers, capable of rapid take-off, would still remain. In the early 1980s, the overall vulnerabilities of the forces on both sides, including submarines in port and airplanes on the ground, were about equal—approximately 35–40 percent.[2] This means that 60–65 percent of each power's strategic forces were capable of surviving and carrying out a retaliatory strike that would be 10–15 times stronger than the level of unacceptable damage estimated by US Secretary of Defense Robert McNamara in the 1960s.

Some prominent US figures suggest that the "window of vulnerability" problem exists only for US ICBMs, because of the Soviet monopoly on highly MIRVed "heavy" (SS-18) ICBMs. However, the fact is that the silo-based ICBMs of both sides are becoming vulnerable to increasingly accurate multiple independently targetable re-entry vehicles (MIRVs), which are appearing on both ICBMs and sea-launched ballistic missiles (SLBMs). As the former chief of the General Staff of the Armed Forces of the USSR, S.F. Akhromeev, noted:

> Strategic offensive forces are now becoming approximately equal. In terms of battle effectiveness there is no difference between the Soviet ICBM and the American "Trident" SLBM. Thus, strategic weapons must be viewed and evaluated as parts of a complex, as a unified whole.

The US ICBM Minuteman III, equipped with the MK-12A re-entry vehicle, has for many years been an effective countersilo weapon that has made some Soviet ICBMs as vulnerable as those of the US. Furthermore, the development of ICBMs and SLBMs with highly accurate and powerful warheads is currently a special priority of the US strategic modernization program. In order to enhance the survivability of its ICBMs against increasing US countersilo capabilities, the Soviet Union began in the early 1980s to deploy mobile ICBMs.

The argument made by proponents of the "window of vulnerability" concept—that the very existence of the vulnerability of ICBMs could be used in a crisis situation for the purpose of obtaining political concessions—does not withstand scrutiny. Such a threat could well result in the threatened side adopting a launch-on-warning posture. If so, the threatened attack would serve only to destroy empty silos and prompt a devastating counterstrike.

Many specialists also note the highly significant technical and operational uncertainties that would be involved in launching a synchronized missile strike calculated to destroy more than 1,000 ICBM silos. Full-scale practice of such an attack is impossible, and success in computer simulations would not be convincing. The fact that the attacking ICBMs would be launched on polar rather than along their usual (east–west) test trajectories would cause additional uncertainties, and the clouds of debris created by the explosions of the first warheads would also affect the remainder of the attack. Such uncertainties increase the stability of the strategic balance. Arms-control agreements such as the Partial Nuclear Test Ban Treaty, which prohibits the testing of nuclear weapons in the atmosphere, preserve some of these stabilizing uncertainties.[3]

If either side attempted a counterforce strike, there would be enormous consequences, including the annihilation of many millions of people as a direct result. The numbers of such deaths have been convincingly estimated to range from 5–34 million people.[4] Such a strike could under no circumstances be regarded as a "surgical" nuclear attack, as some US strategists seem to think. From a political, military, and moral point of view, it would be nothing less than an act of total thermonuclear aggression with all the ensuing catastrophic consequences.

It is important, of course, to take into account possible technological developments in discussions of the future stability of the strategic balance. It might be possible, for example, to develop low-yield ballistic-missile warheads with high accuracies, achieved by maneuvers during the final stage of flight, that could destroy strategic objects with lower civilian casualties. The appearance of such weaponry might give rise to additional illusions regarding the possibility of conducting "limited nuclear war." It is necessary also to bear in mind the possibility of other developments capable of reducing the strength of a retaliatory strike. These include antimissile systems and strategic antisubmarine warfare capabilities.

Any analysis of the problem of strategic stability must also take into account an entire range of military-technical considerations, some of which have been ignored by the majority of US specialists. Thus, for example, the introduction into Western Europe of the Pershing II, with its short flight time (8–12 minutes in comparison to 25–30 min for ICBMs) greatly reduced the time available to Soviet political leaders to make decisions concerning retaliatory strikes. Reduced radar cross sections and infrared emissions by "stealthy" bombers and missiles will similarly increase fears of surprise attacks. And difficulties persist in imposing reliable controls on SLBMs, which increases the risk of their unauthorized launch.

The deployment of partially effective space-based antimissile systems and countermeasures to them would be especially destabilizing. Each side, proceeding on the basis of worst-case analyses, would view the other side's antimissile system as a threat to its retaliatory capabilities. In order to counter this threat, each side would develop its own antimissile system and build up its offensive weapons. Space-based systems would also create strategic instability by virtue of their inherent capability to attack and destroy the other side's antimissile system in a first strike using only an insignificant part of their own military potential. Finally, it would be almost impossible to impose arms control on space-based antimissile systems because their effectiveness would depend heavily on unverifiable technical characteristics (target detection and identification capabilities, the reliability of complex computerized guidance subsystems, etc.).[5]

## PREREQUISITES FOR DEEP REDUCTIONS

We assume that cuts of the Soviet and US strategic forces to 25 percent of their current levels or less will be impossible from either a political or military point if the other nuclear powers (i.e. France, China, and the United Kingdom) do not join in.

We also assume that, in parallel with moves to increase the stability of the strategic nuclear balance, the stability of the conventional military equilibrium must also be greatly increased—especially between the Warsaw Pact and NATO.[6]

Specifically, the force structures in Europe should be reshaped by greatly reducing their capabilities to conduct offensive operations and creating a situation of "defensive dominance." In such a situation, it would become apparent to both sides that the defensive capabilities of the Warsaw Pact were significantly superior to the offensive capabilities of NATO and, conversely, that the defensive capabilities of NATO were clearly superior to the offensive capabilities of the Warsaw Pact.[7]

Also, in order to reduce the possibility of surprise attack, a "third generation" of confidence-building measures must be created, and extended to include the naval and airborne activities of both sides, following on the "second generation" measures agreed to at the 1986 Stockholm conference and the "first generation" agreements in the Helsinki Final Act of 1975.

Assessing conventional stability requires much more complex calculations and investigation than traditional assessments of the military balance in terms of quantitative comparison of numbers of divisions, tanks, warplanes, artillery pieces, missile launchers, etc.[8] In-depth professional work will be required, including a detailed investigation of the military training of the armed forces of both sides.

## EFFECTS OF REDUCTIONS ON STABILITY

On 15 January 1986, General Secretary Mikhail Gorbachev proposed radical reductions of strategic offensive weaponry. This was followed by Soviet–US discussions of such reductions in Reykjavik in October 1986 and then by negotiations on 50-percent reductions in strategic ballistic-missile warheads during the Washington DC summit of December 1987. This focus on reductions requires a detailed analysis of the requirements for maintaining strategic stability during reductions all the way to the ultimate liquidation and prohibition of nuclear weapons. A baseline study of this type was conducted during 1984–87 by a working group of the Committee of Soviet Scientists.[9]

The effect on stability of different force structures and levels of reductions were explored using "AC" dialogue computer models worked out by the Laboratory of Structural Analysis and Modeling of the Institute of US and Canadian Studies of the Soviet Academy of Sciences.[10]

All acceptable force structures were required to meet the following two basic conditions:

♦ Both sides should lack incentives to use nuclear weapons first. In particular, neither side should have the capability for a disarming first strike. If either side is subjected to such an attack, it should retain the potential for a retaliatory strike that would cause comparable and unacceptable damage.

♦ Conditions for unsanctioned and accidental use of nuclear weapons are absent.[11]

The satisfaction of these conditions requires the presence, on each side, of reliable redundant systems for command and communication and for early warning of missile attack. It is possible that cooperative arrangements, involving supplementary systems for joint monitoring and communications, might also be required.

Our calculations show that, in the majority of cases, 50 percent reductions of the Soviet and US strategic arsenals would not significantly alter the stability of the strategic balance as it has taken shape in the latter half of the 1980s. Either side, having suffered a first strike (especially if it received some advanced warning), would retain both the capability to destroy a wide class of military targets in a retaliatory attack and the capability of causing unacceptable damage to population and industry of the attacking country (i.e. to commit an act of "assured destruction"). However, if the attacking country possessed ballistic-missile defenses which could reduce the strength of the retaliatory strike by 50–60 percent, this stability would be reduced in the absence of specific countermeasures by the other side.[12]

The fact that a simple and cost-effective countermeasure against an ABM system would be to increase the number of ICBMs and warheads on the other side shows the incompatibility of ballistic-missile defense (BMD) systems with reductions in offensive forces.[13]

Analytical studies were also conducted of the effects on the stability of the strategic balance of two further levels of reductions:

♦ A cut to approximately 25 percent of late-1980s warhead levels, and

♦ A further cut to approximately 5 percent of late-1980s levels.

Ten possible combinations of force structures were considered at the first level of reductions and seven at the second. The stability of the resulting strategic balances were tested against various scenarios for conflicts.

Analysis of the first-level (75-percent cut) reductions revealed that a retaliatory strike could still destroy a wide range of military objectives, or inflict unacceptable damage to industry and populations. Given an equal division of targets between military and nonmilitary objectives, the damage inflicted on the civilian population, although still enormous, would be somewhat less than would be caused by a retaliatory attack with the 50-percent cut forces.

The use of the "AC" dialogue system made it possible not only to determine the presence or absence of a stable state of the strategic balance, but also to find by "feel" the surfaces between regions of stable and unstable balance in a parameter space involving different strategic forces and weapon-system characteristics. It was found that the stability regions became relatively smaller with assumptions of more accurate strategic nuclear weapons and smaller still when BMD systems were introduced. Specifically, when one or both sides deploy BMD systems, the situation becomes unstable even if the BMD systems have a very low probability of interception (30–50 percent). The situation is not so unstable if one of the countries deploys countermeasures instead of its own BMD system (an asymmetrical response). However, as in the case of 50-percent reductions, the stabilizing influence of such restraint has its limitations.[14]

The strategic equilibrium was found to be even more sensitive to antisubmarine warfare (ASW) capabilities. If deep reductions are to be achieved, it will be necessary to take measures on a joint basis to significantly increase the survivability of ballistic-missile submarines—for example, with the creation of sanctuaries free from the ASW activities of the other side.[15]

It is interesting to note that a 75-percent reduction in strategic warheads would return the number of ballistic-missile warheads to approximately the same levels as in the beginning of the 1970s. However, the qualitative characteristics of these forces would have changed greatly. On the one hand, the accuracy of ballistic-missile warheads has increased greatly along with antisubmarine warfare capabilities. On the other hand, ICBM silos and command-and-control facilities have been hardened, longer-range SLBMs have made it possible for ballistic-missile submarines to be stationed where they are less vulnerable, and some ICBMs have been put on mobile launchers.

On balance, our calculations show that the effects of increased accuracy have not been offset by the hardening of ICBM silos and command-and-control systems. Unfortunately, re-entry–vehicle accuracy is a characteristic that would be extremely difficult, if not impossible, to limit verifiably. However, at the level of 75-percent reductions, severely limiting the number of missiles with accurate warheads capable of destroying hardened targets would enhance strategic stability.

## MOBILE ICBMs

In the scenario involving a 95-percent reduction in the nuclear arsenals of both sides, only several hundred nuclear warheads would remain on Soviet and US strategic delivery vehicles. It was also assumed that tactical nuclear weapons would be eliminated and the nuclear forces of other nations would either be reduced proportionately or entirely eliminated. Furthermore, it was assumed that the ABM treaty would remain in force and there would be other arms-control agreements, including a comprehensive nuclear test ban and a ban on the production of fissile materials for nuclear warheads.

Our examination of different force structures led us to the conclusion that strategic stability would be maximized if each side possessed approximately 600 small, single-warhead ICBMs—some in fixed silos and some on mobile launchers. Heavy bombers armed with either freefalling bombs or air-launched cruise missiles (ALCMs) would be eliminated—as would be SLBMs, sea-launched cruise missiles (SLCMs), and all other nuclear weapons and carriers.

The advantage of land-based missiles over submarine-based missiles comes from the fact that communications with them can be more reliable.

The possibility of accidental launch is thereby reduced. Also, unlike SLBMs, ICBMs would be launched along a relatively predictable trajectories, which would make early-warning arrangements more reliable. In comparison with heavy bombers, land-based missiles have the advantage that they cannot be involved in regional non-nuclear conflicts.

The very fact that each missile carried only a single warhead would be stabilizing since, if the missiles were dispersed, an attack on them would use more warheads than it destroyed.

Silos would be the least expensive basing mode but making some of the missiles mobile would significantly increase their survivability, providing insurance against any secret deployment of additional forces by the other side. The small size and mobility of the single-warhead ICBMs would make it difficult for observation satellites to determine the precise locations of all of them at any one time.

At the same time, silo basing would make it possible to provide enhanced shielding for some of the missiles against high-powered bursts of microwaves produced by "third generation" nuclear weapons which might damage the electronic systems of mobile missiles. The potentially destabilizing impact of such third-generation weapons provides another strong argument for a comprehensive test ban, which would hinder their development.

## THE FINAL ELIMINATION OF NUCLEAR WEAPONS

Although the subject is not, strictly speaking, within the compass of this book, we end this chapter with a few comments on the question of the means and at the rate of the final steps towards the goal of total elimination of nuclear weapons.

In the final steps from several hundreds to zero nuclear weapons, the survivability of the residual forces and the stability of the nuclear balance could be maintained, as before, but the conditions for mutual guaranteed destruction would no longer be fulfilled. At levels of several tens of warheads, the potential losses on each side from nuclear use might become commensurate with the losses in the major non-nuclear wars of the past. At such low levels of the nuclear balance, the restraining influence of the threat of nuclear destruction would be reduced and the military-political balance might become more unstable.

In order to avoid instabilities at such low levels, the final reductions from approximately 600 warheads on each side to zero should therefore be carried out quickly without intermediate stages.

It is assumed that facilities and procedures would be available for the elimination of the nuclear warheads and their carriers. Many of the necessary arrangements would have been worked out in connection with the INF treaty and the 50-percent reductions in ballistic-missile warheads proposed for the START treaty. At the final stage of nuclear-weapon elimination, it would be necessary to have especially rigorous on-site verification arrangements to both guarantee the destruction of existing nuclear weapons and the elimination of the possibility of a secret resumption of their manufacture and deployment.

## NOTES AND REFERENCES

1. *Disarmament—the Call of the Times,* third edition (Moscow: 1984) pp.49–50, 188.

2. *Disarmament and Security*, volume 1, (Moscow: Institute of the World Economy and International Relations, 1987), p.39.

3. A.G. Arbatov, Alexei A. Vasiliev, and Andrei A. Kokoshin, "Nuclear Weapons and Strategic Stability," *USA Economics, Politics, and Ideology* (Moscow: Institute of US and Canadian Studies), 19, 1988, p.19.

4. Frank von Hippel, Barbara G. Levi, Theodore A. Postol, and William H. Daugherty, "Civilian Casualties from Counterforce Attacks," *Scientific American*, September 1988, p.36.

5. Roald Z. Sagdeev and Oleg F. Prilutsky (Committee of Soviet Scientists for Peace and Against the Nuclear Threat), *Strategic Defense and Strategic Stability* (Moscow: Space Research Institute, 1985); Evgeny P. Velikhov, Roald Z. Sagdeev, and Andrei A. Kokoshin, eds., *Space-based Weapons: The Dilemma of Security* (Moscow: Mir Publishers, 1986), p.140–141.

6. Nuclear and conventional forces are interconnected in several ways:
   i) An increasingly large number of weapon carriers—strategic as well as tactical—are "dual purpose" in that they can carry either nuclear or conventional weapons.
   ii) Some kinds of general-purpose conventional forces can be used to destroy strategic nuclear weapons and their command-and-control systems.

iii) The US and NATO envision that they would first use nuclear weapons if they were losing a conventional conflict.
iv) Some types of conventional weapons have nearly the destructive capacity of low-yield nuclear weapons.

7. Andrei A. Kokoshin, *Novaya Vremya*, 33, 1986.

8. Conventional military stability depends also on the ability of command staffs, the organizational structure of the forces, their battle readiness and morale, weapon technology, the strength and nature of the opponent's counter-action, local conditions, weather, and other factors.

9. Working group of the Committee of Soviet Scientists for Peace and Against the Nuclear Threat (Roald Z. Sagdeev and Andrei A. Kokoshin, research leaders), *Strategic Stability Under Conditions of Radical Reductions of Nuclear Weapons* (Moscow, 1987).

10. Input to these models included data and expressions for: the number of bomber bases, the at-sea rates of ballistic-missile submarines and the alert rates of bombers under peacetime and crisis conditions, the probable effectiveness of air-defense and ABM systems, losses of ballistic-missile submarines to antisubmarine warfare, the area-density of mobile-missile launchers, and the maximum number of warheads that can simultaneously be directed at a single target (because of the destruction of later warheads by the effects of the explosions of earlier warheads).

11. It is possible to single out four basic sources of danger of accidental nuclear war:
i) Technical errors in early-warning systems and nuclear-weapon control systems
ii) Failures in communicating information in the decision-making system
iii) Human error in assessment of the strategic situation as a result of inaccurate interpretation of the incoming data
iv) Other human errors or nervous breakdown as a result of fatigue, illness, etc.
Analysis of major disasters in sociotechnical systems (airplane crashes, nuclear reactor accidents, etc.) has shown that the most dangerous situations combine several of these sources of error—for example, when an intense psychological burden is created as a result of a purely technical error. Prevention of the unpremeditated onset of nuclear war requires above all an organization of the decision-making process and technological systems that will minimize such errors.

12. One of the most vulnerable components of a space-based antimissile defense would be its systems for command and communication and its subsystems for target detection and identification. The effectiveness of the entire antimissile system may be substantively degraded—if not entirely eliminated—by electromagnetic countermeasures. Such techniques would therefore be the most cost-effective countermeasures to space-based defenses.

13. Yu.P. Maksimov, "A Reliable Shield," *Novaya Vremya*, 51, 1986, p.13.

14. *Space Weapons: the Dilemma of Security*, pp.128–157.

15. Negotiations to this end were proposed by the Committee of Soviet Scientists for Peace and Against the Nuclear Threat in *The Problem of the Nuclear Freeze* (Moscow, 1984), p.2; see also *Pravda*, 10 October 1985.

# Chapter 3

*Harold A. Feiveson*
*Frank von Hippel*

# Stability of the Nuclear Balance after Deep Reductions II

There appears to be a greater possibility of radical changes in the nuclear policies of the US and Soviet Union today than at any time in the nuclear age. The two countries have concluded the Intermediate-range Nuclear Forces (INF) agreement, eliminating land-based missiles with ranges between 500 and 5,500 kilometers, and, at the time of this writing, were completing a Strategic Arms Reduction (START) agreement, which will reduce the number of warheads carried by US and Soviet strategic ballistic missiles (see table 3.2). They were also completing an agreement to reduce conventional forces in Europe against a dramatic backdrop of radical political change in Eastern Europe and the Soviet Union.

So far, however, none of these developments has changed the basic dynamics of the nuclear arms race. In the case of the START agreement, both sides appear likely to retire their older systems while more modern systems designed to attack hardened military targets continue to be phased into their forces. The US Navy, for example, currently plans to replace a large fraction of its 5,000 relatively inaccurate and low-yield Poseidon (0.04 megaton) and Trident I (0.1 megaton) warheads with a somewhat smaller number of much more accurate and powerful Trident II (0.5 megaton) warheads designed explicitly for attacking Soviet missile silos and underground command posts. The US Air Force has designed its new intercontinental ballistic missiles (ICBMs), the MX and the Midgetman, with warheads much more accurate than those of the previous generation Minuteman II and III ICBMs. The expense of a second new strategic nuclear bomber, the B-2, is being justified by its superior intended ability to penetrate Soviet defenses and attack mobile missiles and command posts.[1] The US Department of Energy is developing earth-penetrating warheads to increase the effectiveness of ballistic missiles against underground targets.[2] Meanwhile, the Soviet Union is replacing its heavy ICBM, the SS-18, with a newer, more accurate and powerful version, the SS-18 mod 5.[3]

The constant pressure to "modernize" nuclear forces derives from each side's efforts to maintain and improve its "counterforce" capabilities to attack the nuclear forces of the other side and to reduce the vulnerability of its own forces. From this competition, worst-case analysts conjure images of a remorseless enemy to justify new armaments. Thus the nuclear weapons of each side help to justify those of the other, each generation of nuclear weapons justifies the next, and nuclear arsenals are rationalized that contain many times more nuclear warheads than required to maintain a stable mutual nuclear-hostage relationship. Although neither side will be able to mount a successful first strike in the foreseeable future, each pushes the other towards accident-prone launch-on-warning strategies at best and pre-emptive–strike strategies at worst.

Concerns about crisis instability have led to efforts to curb the nuclear counterforce competition indirectly—initially through a nuclear-weapon test ban and later primarily through limitations on ballistic-missile tests. However, the weapon developers and militaries have successfully resisted such constraints by arguing that, as long as they are tasked with being ready to fight a nuclear war, it is necessary to have the most capable warheads and delivery systems that science and technology can provide.

The changed political environment resulting from the establishment of independent governments in Poland, Hungary, Czechoslovakia, and East Germany and the expected withdrawal of most Soviet military forces from these countries may, at last, allow the US and the USSR to bring the arsenals into closer correspondence with the reality that was recognized by

Gorbachev and Reagan: "A nuclear war cannot be won and must never be fought."[4]

Below we review the current strategic nuclear arsenals, their capabilities to attack each other, and the consequences for civilians of such attacks. Then we consider the stability of three potential future levels of the strategic forces:

♦ Forces incorporating the reductions being negotiated in the START talks

♦ Forces resulting from further reductions, including the elimination of multiple-warhead ICBMs (START II)

♦ Forces resulting from reductions to 2,000-warhead "finite deterrence" levels based on strategic weapon systems that have already been deployed or are in an advanced stage of development.

We do not consider reductions to levels below 2,000 warheads here because they would require taking into account the British, Chinese, and French nuclear forces.[5] We also do not consider the issues associated with tactical nuclear weapons or "semistrategic" nuclear-armed long-range sea-launched cruise missiles, although such weapon systems are discussed in subsequent chapters on the verification of deep reductions.


## THE CURRENT STRATEGIC NUCLEAR BALANCE

As of September 1989, the US had deployed over 12,500 strategic nuclear warheads on vehicles designed to attack the USSR. For its part, the Soviet Union had deployed about 11,000 strategic nuclear warheads (see table 3.1). In addition, the US and USSR each had more than 10,000 "tactical" nuclear warheads available primarily for use in regional confrontations and deliverable by fighter-bombers, short- to intermediate-range missiles, artillery, and other means.

These arsenals are grossly in excess of what either country requires to hold the other hostage. According to US Department of Defense (DoD) estimates, published in 1968 by former US Secretary of Defense Robert McNamara, the explosion of the equivalent of only 200 one-megaton

**Table 3.1:** Numbers of warheads and destructive power in US and Soviet strategic nuclear arsenals, September 1989

| US FORCES | Launchers | Warheads | MT/warhead | EMT/warhead | Total EMT |
|---|---|---|---|---|---|
| *ICBMs* | | | | | |
| Minuteman II | 450 | 450 | 1.2 | 1.13 | 508 |
| Minuteman III | 200 | 600 | 0.17 | 0.31 | 184 |
| Minuteman IIIA | 300 | 900 | 0.335 | 0.48 | 434 |
| MX | 50 | 500 | 0.3 | 0.45 | 224 |
| total ICBMs | 1,000 | 2,450 | | | 1,351 |
| *SLBMs* | | | | | |
| Poseidon | 224 | 2,240 | 0.04 | 0.12 | 263 |
| Trident I | 384 | 3,072 | 0.10 | 0.22 | 663 |
| total SLBMs | 608 | 5,312 | | | 926 |
| *Bombers** | | | | | |
| B-52 ALCM | 194 | 3,480 | 0.23 *average* | 0.36 *average* | 1,250 |
| B-1B ALCM | 2 | 16 | 0.15 | 0.28 | 4 |
| B-1B | 95 | 1,520 | 0.35 *average* | 0.50 *average* | 760 |
| total bombers | 290 | 5,016 | | | 2,014 |
| TOTAL US | 1,899 | 12,778 | | | 4,291 |
| **SOVIET FORCES** | | | | | |
| *ICBMs* | | | | | |
| SS-11/13 | 409 | 409 | 1.0 | 1.0 | 349 |
| SS-17 | 108 | 432 | 0.5 | 0.63 | 272 |
| SS-18 | 308 | 3,080 | 0.5 | 0.63 | 1,941 |
| SS-19 | 320 | 1,920 | 0.5 | 0.63 | 1,210 |
| SS-24 | 58 | 580 | 0.5 | 0.63 | 365 |
| SS-25 | 171 | 171 | 0.5 | 0.63 | 108 |
| Total ICBMs | 1,374 | 6,592 | | | 4,245 |
| *SLBMs* | | | | | |
| SS-N-6/8/17 | 538 | 538 | 0.75 | 0.83 | 446 |
| SS-N-18 | 224 | 1,568 | 0.2 | 0.34 | 537 |
| SS-N-20 | 100 | 1,000 | 0.1 | 0.22 | 220 |
| SS-N-23 | 80 | 320 | 0.1 | 0.22 | 70 |
| Total SLBMs | 942 | 3,426 | | | 1,273 |
| *Bombers* | | | | | |
| Bear | 85 | 310 | 1.0 | 1.0 | 310 |
| Bear-H ALCM | 75 | 450 | 0.15 | 0.28 | 126 |
| Blackjack | 10 | 120 | 0.15 | 0.28 | 44 |
| Total bombers | 170 | 880 | | | 480 |
| TOTAL USSR | 2,486 | 10,898 | | | 5,998 |

* Each B-52G is assumed to carry eight air-launched cruise missiles (ALCMs, yield = 0.15 megatons) externally; each B-52H is assumed to carry 12 wing-mounted ALCMs. About one half the weapons on the B-52s are ALCMs and one half bombs and SRAMs. Each non-ALCM B-1B is assumed to carry 16 weapons internally, and each B-52 is assumed to carry eight weapons internally. In each case, the internal weapons are assumed to be half short-range attack missiles (SRAMs, yield = 0.17 megatons) and half bombs (average yield = 0.5 megatons).

Sources: Launchers and warhead numbers from "Strategic Nuclear Forces of the United States and the Soviet Union," (Washington DC: Arms Control Association Fact Sheet, September 1989). Warhead yields from Thomas B. Cochran, William M. Arkin, and Robert Norris, *The Bomb Book* (Washington DC: Natural Resources Defense Council, updated data for the Nuclear Weapon Databook series, December 1987), pp.13–14, except that we find implausible the Soviet ALCM warhead yield of 0.25 megatons and have reduced it to the same as for the US ALCM, 0.15 megatons.

warheads (200 "equivalent megatons" or EMT)* over the largest Soviet cities could kill over 20 percent of the Soviet population and destroy over 70 percent of Soviet industry.[6] As figure 3.1 shows, when the potential effects of nuclear-explosion–ignited "superfires" are taken into account, as little as 50 EMT exploding over US or Soviet urban areas would kill 20 percent of their respective populations directly.[7]

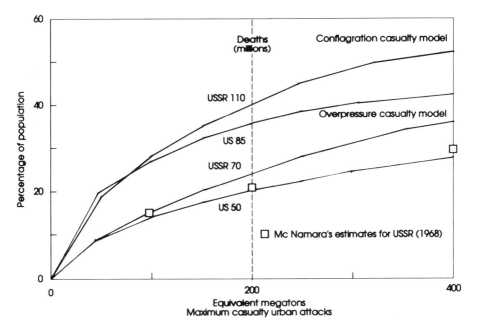

**Figure 3.1:** Measuring the US–USSR mutual nuclear-hostage relationship. Estimates of the number of people who could be killed in either the US or USSR by the direct effects of 1-megaton nuclear explosions are shown as a function of the cumulative number of such explosions. The "McNamara Points" are estimates published by former US Secretary of Defense Robert S. McNamara in 1968 for an attack on the Soviet Union. The curves represent the results of calculations for mid-1980s Soviet and US population distributions. The curves labeled "overpressure model" were obtained using methods similar to those that were used to make the calculations for McNamara. The curves labeled "conflagration model" take into account the lethal effects of the huge "superfires" that might be ignited by large nuclear bursts over urban areas.

---

* Equivalent megatonnage is calculated by raising the yield of a warhead, measured in megatons, to the two-thirds power. A one-eighth-megaton warhead, for example, would then carry one quarter of an EMT of destructive power. The scaling law is based on the fact that the size of the area that can be subjected to overpressures greater than a certain level increases as the two-thirds power of the warhead yield. (See, for example, Samuel Glasstone and Philip J. Dolan, eds., *The Effects of Nuclear Weapons*, 3rd edition [Washington DC: US Government Printing Office, 1977], pp.100–102). This scaling law does not apply exactly, however, to the other destructive effects of nuclear explosions such as fire.

After seeing the havoc caused in the USSR by the 1986 Chernobyl nuclear accident, which caused only 31 short-term deaths, the idea that as much as 200 EMT might be required to hold either the US or USSR hostage seems ludicrous. Yet, in 1989, the US and Soviet strategic arsenals each contained about 5,000 EMT.

## Stability

Figure 3.2a shows estimates of the numbers and total destructive powers of the strategic warheads that would be expected to survive all-out attacks by the US and USSR on each other's strategic forces (at 1989 levels) in the absence of warning—i.e. with the forces on ordinary peacetime alert.

Nearly 4,500 US strategic nuclear warheads with a combined destructive power of over 1,100 EMT would be expected to survive a Soviet attack. About 1,700 Soviet strategic warheads with a destructive power of nearly 1,000 EMT would survive a US attack.

Under crisis conditions, the fraction of bombers on alert and dispersed and the number of ballistic-missile submarines at sea would increase and a larger fraction of both forces would survive (see figure 3.2b).*

If we measure stability by the survivability of an assured retaliatory capability, therefore, the current nuclear balance is very stable. However, since both sides have counterforce strategies, crisis instability arises from pressures on nuclear-weapon policy makers to use their nuclear-counterforce capabilities before losing them.

## Civilian Casualties from Counterforce Attacks

The abstractness of calculations such as those whose results are displayed in figure 3.2 tends to obscure the fact that the use of thousands of nuclear warheads on military targets would almost inevitably result in tens of millions of civilian casualties. Modern strategic nuclear warheads are so powerful that, almost independently of their targets, they are weapons of mass destruction. Thus, for example, the blast and fire from the explosion

---

* The detailed assumptions and calculations behind the results in figures 3.2 are given in the appendix.

**Warheads before/after exchange**
1989 forces before exchange (rear)
1989 forces, after exchange: generated alert (center) peacetime (front)

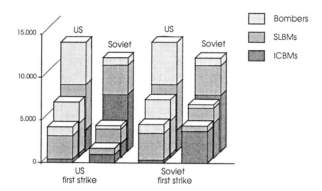

**EMT before/after exchange**
1989 forces before exchange (rear)
1989 forces, after exchange: generated alert (center) peacetime (front)

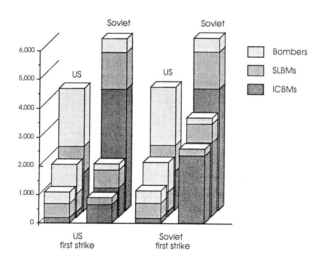

**Figure 3.2:** The strategic nuclear balance before and after a counterforce exchange, 1989. The US and Soviet Union each have in their strategic nuclear arsenals over 20 times the 200 equivalent-megatons that McNamara estimated was adequate for an "assured destruction" capacity in 1968. Even after a worst-case first strike by one side on the forces of the other, the attacked side would still have about 10 times the destructive power required to destroy its attacker in a retaliatory strike. These estimates have been made for both a case in which the first strike would occur without warning—so that the forces of both sides would be at peacetime levels of alert, and for a case in which the forces would be on generated alert. The estimates assume that ballistic missiles would not be launched under attack and that the first response of the attacked side would be to use 200 of its surviving ICBM warheads to destroy the in-port ballistic-missile submarines, nonalert bombers and command-and-communications system of the attacking side.

of a half-megaton strategic warhead will destroy an area of approximately 100 square miles (250 square kilometers). Attacks on air bases, command centers and ports would therefore exterminate the substantial population living near such targets. Large numbers of warheads exploding over "hardened" missile silos—necessarily low enough for the fireball to contact the ground—would result in fallout lethal to unprotected civilians up to 1,000 kilometers downwind from the target area.[8] And tens of millions of additional deaths could be expected from exposure, famine, and disease among the surviving populations as a result of the economic and environmental effects of the attacks.[9]

In view of these enormous consequences, it seems foolhardy to expect that counterforce attacks involving the use of thousands of nuclear warheads would be significantly less likely than deliberate attacks on civilian populations to trigger a devastating retaliatory attack.

It is possible to imagine scenarios for counterforce attacks with tens or hundreds of warheads that would not have such devastating consequences but such scenarios cannot be used to justify the existence of over 10,000 strategic warheads that the US and USSR each possesses today.


## STABILITY AFTER REDUCTIONS

### Levels of Reductions

To give our analysis concreteness, we focus here principally on two hypothetical levels for US and Soviet strategic nuclear forces after reductions:

♦ The forces after the START-mandated cuts

♦ "Finite deterrence" levels at which the US and USSR would each have only 2,000 strategic warheads in forces designed primarily for survivability rather than for counterforce capabilities.

In practice, at least one intermediate stage of strategic reductions after START might be negotiated before finite-deterrence levels were reached. At such a stage, land-based multiple-warhead ("MIRVed") ICBMs, heavy sea-launched ballistic missiles (SLBMs) such as the Trident II, and nuclear-armed sea-launched cruise missiles (SLCMs) might be banned. Below, we will therefore also briefly discuss the possibility of such an intermediate "START II" level of reductions.

Our hypothetical START and finite-deterrence forces are shown in table 3.2. This table shows both the actual total number of warheads and, where different, the total number counted under the special counting rules of START (in parentheses). Below, we briefly discuss the design of these forces.

*START Forces*

Although several issues were still unresolved at the time of this writing, the US and the Soviet Union were at a relatively advanced stage of negotiating a START treaty. This agreement will set a 6,000-warhead ceiling on the total number of *counted* warheads on strategic ballistic missiles and long-range bombers in each country's arsenal and a 4,900-warhead subceiling on the total number of ballistic-missile warheads. The agreement will also limit the number of strategic delivery vehicles to 1,600 and the number of warheads permitted on "heavy missiles" (the Soviet SS-18) to 1,540.[10]

The agreement will definitely reduce the number of strategic warheads on deployed ballistic missiles. However, according to agreed and proposed START counting rules, bomber-delivered warheads will be undercounted and could even increase in number. Most important, strategic bombers other than those armed with long-range air-launched cruise missiles (ALCMs) will be counted as carrying only one nuclear warhead each.[11]

Under these counting rules, 100 non-ALCM bombers, each carrying 16 bombs and short-range-attack missiles will count as carrying only 100 warheads—although they will actually carry 1,500 more than that.[12] The deployment of such heavily "MIRVed" non-ALCM bombers on each side in combination with the deployment of hundreds of additional nuclear warheads on long-range nuclear-armed SLCMs—which are to be counted outside the START totals—will result in START-limited arsenals containing thousands more than the advertised 6,000 strategic warheads on each side.

Table 3.2 shows what the US and Soviet forces might look like in the mid-1990s under a START agreement, assuming continued deployment of nuclear weapon systems currently in production and a compensating phase-

**Table 3.2:** US and Soviet strategic nuclear arsenals in late 1989 and hypothetical START and finite-deterrence forces in the mid-1990s

| | Delivery vehicles | | | Warheads/totals | | |
|---|---|---|---|---|---|---|
| | 1989* | START | Finite deterrence | 1989* | START† | Finite deterrence |
| **US FORCES** | | | | | | |
| *ICBMs* | | | | | | |
| Minuteman II | 450 | 0 | 0 | 450 | 0 | 0 |
| Minuteman III/IIIA | 500 | 200 | 0 | 1,500 | 600 | 0 |
| MX | 50 | 50 | 0 | 500 | 500 | 0 |
| Midgetman | 0 | 344 | 492 | 0 | 344 | 492 |
| Total ICBMs | 1,000 | 594 | 492 | 2,450 | 1,444 | 492 |
| *SLBMs* | | | | | | |
| Poseidon | 224 | 0 | 0 | 2,240 | 0 | 0 |
| Trident I/II | 384 | 432 | 126 | 3,072 | 3,456 | 1,008 |
| Total SLBMs | 608 | 432 | 126 | 5,312 | 3,456 | 1,008 |
| Missile totals | 1,608 | 1,026 | 618 | 7,762 | 4,900 | 1,500 |
| *Bombers* | | | | | | |
| B-52 ALCM | 194 | 90 | 0 | 3,480 | 1,800 ( 900) | 0 |
| B-1/2 | 97 | 150 | 125 | 1,536 | 2,400 ( 150) | 500† |
| Strategic totals | 1,899 | 1,266 | 743 | 12,778 | 9,1000 (5,950) | 2,000 |
| **SOVIET FORCES** | | | | | | |
| *ICBMs* | | | | | | |
| SS-11/13 | 409 | 0 | 0 | 409 | 0 | 0 |
| SS-17 | 108 | 0 | 0 | 432 | 0 | 0 |
| SS-18 | 308 | 154 | 0 | 3,080 | 1,540 | 0 |
| SS-19 | 320 | 0 | 0 | 1,920 | 0 | 0 |
| SS-24 | 58 | 80 | 0 | 580 | 800 | 0 |
| SS-25§ | 171 | 336 | 1,020 | 171 | 336 | 1,020 |
| Total ICBMs | 1,374 | 570 | 1,020 | 6,592 | 2,676 | 1,020 |
| *SLBMs* | | | | | | |
| SS-N-6/8/12/17 | 538 | 0 | 0 | 538 | 0 | 0 |
| SS-N-18 | 224 | 0 | 0 | 1,568 | 0 | 0 |
| SS-N-20 | 100 | 120 | 0 | 1,000 | 1,200 | 0 |
| SS-N-23 | 80 | 256 | 120 | 320 | 1,024 | 480 |
| Total SLBMs | 942 | 376 | 120 | 3,426 | 2,224 | 480 |
| Missile totals | 2,316 | 946 | 1,140 | 10,018 | 4,900 | 1,500 |
| *Bombers* | | | | | | |
| Bear | 85 | 0 | 0 | 310 | 0 | 0 |
| Bear-H ALCM# | 75 | 87 | 75 | 450 | 522 ( 696) | 300 |
| Blackjack | 10 | 50 | 50 | 120 | 600 ( 400) | 200† |
| Strategic totals | 2,486 | 1,083 | 1,265 | 10,898 | 6,022 (6,544) | 2,000 |

* The numbers for 1989 are from "Strategic Nuclear Forces of the United States and the Soviet Union" (Washington DC: Arms Control Association Fact Sheet, September 1989).

† Numbers in parentheses assume a START counting rule of 10 warheads each for US ALCM bombers and eight each for Soviet ALCM bombers and the counting rule agreed at the December 1987 Washington summit of one warhead for each non-ALCM bomber. According to information released by the US Air Force on 16 October 1989, a standard load for the B-2, like the B-1, would be eight SRAMs and eight B-61 gravity bombs (Rowan Scarborough, "Air Force Defends B-2, Disputes Report," *Washington Times*, 17 October 1989, p.1). A full loading on a B-52G (-H) is eight (12) ALCMs plus eight SRAMs and bombs. We assume that all the remaining B-52s under START are B-52Hs.

† Four ALCMs per long-range nuclear bomber.

§ For the finite-deterrence case, we assume that the SS-25 or a follow-on single-warhead missile will be put into either a hardened-mobile launcher or silo.

# We assume that the Bear-H can carry only six ALCMs in an internal rotary launcher (see chapter 9).

out of older systems.

We have assumed that the US will retire its remaining 26 (as of September 1989) 1960s-vintage 16-launch-tube ballistic-missile submarines, which carry Poseidon and Trident I missiles, and continue to build up its fleet of Trident submarines from 8 to 18.* Each Trident submarine can carry 24 eight-warhead Trident I or Trident II missiles.

We have assumed similarly that the Soviet Union will retire its older Yankee and Delta I, II, and III ballistic-missile submarines and build up from 5 to 16 its fleet of Delta IVs, each carrying 16 launchers for the four-warhead SS-N-23 missile,[†] and from 5 to 6 its fleet of Typhoon submarines, each of which carries 20 ten-warhead SS-N-20 missiles. US and Soviet ballistic-missile submarines in the START force would therefore be reduced to approximately one half and one third of their 1989 numbers respectively.[13]

It will be seen that these hypothetical forces approximately saturate the total 6,000-"warhead" limit and 4,900 ballistic-missile–warhead sublimit but not the 1,600 delivery-vehicle limit currently envisioned in the START agreement. From the point of view of stability, it would be desirable to reduce the average number of warheads per ICBM by increasing the number of ICBMs to the limit allowed by START while keeping the number of their warheads constant. This could be done by increasing the number of single-warhead ICBMs on each side while eliminating an equal number of warheads from MIRVed ICBMs. The extra single-warhead missiles could be either mobile or fixed.

If the US and USSR wished to maintain larger fleets of ballistic-missile submarines under START, they could reduce the number of launch tubes in each submarine in a way that could be verified nonintrusively—for

---

* The US Government may actually be arguing that it be allowed 21 Trident submarines under START by exempting from the START limits the three submarines that would be undergoing overhaul at any particular time. See *Arms Control Reporter* (Brookline, Massachusetts: Institute for Defense and Disarmament Studies, 1989) p.611.B.556.

† It might also be possible to build up the number of SS-N-23 missiles by backfitting them into some of the 14 Delta III submarines (*Hearings on the Fiscal Year 1988–89 DOD Budget* before the Subcommittee on Seapower and Strategic and Critical Materials of the House Armed Services Committee, pp.11–12).

example, by random challenge inspections just before the submarines left port.

*START II*

Since the step from the START levels to the finite-deterrence levels is quite large (see table 3.2), it might be considered desirable to introduce an intermediate level of reductions, which we call "START II."

An obvious objective of START II would be to complete the dismantlement of all land-based MIRVed missiles. Such missiles are destabilizing by virtue of their ability to threaten multiple targets in a first strike and their attractiveness as targets of a first strike. In the case of the hypothetical START forces shown in table 3.2, this would mean that the Soviet Union would dismantle its remaining 10-warhead silo-based SS-18 missiles and its 10-warhead silo and rail-mobile SS-24s while the US would dismantle its remaining silo-based three-warhead Minuteman III and its silo- or rail-mobile 10-warhead MX missiles. Since this would be an unequal trade of more than 2,300 Soviet for 950 US warheads, some additional US warheads would have to be traded in. These would presumably be predominantly SLBM warheads. If the US still had a very much larger number of bomber warheads than the USSR, that disparity could also be reduced.

Among the other items that might be considered for inclusion in a START II package would be a freeze on the development and deployment of more counterforce-capable SLBMs. Here the Trident II and SS-N-20 stand out among SLBMs in the same way that the SS-18 stands out among ICBMs. These missiles have a launch weight of approximately 60,000 kilograms—about twice the launch weight of the Trident I (30,000 kilograms) and 50 percent greater than the estimated launch weight of the SS-N-23, the most modern MIRVed Soviet SLBM.[14] Both sides could also agree to include in START II a ban on nuclear-armed SLCMs and/or whole classes of tactical nuclear weapons. Both SLCMs and tactical nuclear weapons lower the threshold to nuclear war by mingling nuclear weapons with conventional multipurpose forces.[15]

*Finite-deterrence Levels*

In table 3.2, we show how US and Soviet finite-deterrence forces could be constructed—once again using strategic weapons that have already been deployed or are in an advanced stage of development. We have assumed that the US would continue to favor SLBMs and keep one half of its strategic warheads on ballistic-missile submarines and that the Soviet Union would similarly continue to favor ICBMs. The total destructive powers of the US and the Soviet finite-deterrence forces would each be about 900 EMT.* Although there would be less than 40 percent as many ICBM warheads in these finite-deterrence forces than in the corresponding START forces, the numbers of ICBMs would be about the same in the case of the US and double in the case of the Soviet ICBM forces because all the ICBMs would have only a single warhead. Stability would be enhanced because each country would have many fewer warheads per strategic target on the other side.

The hypothetical finite-deterrence forces in table 3.2 have many fewer SLBMs than the START forces because basically the same MIRVed missiles are assumed to be used. However, the survivability of the SLBM force does not depend only on the number of SLBMs, although it depends to some extent on the number of ballistic-missile submarines. If the number of launch tubes per ballistic-missile submarine were reduced to six, the US would be able to keep 21 Trident submarines and the USSR would be able to keep six Typhoon and 24 Delta submarines—more in each case than under the hypothetical START agreement.†

We also assume that a finite-deterrence force might contain 125 long-range nuclear bombers equipped with ALCMs. This is similar to the number of ALCM bombers that we assumed for the START force but the number of warheads carried by each bomber in the finite-deterrence force

---

* We assume that the yields of the warheads on both the US and Soviet single-warhead ICBMs are 0.5 megatons, that the yields of the warheads on their SLBMs average respectively 0.3 and 0.1 megatons, and that the yields of the warheads on their ALCMs are 0.15 megatons.

† As has already been noted, an alternative way to reduce the number of warheads per ballistic-missile submarine is being explored in the START negotiations: reducing the numbers of warheads carried by the existing missiles. In the longer term, smaller missiles carrying smaller numbers of warheads could be put in the launch tubes, or submarines with fewer launch tubes per submarine could be deployed.

is reduced to four. As the carrying capacity of current strategic bombers could not be verifiably limited to this level, verification of the ALCM limits would have to be accomplished through separate limits on the nuclear-armed missiles themselves, as is discussed later in the book.[16]

## Effects of Reductions on Stability

There are two different types of stability of concern in the design of any arms-control proposal: *crisis stability*, to ensure that neither side would perceive an advantage in striking first during a crisis; and *breakout and arms-race stability*, to ensure that neither side would perceive an advantage in breaking out from treaty restrictions by either rapidly deploying more warheads than permitted by the arms-control agreement or by developing new types of weapon systems.

The primary determinants of both types of stability are the vulnerabilities of the forces and effectiveness of the defenses of each side. Some of the determinants of stability with today's nuclear forces are listed in table 3.3. It will be seen that stability can be improved by both unilateral and cooperative measures.

### Crisis Stability

Figure 3.3 shows our estimates of the sizes of the hypothetical US and Soviet START* and finite-deterrence arsenals before and after counterforce exchanges—assuming both peacetime and crisis alert levels. Those numbers include losses by the attacker to a counterstrike with 200 ICBM warheads against its ballistic-missile submarines that are not at sea and its bombers that are not on alert.†

---

* For simplicity, we have neglected the SLCMs in our calculations of the results of first strikes on the START forces.

† The assumptions used in these calculations are given in the appendix. It is assumed that, for a "bolt from the blue" attack under peacetime conditions, the attacking nation would also have its ballistic-missile submarine and bomber forces at peacetime alert levels so as not to provide warning to the other side. The US and USSR monitor each others' deployments and military communications so closely that any substantial increase in alert rates would be likely to be detected by the other side, causing an increase in its own state of alert.

**Table 3.3:** Some key determinants of stability for strategic systems

| Launcher | Determinants of stability |
|---|---|
| **Missile silos** | Warheads per silo |
| | Hardness of silos |
| | Number, accuracy, and yield of the ballistic-missile warheads of the other side |
| | Effectiveness of ballistic-missile defenses of the other side |
| **Mobile land-based missile launchers** | Warheads per launcher |
| | Hardness of launcher |
| | Area over which launchers can scatter between warning and arrival of warheads |
| | Number and yields of submarine-launched ballistic-missile warheads available to the other side for a barrage attack from near-coastal waters |
| | Possibility of the use of submarine-launched ballistic-missile (SLBMs) on short-flight-time depressed trajectories |
| | Effectiveness of ballistic-missile defenses of the other side |
| **Ballistic-missile submarines** | Warheads per submarine |
| | Fraction of submarines at sea |
| | Quietness of submarines |
| | Deployment and detection technologies of the antisubmarine-warfare forces of the other side |
| | Effectiveness of ballistic-missile defenses of the other side |
| **Long-range bombers** | Warheads per base |
| | Closeness of bomber and tanker bases to the coast |
| | Area over which aircraft can scatter between warning and arrival of warheads |
| | Number and yields of SLBM warheads available to the other side for barrage attack |
| | Possibility of the use of SLBMs on short-flight-time depressed trajectories |
| | Effectiveness of air defenses of the other side |

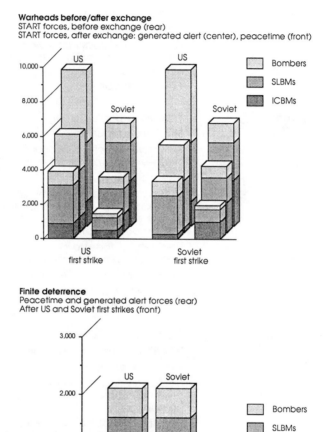

**Figure 3.3:** US and Soviet START and finite-deterrence arsenals before and after counterforce exchanges. The figures show the estimated number of strategic warheads on each side, before and after strategic counterforce exchanges, assuming peacetime or crisis alert conditions.

It will be seen that there is no case in which either side could realistically hope to remove itself from the mutual-nuclear-hostage relationship with a first strike. Even in the case of a "bolt from the blue" attack on one of the finite-deterrence forces, approximately one half of the weapons in the attacked force, carrying about 500 EMT of destructive power, would survive.

In estimating the effects of first strikes, we have considered the possibility of "barrage attacks" on the areas around bomber and airborne-tanker bases or mobile-missile bases during the minutes when bombers or missile carriers escaping these bases would still provide relatively concentrated targets. We found barrage relatively effective against US B-52 bombers and tankers located at bases near the US east and west coasts and against bases for soft mobile ICBM carriers such as the trucks carrying the Soviet SS-25 or railroad trains carrying the Soviet SS-24—or possibly in the future, the US MX.* However, we have assumed that by the time finite-deterrence levels were achieved, bombers and tankers would all be based inland and all mobile ICBMs would be on cross-country-capable hardened launchers.

Barrage would become a much more serious concern if used in conjunction with depressed trajectories on which SLBM warheads could travel 2,000–3,000 kilometers in about 10 minutes instead of the 15–18 minutes assumed in our calculations.[17] Fortunately, neither the US nor USSR seems to have tested its SLBMs on such trajectories.

If depressed-trajectory attacks became a serious possibility, missile launchers would not be able to survive simply through their ability to scatter on warning of attack. The warning time would be too short. Most mobile launchers would have to remain dispersed, moving frequently enough so that they could not be targeted. If this was impractical, it might be better simply to base the missiles in separate hard silos. If desired, such a scheme could be elaborated by a deceptive basing scheme similar to that proposed for the MX by the Carter Administration. Such a scheme would be much less costly and environmentally damaging for a small single-warhead missile than for the MX.

---

\* It must be remembered, however, that the ICBMs could be launched on warning.

Some analysts believe that command-and-control vulnerability and the resulting "use it or lose it" pressures for commanders anticipating the loss of their ability to order coordinated attacks on "time urgent" counterforce targets are the single most important source of instability in the nuclear balance today.[18] The US command-and-control system is currently in the midst of a hardening program, which, it is believed, will increase the number of warheads required to incapacitate it from the tens to the hundreds—but still far fewer than the thousands of warheads that would be required to destroy a significant fraction of US strategic weapon systems.[19] The Soviet command-and-control system has redundant underground bunkers and communication systems but is also highly centralized and may be as vulnerable as that of the US.

Attacks on command-and-control systems would require a relatively small fraction of the warheads in today's strategic arsenals or in the proposed START arsenals but they would require a more substantial fraction of the finite-deterrence arsenals. Therefore, this source of crisis instability would be reduced at finite-deterrence levels.

Crisis instability due to the vulnerability of command-and-control systems would be reduced much more, however, by the different attitude toward retaliation implicit in a finite-deterrence posture. Since the finite-deterrence forces would not be targeted on each other, there would be less pressure for a pre-emptive attack. If one side *did* attack, the attacked side could ride out the attack and the surviving members of the command structure could take time to regroup, decide what to do, establish communication with the surviving nuclear forces, and give orders for whatever retaliation was considered appropriate. Days or weeks could be spent in this process rather than the 10-minute or shorter response time that is the fundamental source of instability in today's arrangements for launch-on-warning counterforce attacks. The delay would not significantly reduce the deterrent value of potential retaliation.[20]

*Breakout and Arms-race Stability*
A critical design requirement for any arms-control agreement is to make each side confident that the other side could gain no significant military advantage by a sudden breakout from the arms-control restrictions. Below,

we consider two breakout scenarios: a rapid increase in the number of deployed warheads and a rapid deployment of ballistic-missile defenses.

*Additional Warheads.* Concerns about the possibility of a rapid increase in the number of deployed ICBM Soviet warheads in a breakout from a START agreement were raised in a report published by the US House Armed Services Committee in 1988, and strongly rebutted in a Report to Congress in September 1988 by Frank Carlucci, then Secretary of Defense.[21] Although the US and the Soviet Union currently possess numbers of test and spare ICBMs comparable to the number deployed, the report of the Secretary of Defense regarded as "improbable" the possibility that the Soviets could suddenly deploy these missiles on soft launch pads preparatory to a first strike. The report judged that such a scenario "would take years to accomplish."

In any case, even a doubling of the number of deployed ICBM warheads would not significantly affect the survivability of submarines at sea because their locations would be too uncertain for them to be subject to barrage by any credible number of warheads.[22] Such a doubling would also not greatly increase the barrage threat to alert bombers and mobile missiles because of the relatively long warning times for attacks by ballistic-missile warheads fired from intercontinental distances.

The barrage threat would be more sensitive, however, to a great increase in the number of SLBM warheads available for launch from near-coastal waters. Table 3.4 presents our estimates of the number of extra 100-kiloton SLBM warheads that would be required to mount successful barrage attacks against inland bomber bases hosting modern bombers or bases for hardened mobile-missile carriers. It will be seen that an estimated 13,000–30,000 one-hundred-kiloton SLBM warheads would be required to barrage 13 bomber bases and an additional 7,500–15,000 warheads to attack 250 mobile-missile bases, given SLBM flight times of 15–18 minutes.[23] For comparison, the hypothetical US and Soviet START forces in table 3.2 contain 3,456 and 2,224 SLBM warheads respectively.

The only possibility for a rapid increase in the number of SLBM warheads under the hypothetical START regime would be an increase in the number of re-entry vehicles on the SLBMs from their treaty-limited

**Table 3.4:** Numbers of close-in 100-kiloton SLBM warheads required for barrage attacks on inland bomber or mobile-missile bases

|  | Number of bases | Warheads required to barrage each[*] | Warheads required to barrage all |
|---|---|---|---|
| Bomber bases (modern bombers)[†] | 13 | 1,000–2,400 | 13,000–30,000 |
| Hard mobile-missile carriers | 250 | 30–60 | 7,500–15,000 |

[*] Assuming a total of 15–18 minutes warning time and that 8 of these minutes are used up before the mobile-missile carriers are actually put into motion. See appendix—especially table 3.A.2.

[†] A US B-1, B-2, or Soviet Blackjack bomber. A US B-52 or Soviet Bear bomber base would be vulnerable to a somewhat smaller barrage because their takeoff is somewhat slower and they are assumed to be somewhat more vulnerable to blast. See appendix, table 3.A.2.

values to their maximum numbers, which we assume to be from 8 to 12 for the Trident I/II, and from 4 to 10 for the Soviet SS-N-23.[*] This would increase the number of warheads on US and Soviet START SLBM forces by 1,728 and 1,536 respectively. If we assume that it is not credible that more than half of the ballistic-missile submarines of either side could or would be secretly brought close in to the other country for a barrage attack, the maximum number of SLBM warheads available to the US and USSR for close-in barrage would be about 1,700 and 1,100 respectively under START and 2,500 and 1,800 respectively after a breakout from START. These numbers are all low relative to those required for a barrage attack.

We have assumed in our hypothetical finite-deterrence force that the submarines would be of the same types and approximate number as today but with the number of launchers reduced to six per submarine. If the elimination of launch tubes were rapidly reversible, then the breakout-barrage threats would be comparable to those for the hypothetical START postures. It would, of course, be desirable to make the launch-tube elimination as irreversible as possible and ensure that movement towards its reversal would be detected soon enough so that countermeasures such as

---

[*] It is assumed that the 10 re-entry vehicles with which the SS-N-20 is credited under the START counting rules already put it at its maximum loading.

further dispersal of bombers and mobile missiles could be undertaken.

Concern about this breakout scenario could be reduced further if each side agreed to store its spare and test missiles at a few central locations where the other side would have a permanent inspection force controlling the perimeter and was able periodically to check the missiles for nuclear warheads.

*Strategic Ballistic-missile and Air Defenses.* The US and USSR agreed to limit their defenses against ballistic missiles to trivial levels in the 1972 ABM treaty, in part because of concerns about launching a wasteful offensive–defensive arms race. Some officials were also concerned that defenses would increase incentives to attack first in a crisis because the defenses might inspire the hope that it would be possible to protect one's country against a "ragged" retaliatory attack. Undoubtedly the most persuasive argument, however, was that the defenses could easily be overwhelmed or otherwise rendered ineffective. One effective demonstration of the ease with which defenses could be overwhelmed was the development of MIRVed ballistic missiles.

One might worry that a reduction in the number of ballistic-missile warheads from about 10,000 today to 1,500 in a finite-deterrence regime might make the idea of ballistic-missile defense (BMD) credible enough to tempt one side or the other to break out of the ABM treaty. However, in the absence of an ability to destroy ballistic missiles in their boost phase, it is well understood that BMD systems could be overwhelmed if each ballistic-missile re-entry vehicle containing a real warhead were accompanied by a dispersed swarm of decoys made indistinguishable from the warhead through simulation or antisimulation.*

If the decoys were meant to be indistinguishable from re-entry vehicles only above the atmosphere, the total weight of 100 of them might be no more than that of one real re-entry vehicle. The insubstantiality of such decoys would be revealed only when they entered the atmosphere during

---

* In simulation, each decoy would be made to appear like a warhead. In anti-simulation, the warheads would be made to appear like decoys by, for example, surrounding them with balloons.

the last 20 seconds or so before impact.[24]

The same kind of logic applies to the competition between strategic air defenses and strategic bombers. During the 1970s, improvements in Soviet air defenses were used to justify the "MIRVing" of US bombers with air-launched cruise missiles. Just as with ballistic-missile MIRVs and exo-atmospheric decoys, this approach and the recent development of "stealthy" cruise missiles seems to have effectively neutralized area defenses for the indefinite future. However, just as MIRVing shifted the focus of BMD interest to the boost phase before fractionation occurs, the MIRVing of bombers has shifted attention of the designers of strategic air defenses to the possibility of attacking bombers thousands of kilometers from their targets, before they have a chance to launch their cruise missiles. The obvious means of attack are ballistic missiles carrying homing warheads. In this scheme, targeting could be done with over-the-horizon radars and space-based sensors. Such systems for attacking aircraft at long ranges could also pose a serious threat to airborne command-and-communications relay aircraft.

It would therefore be important to complement an accord on reductions to finite-deterrence levels with a set of agreements to limit the long-lead-time items for systems designed to attack aircraft at long ranges. Such limits could include limitations on the number and power of over-the-horizon and space-based radars, and a ban on tests of ballistic-missile re-entry vehicles capable of maneuvering to home in on aircraft.* With long–lead-time items limited, the temptation to deploy such systems would be dampened by the realization that additional cruise missiles and cruise-missile carriers could be deployed more rapidly than the defenses.

*Antisubmarine-warfare Systems.* There are currently no agreed constraints on the development of strategic antisubmarine-warfare (ASW) systems. However, there is also no current prospect of a breakthrough in detection technology that cannot easily be neutralized.[25]

Nevertheless, agreements to limit strategic ASW would be desirable if

---

* This would probably require an agreement that ballistic-missile tests would involve re-entry only in areas observable by the other side's monitoring systems.

they would help damp the arms race in this area. One agreement that would be relatively easy to verify would be for the US and USSR to limit the number of their nuclear-powered attack submarines.[26]

*Third-generation Nuclear Warheads.* Currently, US and possibly also Soviet nuclear-weapon laboratories are investigating new types of warhead such as the nuclear-explosion-powered x-ray laser for use against targets in outer space, nuclear-powered microwave generators to burn out the electronics of mobile missiles and mobile command posts over large areas, and earth-penetrating warheads designed to increase the shock to underground facilities for a given weight warhead.[27]

It is unlikely that any of these weapons, even if they could all be successfully developed would have a fundamental impact on the stability of the arms race. However, because their inherent characteristics make them more relevant to a first strike than to a retaliatory strike, their deployment would make assessment of the balance that much more complicated, and in some cases it would be necessary to develop and deploy countermeasures. A ban or very-low–threshold limit on underground nuclear testing would therefore be desirable to prevent the additional complications and waste of talent and resources.[28]

## VERIFICATION REQUIREMENTS

Verifiability is typically one of the most contentious issues in the US debate over the ratification of proposed arms-control treaties—not because of its inherently overwhelming importance but because opponents of arms control use verification issues as an apparently objective reason for their opposition to agreements. Unfortunately, it is very seldom that anyone participating in these debates offers a definition of how verifiable is verifiable enough.

An unarguable minimal requirement for verifiability is that no clandestine activity remain undetected long enough for the US or Soviet Union unilaterally to release itself from the mutual-hostage relationship before the other side could take effective countermeasures. Based on the analysis above, the mutual-hostage relationship cannot be broken as long as:

♦ Ballistic-missile submarines remain substantially invulnerable

♦ Intercontinental-range weapons that can home in on aircraft and mobile-missile launchers are not developed

♦ Strategic defenses are prevented from developing significant effectiveness.

Of course, the *political* standards of adequacy of verifiability would be more stringent.

Many of the verification tasks under finite deterrence can be handled by procedures established under past and existing arms-control agreements. In particular:

♦ Monitoring of numbers and types of deployed ballistic-missile submarines, ICBM silos, and strategic bombers can be accomplished principally through "national technical means" (imaging satellites). This was established with the SALT I and SALT II agreements, which also included agreements that the parties not use deliberate concealment measures to interfere with the verification.

♦ Monitoring of ballistic-missile flight tests to determine the number of re-entry vehicles released by a missile and the overall missile characteristics can also be accomplished largely through national technical means. SALT II depended on such monitoring to verify restrictions on the development of new ballistic missiles, the number of warheads permitted to be flight tested on each ballistic missile, and missile throw-weight.

♦ On-site inspection procedures to monitor numbers of deployed and non-deployed missiles and launchers during the phase-out period and to monitor production bans on prohibited systems were established by the INF treaty.

Therefore, many of the arrangements required to verify reductions to finite-deterrence levels are already in place. The principal new tasks essential to finite deterrence (including those being discussed, but not yet worked out under START) would be to verify the elimination of ballistic-missile submarine launch tubes, and limits on: numbers of mobile-missile launchers, numbers of warheads deployed on each type of ballistic missile, the number of nuclear-armed cruise missiles, and on the overall availability of warheads and the fissile material required for their manufacture.

These tasks will generally require cooperative arrangements and they are discussed in the remainder of this book.

## NOTES AND REFERENCES

1. See, for example, Senator William Cohen, "The B-2: Mission Questionable, Cost Impossible," *Arms Control Today*, October 1989, pp.3–8.

2. See, for example, C.P. Robinson, then Director of National Security Programs, Los Alamos National Laboratory, in *Review of Arms Control and Disarmament Activities*, Hearings before the House Armed Services Committee, 1985, pp.140–142.

3. *Soviet Military Power 1989* (Washington DC: US Government Printing Office, 1989), p.45.

4. Gorbachev–Reagan Joint Summit Statements: Geneva, 21 November 1985; Washington, 10 December 1987; and Moscow, 1 June 1988.

5. These countries should, however, be involved in the nuclear-arms–reduction negotiations well before finite-deterrence levels are reached—if only informally. Under current modernization plans, the French, British, and Chinese forces are expected by the late 1990s to have a combined total of about 1,500 strategic warheads—approximately the same as the number of warheads assumed for each of the "superpower" finite-deterrence forces. Richard Garwin has proposed that, in the context of a US–Soviet agreement to reduce their arsenals to 1,000 warheads each, France, Britain, and China agree to accept a ceiling of 200 warheads each. (Richard L. Garwin, "A Blueprint for Radical Weapons Cuts," *Bulletin of the Atomic Scientists*, March 1988, pp.10–13.)

6. Secretary of Defense, Robert S. McNamara, *The Fiscal Year 1969–73 Defense Program and the 1969 Defense Budget* (Washington DC: Department of Defense, 1968) p.57.

7. See Frank von Hippel, Barbara G. Levi, Theodore A. Postol, and William H. Daugherty, "Civilian Casualties from Counterforce Attacks," *Scientific American*, September 1988, pp.36–42, and articles cited there.

8. Ibid.

9. See, for example, M.A. Harwell, T.C. Hutchinson et al., *Environmental Consequences of Nuclear War, II: Ecological and Agricultural Effects* (New York: John Wiley & Sons, 1985).

10. US–Soviet Joint Summit Statement, Washington DC, 10 December 1987.

11. The US and the Soviet Union have agreed on a START counting rule that would credit US ALCM bombers with 10 warheads each and Soviet ALCM bombers with eight warheads each—considerably fewer than these bombers can actually carry.

12. The US currently has 95 B-1s which are configured as "penetrating" (non-ALCM) bombers. In addition, the US intends to produce 75 B-2 ("Stealth") aircraft to be deployed as penetrating bombers. As the B-2 is phased in, the B-1s may be refitted as ALCM carriers. The Soviet Union has produced a small number of Blackjack bombers that the US Department of Defense believes have capabilities comparable to the B-1. Both countries have equipped many of their older types of bomber with ALCMs (the B-52G and -H in the case of the US and the Bear-H in the case of the USSR).

13. Our numbers for US and Soviet submarines in 1989 are based on *Strategic Forces of the United States and the Soviet Union* (Washington DC: Arms Control Association, September 1989).

14. Thomas B. Cochran, William M. Arkin, and Milton M. Hoenig, *Nuclear Weapons Databook Volume I: U.S. Nuclear Forces and Capabilities* (New York: Ballinger, 1989), pp.121, 145; Cochran, Arkin, Robert S. Norris, and Jeffrey I. Sands, *Nuclear Weapons Databook Volume 4: Soviet Nuclear Weapons*, pp.150, 151.

15. US nuclear SLCMs and sea-based tactical nuclear weapons are also not equipped with permissive-action links that would physically prevent their unauthorized use.

16. See chapters 4 and 9.

17. For ballistic-missile flight times, see, for example, Harold Feiveson and Frank von Hippel, "The Freeze and the Counterforce Race," *Physics Today*, January 1983, figure 3.

18. See, for example, Bruce Blair, *Strategic Command and Control* (Washington DC: Brookings Institution, 1985).

19. Charles A. Zraket, "Strategic Command, Control, Communications and Intelligence," *Science* **224**, (1984), p.1306.

20. This argument is made at length in Morton Halperin, *The Fallacy of Nuclear Weapons* (New York: Ballinger, 1987).

21. *Breakout, Verification and Force Structure: Dealing With the Full Implications of START* (see especially Annex 2); Frank Carlucci, "Report to the Congress on Verification of START," September 1988. See also James P. Rubin, "Carlucci Responds to Critics of Proposed START Treaty", *Arms Control Today*, November 1988, p.28.

22. Consider, for example, a situation in which a significant fraction of Soviet submarines were deployed in one half of the area of the Arctic Ocean (seven million square kilometers). Assuming that a single 1-megaton underwater explosion could destroy a submarine within an area of 100 square kilometers (5.6-kilometer radius, Richard L. Garwin, "Will Strategic Submarines be Vulnerable?," *International Security*, Fall 1983, p.52), it would take 70,000 EMT to barrage the area. (In deep water, the shock wave from a 1-megaton explosion would have a peak pressure of about 1,000 psi for about 0.1 seconds at a radius of 5.6 kilometers [Samuel Glasstone and Philip J. Dolan, *The Effects of Nuclear Weapons*, 3rd Edition, pp.268–271]). One thousand psi corresponds to the static pressure at a depth of about 700 meters. However, the shock overpressure at the surface will be canceled by the reflected rarefaction wave and will be partially canceled at shallow depths. Since a submarine would become detectable at the surface, Garwin suggests that the optimum depth for survivability might be 100 meters.

23. There are currently 17 strategic bomber bases in the US, of which six can be considered inland (*Air Force Magazine*, May 1988, pp.194–202; Frank Carlucci, Secretary of Defense, *Annual Report to the Congress FY 1989*, p.237). The one specific DoD design for basing of hardened Midgetman mobile launchers at fixed sites ready to scatter on warning would have 500 launchers located in pairs at 250 Minuteman missile silos. (See, for example, Art Hobson, "Survivability of Mobile Midgetman" in Barbara Levi, Mark Sakitt and Art Hobson, eds., *The Future of Land-Based Missiles* [New York: American Institute of Physics, 1989].)

24. In the case of long-range SLBMs, the intercept time could be reduced to less than 10 seconds if the attacking missile and decoys were fired at shorter range in a "lofted" trajectory. The radar cross sections of modern re-entry vehicles are so small and the capabilities of air-defense radars for detecting them at the required distance of about 100 kilometers so marginal that it would also.be easy to neutralize them with lightweight jammers carried by the ballistic missiles. (For a good discussion of countermeasures to ground-based ballistic-missile defense systems, see, for example, the statement by Theodore A. Postol in *Special Panel on the Strategic Defense Initiative* [a hearing before the Strategic Defense Initiative Panel of the House Committee on Armed Services, 20 April 1988] p.199 and attachments.) An early published discussion of decoys may be found in Richard L. Garwin and Hans A. Bethe, "Anti–Ballistic-Missile Systems," *Scientific American*, March 1968, pp.21–31.

Maneuvering re-entry vehicles have already been tested that could be programmed to carry out sudden changes of trajectory on their ways down through the atmosphere to their targets and render interception even more difficult. The US has tested three Maneuvering Re-entry Vehicles: the Mark 500 Evader, an alternative re-entry vehicle for the Trident I SLBM, which has a bent nose whose azimuth relative to the axis of the re-entry vehicle can be controlled by moving an internal weight; the Advanced Maneuvering Re-entry Vehicle, an alternative re-entry vehicle for the Minuteman or MX, which is maneuvered with two flaps; and the Pershing II re-entry vehicle, which is maneuvered with four flaps and has a radar in the side of its nose for recognizing the area around its target, allowing it to maneuver for a more accurate impact. See for example Matthew Bunn, "Technology of Ballistic Missile Re-entry Vehicles" in Kosta Tsipis and Penny Janeway, eds., *Review of U.S. Military Research and Development, 1984*, (McLean, Virginia: Pergamon-Brassey's International Defense Publishers, 1984), pp.67–116.

25. See, for example, "Will Strategic Submarines be Vulnerable?"

26. This suggestion has been made by Mark Sakitt in *Submarine Warfare in the Arctic: Option or Illusion* (Stanford University, Center for International Security and Arms Control, 1988), p.62.

27. See, for example, Robinson, op. cit.

28. Frank von Hippel, Harold A. Feiveson, and Christopher E. Paine, "A Low-Threshold Nuclear Test Ban," *International Security*, Fall 1987, pp.135–151; Dan Fenstermacher, "The Effects of Nuclear Test-ban Regimes on Third-generation–weapon Innovation," *Science & Global Security*, 1, 3–4 (1990), p.187.

*Harold A. Feiveson*
*Frank von Hippel*

# Calculations of Counterforce Exchanges

**F**igures 3.2 and 3.3 in chapter 3 show the calculated results of strategic counterforce exchanges—both for "bolt from the blue" attacks—when bombers, ballistic-missile submarines and land-mobile ballistic missiles on both sides are assumed to be on normal peacetime alert—and attacks during crises when both sides' strategic forces are assumed to be on "generated alert."

In arriving at these results, it has been assumed that all submarines at sea would be invulnerable to attack and that all submarines not at sea, all bombers and mobile missiles not on alert, and all rail-mobile missiles not dispersed would be destroyed.

It has also been assumed, on the basis of the barrage calculations to be described below that:

♦ All hardened truck-mobile missiles on alert (for example, the Midgetman) would survive

♦ All *dispersed* rail-mobile missiles (the US MX and the Soviet SS-24) would survive

♦ Alert bombers based away from coasts would survive, but US bombers based on the west and east coasts would be destroyed by a small number of ballistic missiles launched from off-shore submarines before they had been able to travel far from their bases. One third of all US bombers (and we assume a corresponding share of US alert bombers) are located at coastal bases (as they were in 1989). It is assumed that this situation would be rectified for the US finite-deterrence force by basing all bombers away from the coasts.

## DEPLOYMENT AND ALERT RATES

For forces at 1989 levels, it is assumed that, under a generated alert, the US would increase its fraction of ballistic-missile submarines at sea from 60 percent (peacetime rate) to 75 percent and its fraction of bombers on alert from 30 percent to 75 percent. The corresponding increases for the Soviet Union are assumed to be from 20 percent to 50 percent for the at-sea rates and zero to 50 percent for the fraction of bombers on alert. As shown in table 3.A.1, both the US and USSR are assumed to maintain higher alert rates for their strategic forces after reductions. For the finite-deterrence force we assume, for simplicity, that US and Soviet alert rates are the same and will be unchanged from peacetime to generated-alert conditions.

With respect to mobile missiles, it is assumed in all cases that 80 percent of both US and Soviet truck-mobile missiles would be on continuous alert. It is also assumed that the percentage of US and Soviet rail-mobile missiles dispersed would increase from zero in peacetime to 80 percent in a crisis.

Since it is assumed that ICBMs in silos would not be launched before the arrival of the attacking warheads, their survival rate would depend upon the hardness of the silos and the number, accuracy, and yields of the attacking warheads. It has been assumed that, in a first strike, each fixed silo would be targeted by two warheads and that a silo so attacked would have only a 20 percent probability of surviving. In all cases except that of a US first strike with 1989 forces, we assume that the attacks on silos will be made with ICBM warheads exclusively. For the one excepted case, we assume that half of the warheads used against the fixed silos of non-MIRVed missiles will be SLBM warheads. It has also been assumed that, regardless of which side attacked first, each side would expend 200 ICBM warheads attacking the other's bomber and submarine bases, garrisoned mobile missiles, and command-and-communications facilities.

**Table 3.A.1:** Assumed alert and dispersal rates for strategic forces in alternative postures *percent*

|  | 1989 | | START | | Finite deterrence | |
|---|---|---|---|---|---|---|
|  | peacetime | crisis | peacetime | crisis | peacetime | crisis |
| *Ballistic-missile submarines at sea* | | | | | | |
| US | 60 | 75 | 65 | 80 | 50 | 50 |
| USSR | 20 | 50 | 33 | 66 | 50 | 50 |
| *Bombers and tankers on alert* | | | | | | |
| US | 30 | 75 | 30 | 75 | 50 | 50 |
| USSR | 0 | 50 | 20 | 60 | 50 | 50 |
| *Truck-mobile ICBMs on alert* | | | | | | |
| US *Midgetman* | na | na | 80 | 80 | 80 | 80 |
| USSR *SS-25* | 20 | 80 | 50 | 80 | 80 | 80 |
| *Rail-mobile ICBMs dispersed* | | | | | | |
| US *MX* | na | na | 0 | 80 | na | na |
| USSR *SS-24* | 0 | 80 | 0 | 80 | na | na |

na = not applicable

Source: 1989 alert rates for bombers and at-sea rates for ballistic-missile submarines from "Strategic Nuclear Forces of the United States and the Soviet Union" (Washington DC: Arms Control Association factsheet, January 1989). All other numbers are authors' assumptions.

## BARRAGE-ATTACK CALCULATIONS

In principle, the escape routes for the strategic bombers and their associated refueling aircraft ("tankers") and the deployment areas of mobile missiles could be "barraged" by large numbers of nuclear explosions. The feasibility of such a barrage would depend upon the areas over which the aircraft or missile launchers had scattered by the time the attacking missiles arrived and the "hardness" or resistance to blast and other nuclear effects of the aircraft and missile carriers.

We have assumed in the barrage calculations that the alert bombers and mobile missiles would begin to scatter 8 minutes after the detection of the launch of the attacking ballistic missiles and that:

◆ The modern bombers (B-1, B-2, and Blackjack) would scatter in all directions surrounding the air base and they would reach a constant radial airspeed of 12.5 kilometers per minute 3 minutes after brake release. We assume that a peak blast overpressure of 0.2 atmospheres would be required to destroy the bombers. The older (unhardened) bombers and tankers would similarly scatter in all directions. These aircraft would reach a constant radial airspeed of 10 kilometers per minute 3 minutes after brake release and would be vulnerable to a peak blast overpressure of 0.1 atmospheres.[1]

◆ Because of constraints imposed by the terrain and the existing road system, we assume that the average truck-carried mobile missile could scatter only into an area equivalent to a semicircle around its base point. Average radial speed could be as high as 1 kilometer per minute. The hardened mobile launchers would be able to survive up to 2.0 atmospheres peak blast overpressure.[2]

It is assumed that the weapons attacking the aircraft flyout routes are exploded at an altitude of three kilometers. Because of reflection of the blast wave from the ground, the lethal volume corresponding to a given blast overpressure has the approximate shape of a cylinder with a hemisphere of equal radius on top with the radii being approximately the horizontal radius that would be calculated for a free-air explosion (2.2 and 3.8 kilometers respectively for a 0.2-atmosphere overpressure from a 100-kiloton and 500-kiloton explosion respectively).[3]

The lethal areas of destruction of a 100-kiloton weapon against the modern [older] bombers and hard mobile-missile carriers would be about 15 [35] and 2.5 square kilometers respectively; for a 500-kiloton weapon, these lethal areas would be about 44 [100] and 7 square kilometers respectively.

The approximate number of 100-kiloton warheads required to blanket the scatter area as a function of time ($t$ in minutes) after detection of the attacking missiles can then be expressed

for hardened bombers:     $32(t - 9.5)^2$ for $t > 11$
                          $0.9(t - 9.5)^4$ for $11 > t > 8$

for mobile missiles:     $0.6(t - 8)^2$

**Table 3.A.2:** Susceptibility of various types of aircraft and mobile-missile bases to barrage attack

| Base type | Scatter area* $km^2$ | Hardness atmospheres | 100-kiloton area of kill $km^2$ | 100-kiloton warheads required for barrage* |
|---|---|---|---|---|
| Refueling aircraft or older bomber[†] | $315(t - 9.5)^2$ | 0.1 | 35 | $9(t - 9.5)^2$ |
| Modern bomber[‡] | $490(t - 9.5)^2$ | 0.2 | 15 | $32(t - 9.5)^2$ |
| Missile carrier soft[§] | $1.5(t - 8)^2$ | 0.3 | 30 | $0.05(t - 8)^2$ |
| hard[□] | $1.5(t - 8)^2$ | 2.0 | 2.5 | $0.6(t - 8)^2$ |

\* $t$ is time in units of minutes after the attacking missiles are detected. It is assumed that the aircraft or missile carriers begin to scatter when $t = 8$ minutes. The formulas presented in this column assume $t > 11$ minutes

† Refueling aircraft, US B-52 or Soviet Bear

‡ US B-1 or B-2 or Soviet Blackjack bomber

§ Rail-mobile US MX or Soviet SS-24 or Soviet truck-mobile SS-25

□ US Midgetman

For 500-kiloton warheads, these numbers would be one-third as large. Table 3.A.2 summarizes these results and the corresponding results for older bombers and refueling aircraft and for soft mobile-missile launchers.

In a barrage attack, there would obviously be a premium on short warning times. The quickest-arriving warheads would be from quiet ballistic-missile submarines lying 1,000 kilometers or so off the coasts of the US or USSR. For bases near the coasts, the flight time of these warheads might be as short as 10 minutes.[4] Existing or likely locations of Soviet and US mobile-missile bases and two-thirds of US bomber bases (and we assume all bomber bases in our finite-deterrence force) are located away from the ocean coasts, however, where distances from the closest potential SLBM launching areas would be 2,000-3,000 kilometers away. At these distances, ballistic-missile flight times on minimum-energy trajectories would be more like 15–18 minutes.

## NOTES AND REFERENCES

1. We would like to thank David Ochmanek of RAND for a discussion of these assumptions. See also Alton H. Quanbeck and Archie L. Wood, *Modernizing the Strategic Bomber Force: Why and How* (Washington DC: The Brookings Institution, 1976).

2. See Art Hobson, "Evaluation: Land-mobile Midgetman," in Barbara G. Levi, Mark Sakitt, and Art Hobson, eds., *The Future of Land-Based Strategic Missiles* (New York: American Institute of Physics, 1989), p.51.

3. See Samuel Glasstone and Philip J. Dolan, *The Effects of Nuclear Explosions*, 3rd edition, (Washington DC: US Departments of Defense and Energy, 1977), figure 3.72 and associated altitude corrections. We make the pessimistic approximation that any aircraft within the horizontal destruction radius of an explosion is destroyed, no matter what its altitude. For the barrage attack against mobile-missile launchers, we assume that the explosion occurs at a height that maximizes the area covered by overpressures exceeding the lethal level for the launcher. See ibid., figure 3.73b and associated scaling factors.

4. For ballistic-missile flight times, see, for example, Harold A. Feiveson and Frank von Hippel, "The Freeze and the Counterforce Race," *Physics Today*, January 1983, p.36.

# Verifying Reductions of Nuclear Warheads

Thus far, nuclear arms control has imposed controls on nuclear warheads only indirectly through controls on missile launchers, missiles, and bombers. This limited coverage has not been a major issue in an era of huge excesses of nuclear weapons. But it is likely to be insufficient for monitoring agreements that make deep reductions in the strategic arsenals.

Such agreements would be greatly strengthened by a comprehensive system of controls over the availability of nuclear warheads. These controls would also make it possible to extend nuclear-arms control to limit the nuclear capabilities of "dual capable" systems such as fighter-bombers and artillery.

Controls over nuclear warheads would also provide a method for dealing with concerns about the possibility of a "breakout," where one side suddenly increased the numbers of warheads carried by missiles or bombers, or converted conventional cruise missiles into nuclear-armed ones.

A logical first step towards a nuclear-warhead–control system would be to verifiably cut off the production of fissile materials for weapons. This would require capabilities both to safeguard fissile material being used for nonweapon purposes and to detect significant clandestine fissile-material production activities.

Systems to safeguard civilian fissile material have already been established to verify the compliance of non–nuclear-weapon states with their commitments under the Non-Proliferation Treaty (NPT). Civilian fissile materials in those countries are safeguarded and tracked using seals, audits, and various other arrangements overseen by inspectors from the International Atomic Energy Agency (IAEA). Similar arrangements could be applied to civilian nuclear activities in the nuclear-weapon states and also to two military nuclear activities that have not previously been dealt with by the IAEA safeguards system: the production of naval propulsion-reactor fuel and the production of tritium for warheads.

The task of verifying that the other side was not producing weapon-useable fissile materials on a significant scale at clandestine facilities would of necessity be left to national intelligence capabilities, supplemented by challenge inspections.

Once the flow of new fissile material into the nuclear-weapon establishments had been halted, the amounts of weapon-useable fissile material available to those establishments could then be reduced by verifiable transfers to safeguarded nonweapon uses.

One way to accomplish such transfers would be as part of the verified dismantlement of nuclear warheads. These warheads would be dismantled by technicians of the owning government in a facility whose perimeter was monitored by the other government's inspectors. The recovered fissile material would then be transferred to safeguarded nonweapon facilities for use as reactor fuel or disposal. Between dismantlement campaigns, the inside of the dismantlement facility could be checked to verify that the warheads had indeed been destroyed and that all their fissile material had been placed under safeguards.

The only difficult part of these arrangements is to design the methods by which the inspectors monitoring the delivery of the warheads to the dismantlement facility would be able verify that they were indeed intact warheads of the agreed types. As is discussed in chapter 5, there are various plausible ways to "fingerprint" warheads without revealing sensitive design information but the detailed elaboration of a mutually satisfactory set of arrangements will require inputs from a broad range of experts and officials from each side.

As reductions proceeded, uncertainties in estimates of the initial sizes of the stockpiles of unsafeguarded fissile materials might loom larger. In chapter 4 we therefore propose that the sizes of these initial stockpiles be declared and that there be an exchange of historical records concerning the production and disposition of fissile materials by each country, protecting only details that are considered sensitive, such as perhaps the amounts of fissile material in individual warheads and in individual naval-propulsion reactors. Related records, historical intelligence data, and various types of physical data obtainable from on-site measurements could be used to check the authenticity of the production and disposal records.

Of course, the objects of ultimate concern are nuclear warheads and not fissile materials, and the amount of fissile materials in warheads can vary greatly. The US and USSR should therefore also declare numbers of warheads of each type in their nuclear arsenals and their locations. Since the two sides have already agreed to exchange information about the locations of all treaty-limited strategic-missile launchers, bombers, missiles, and deployed ballistic-missile warheads under the START treaty, the additional information exchange would not create a significant new vulnerability for strategic weapons and would extend the arms-control regime to cover tactical nuclear weapons as well. As with the verification of the numbers and locations of missile launchers and missiles under the INF treaty, declarations of warhead locations could be verified with spot checks with whatever special arrangements are required to safeguard sensitive information at the selected sites.

If it would facilitate progress, warhead stockpile declarations could initially be limited to warheads for delivery systems that have been limited by agreement. However, the full value of an exchange of information on warheads would only be realized if the exchange covered *all* warheads including warheads for dual-purpose weapon-delivery systems. The detection by any means of a nuclear warhead entering or leaving a nondeclared site would then indicate a violation of the agreement.

Given exchanges of information on both fissile material and complete warhead stockpiles, an obvious cross-check would become available if it were possible to measure the average amount of fissile material per warhead on each side. This could be done through the warhead dismantlement arrangements discussed above. If it were found that the amount of fissile material outside warheads exceeded that required for a working inventory in the warhead production complex, arrangements could be made to have the excess placed under safeguards pending negotiated reductions.

The uncertainties in the above procedures should not be minimized. Even if they were carried out in good faith, irreducible measurement errors, losses, etc., would probably make it impossible to rule out the existence of a clandestine stockpile of fissile material or warheads equal to perhaps 10 percent of the cumulative historical production. Controls on fissile material and warheads cannot therefore be represented as so effective as to make controls on delivery vehicles unnecessary. However, the combination of the two systems of control would be much stronger than either alone.

Finally, we discuss in this section the problem of verifying limits on the numbers of warheads deployed on multiple-warhead missiles. Under the SALT II agreement, this problem was finessed by agreeing on "counting rules" that generally credited every missile of a particular type with the maximum number of warheads with which it had been tested. In the 10 December 1987 Joint Summit Statement, however, it was agreed that, under a START agreement, ballistic missiles may be counted as carrying fewer than this maximum number of warheads. It was also agreed that

> procedures will be developed that enable verification of the number of warheads on deployed ballistic missiles of each specific type.

Chapter 6 describes a range of possible procedures that could be used for this purpose. The simplest—and most likely—would be simply to remove the nose cone of a random deployed missile to make the warheads visible for direct counting—either at the deployment site or at a special facility. If such exposure were found, for any reason, to be unacceptable, however, the number of warheads mounted on a missile could also be determined without removal of the shroud by using radiographic or induced-fission techniques.

# Chapter 4

*Frank von Hippel*

# Warhead and Fissile-material Declarations

U ntil recently, arms control agreements were limited by the fact that the only available verification capabilities were "national technical means," which involved instruments in space or beyond national borders. As a result, the SALT II treaty constrained only the construction of large missile silos, ballistic-missile submarines and long-range bombers—and limited the flight testing of long-range ballistic missiles. Recently, however, on-site verification has been accepted, making it possible in the INF treaty to extend controls to small mobile missiles and their launchers. And the START negotiators have worked out arrangements for verifying limits on the numbers of nuclear warheads deployed on ballistic missiles.

However, in a deep-cuts regime, the possible existence of thousands of undeployed warheads would create the basis for concerns about the possibility of a "breakout" from the limitation agreements. For example, extra nuclear warheads might be rapidly installed on strategic missiles and bombers; or conventional cruise missiles, fighter-bombers, etc., might be rapidly converted into nuclear delivery vehicles. If nuclear arms control is to deal with these problems, there appears to be no alternative to agreements directly limiting nuclear warheads.

This chapter therefore outlines a comprehensive system of verifiable limits on nuclear warheads. We discuss in some detail the verifiability of a halt in the production of fissile materials for nuclear warheads, the verifiability of declarations of the amounts of fissile material produced for warheads prior to the production cutoff, and the establishment of a verifiable accounting system for the numbers and types of nuclear warheads possessed by each side. Arrangements for the verifiable dismantlement of agreed numbers of nuclear warheads and transfer of their fissile materials to safeguarded nonweapon purposes are discussed in chapter 5. Figure 4.1 shows the relationship of the various activities and stockpiles that will be discussed in these two chapters.

## A CUTOFF OF THE PRODUCTION OF FISSILE MATERIAL FOR WEAPONS[1]

The idea of cutting off the production of fissile material for nuclear weapons as a way to control the proliferation of nuclear weaponry dates back to the first nuclear-arms–control discussions immediately after World War II.[2] Its validity depends upon the fact that nuclear warheads cannot be made without multikilogram quantities of concentrated fissile material— either uranium enriched to typically over 90 percent in uranium-235 or artificial chain-reacting isotopes such as plutonium-239 or uranium-233.* Such materials do not occur in nature (natural uranium contains only 0.7 percent uranium-235) and are still relatively difficult to produce. The 1970 Non-Proliferation Treaty (NPT) therefore required signatory non–nuclear-weapon states to place under international safeguards all significant quantities of enriched or artificial fissile material and all facilities process-

---

* Plutonium-239 and uranium-233 are created by neutron absorption in the natural "fertile" isotopes uranium-238 and thorium-232 and subsequent transformation of neutrons in the resulting unstable nuclei to protons through radioactive decay. Both plutonium-239 and uranium-233 have significantly smaller "critical masses" than uranium-235 and are therefore useful for the production of compact fission explosives. However, plutonium has been overwhelmingly preferred to uranium-233 by both the US and USSR—probably because uranium-233 is accompanied by trace quantities of uranium-232, whose decay products are intense emitters of penetrating gamma rays.

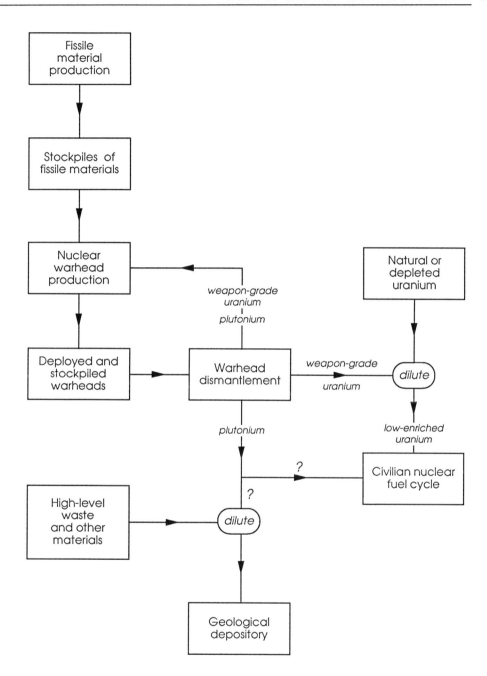

**Figure 4.1:** Warhead production and elimination activities. In the regime described here fissile-material production would be halted, and highly enriched uranium from warheads being retired without replacement would be placed under safeguards and either used as fuel for nuclear-power reactors or (in the case of plutonium) entombed in a geological depository with high-level radioactive waste.

ing or capable of producing significant quantities of such material.*

An agreement to halt the production of fissile material for nuclear weapons would require that such safeguards be applied to the weapon states' nuclear complexes as well. The safeguards could be administered by the IAEA. (All of the five acknowledged nuclear-weapon states have already offered at least some of their civilian nuclear facilities to IAEA safeguards.) The associated safeguards task, which would be comparable to the entire current IAEA safeguards load in the non–nuclear-weapon states,[3] and the personnel and equipment would probably have to be provided initially by the weapon states. However, some general IAEA supervision of such arrangements would make it easier to go beyond a bilateral US–USSR cutoff and include other weapon states later on. IAEA supervision would also dilute the idea of a nuclear-weapon "club" at a crucial time for the NPT—it comes up for extension in 1995.

A fissile production cutoff would not prevent the production of new nuclear warheads, which could be produced with the fissile material recovered from obsolete warheads. Indeed, the US has been producing most of its nuclear warheads with recycled fissile material since the 1960s. However, cutting off the supply of new fissile material for warhead production is an essential precondition to verifiable reductions.

The longer a fissile-cutoff agreement is delayed, the more difficult it will be to reconstruct past production and therefore the more difficult it will be to set low limits on the amounts of fissile material that may have been hidden away. And the larger the potential hidden stockpiles, the more difficult it will be to achieve very deep cuts in the declared arsenals.

Verification arrangements could not realistically ensure that *no* new fissile material was flowing into the nuclear-weapon complexes, but they could impose real restraints on the activities of both sides. We believe that ensuring that the annual diversion or clandestine production of fissile material for weapons was less than 1 percent of current inventories would be a worthwhile constraint.

---

* The NPT contains a loophole, however, in that fissile material may be exempted from safeguards if it is used for nonexplosive military purposes—e.g. to fuel a naval propulsion reactor. Thus far, no nonweapon state that is an NPT signatory has exempted fissile material from safeguards on this basis.

The US and USSR have each produced about 100,000 kilograms of weapon-grade plutonium for nuclear weapons, and the US has produced about 500,000 kilograms of weapon-grade uranium for nuclear weapons.[4][*] (Information that would make possible an estimate of the quantity of highly enriched uranium that has been produced for weapons by the USSR is not publicly available.)

One percent of these quantities is 1,000 kilograms of plutonium and 5,000 kilograms of weapon-grade uranium. These are huge quantities of weapon-grade fissile materials—enough to provide a non–nuclear-weapon state with a nuclear arsenal containing hundreds of warheads; however, the clandestine production or diversion of this amount of fissile material by the US or USSR would have a negligible impact on their current nuclear balance. If undetected for many years, the cumulative clandestine production from sub-threshold violations would mount. However, the likelihood of detection would also grow, making prolonged efforts increasingly implausible.

Having set this context, the problem of verifying a fissile cutoff is discussed below in two parts: safeguards at declared facilities and detection of any clandestine production facilities.

## Safeguards at Declared Facilities

In the years immediately before the consummation of the Non-Proliferation Treaty in 1970, the idea of an agreement by the US and USSR to end their own production of fissile material for nuclear weapons in exchange for an agreement by the nonweapon states to abstain from such production was a subject of intense discussion.[5] The US favored such an agreement and suggested in 1969 that the same IAEA safeguards that were to be used to monitor compliance with the NPT in nonweapon states could be applied to

---

* "Weapon grade" uranium is defined in the US as approximately 94 percent uranium-235, and "weapon grade" plutonium is defined as about 94 percent plutonium-239 (Thomas B. Cochran, William M. Arkin, and Milton M. Hoenig, *Nuclear Weapons Databook Volume 1: US Nuclear Forces and Capabilities* [New York: Ballinger, 1984], pp.23–24). The Soviet definitions are almost identical (interview with Evgeny I. Mikerin, head, Main Department of Manufacturing and Technology, USSR State Committee for the Utilization of Atomic Energy, 7 July 1989).

verify that no nuclear-weapon materials from US and Soviet civilian nuclear facilities had been diverted.[6]

Under such an agreement, the two countries would each declare the locations of all the plants at which they had produced, processed, or stored significant quantities of fissile materials other than uranium-235 in natural $U_3O_8$. Such plants include: most civilian nuclear-reactor fuel-cycle facilities; shut-down facilities previously used to produce fissile material for nuclear weapons; and the fuel-cycle facilities for naval propulsion reactors and tritium-production reactors. Below, we discuss the safeguards task at each of these classes of facility.

*Civilian Reactor Fuel-cycle Facilities*

The nuclear-reactor fuel cycle is shown in figure 4.2. The design objectives of the IAEA safeguards are to detect in a timely manner the diversion of a "significant quantity" of fissile material, defined to be more than 8 kilograms of plutonium or uranium-233, or 25 kilograms of uranium-235 in uranium enriched to 20 percent or more.[7] These quantities are less than one ten thousandth of the amounts of these materials currently available to the US and Soviet nuclear establishments. Smaller diversions would therefore be totally insignificant in the context of a bilateral fissile-production cutoff agreement.[*]

It is well known, however, that the IAEA goals cannot be achieved with facilities in which very large quantities of fissile materials are handled in bulk in gaseous, liquid, or powder form. For such facilities, which include large uranium-enrichment plants, fuel-fabrication plants, and fuel-reprocessing plants, accounting methods can achieve only about 1-percent (one standard deviation) accuracy.[8] With such measurement accuracy, if one sets

---

[*] In this same context, most research reactors would not have large enough inventories or fissile-material production capabilities to be a significant concern. Very few research reactors have more than 25 kilograms of highly enriched uranium-235 in their fuel. A 10-thermal-megawatt reactor could produce no more than about 3.5 kilograms of plutonium per year (i.e. about 1 gram per thermal megawatt-day). According to a 1986 IAEA survey, of 331 operating research and production reactors, only 20 percent had thermal-power ratings greater than 10 megawatts [*Nuclear Research Reactors: Status and Trends* (Vienna: International Atomic Energy Agency, 1986)]. Large power reactors have ratings of about 3,000 thermal megawatts.

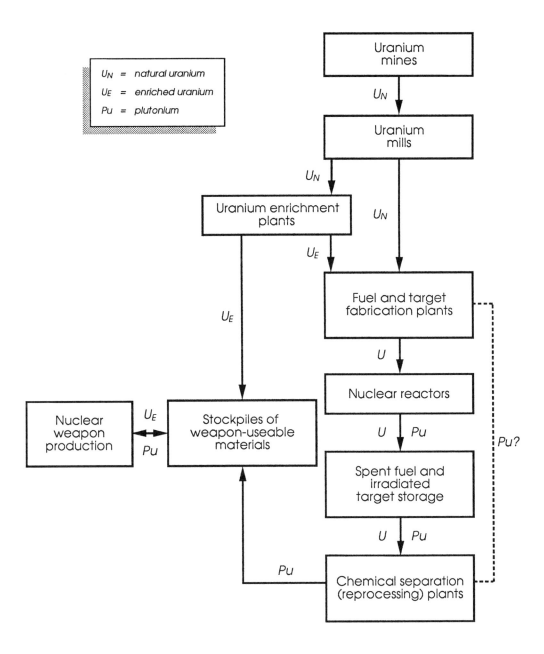

**Figure 4.2:** Flows of fissile material in a nuclear-weapon state

the false-alarm probability at 5 percent, there will be a 95-percent chance of detecting a diversion of 3.3 percent of a specific batch of fissile material passing through the a facility.

Fortunately, the US and Soviet civilian nuclear complexes are not so large that a 3-percent diversion would be significant on the scale of their already existing nuclear-weapon materials stockpiles.[9] Furthermore, almost all civilian power reactors in the US use a "once through" fuel cycle in which the uranium in the fresh fuel contains only a few percent uranium-235 and the spent fuel is not reprocessed. Weapon-grade fissile material could be obtained from this fuel cycle only by reconfiguring some of the enrichment capacity to produce highly enriched uranium or by diverting some of the spent fuel to a clandestine reprocessing plant to recover its plutonium—actions that would be difficult to conceal.*

Concern is sometimes expressed that it will be particularly difficult to effectively safeguard the fuel of Soviet graphite-moderated production reactors, because they are refueled continuously rather than once a year or so, as are the more common light-water–moderated power reactors. However, effective IAEA safeguards have been designed for Canadian heavy-water–moderated power reactors, which are also continuously refueled.

### Facilities Previously Dedicated to the Production of Fissile Materials for Weapons

Under a fissile-material–cutoff agreement, each signatory would have to

---

* The Soviet Union has reprocessed a significant amount of power-reactor fuel but does not currently intend to recycle the recovered plutonium. About 20,000 kilograms of plutonium have been recovered from the spent fuel of Soviet 440-megawatt-electric power reactors (Mark Hibbs and Ann MacLaughlan, "Soviet Union Postpones Completion of Siberian Reprocessing Plant," *Nuclear Fuel*, 16 October 1989, p.1). Both the US and the USSR use highly enriched uranium in research and certain prototype reactors. They also use plutonium in facilities associated with their plutonium-breeder reactor development programs. However, the total quantities of weapon-grade fissile material involved are a few thousand kilograms, and almost total diversion would be required for the amount of diverted material to add more than one percent to the amounts already in the arsenals. The amount of highly enriched (more than 70 percent uranium-235) uranium fuel supplied by the US for research reactors in 1982 was about 1,000 kilograms—about one half for US and one half for foreign research reactors. (J.E. Matos, *RERTR [Reduced Enrichment and Test Reactor] Program Reactor Summary—September 1982* [Argonne, Illinois: Argonne National Laboratory, 22 September 1982], tables 1 and 2.)

identify facilities, such as dedicated plutonium-production reactors and their associated reprocessing plants, to be shut down and perhaps dismantled. It would be relatively easy to verify that these facilities had indeed been shut down—perhaps even remotely, using imaging satellites sensitive to thermal infrared radiation. For extra assurance, however, facilities that were not dismantled could be sealed and checked periodically by inspectors.[10]

*Tritium-production Reactors*

Tritium is an artificial heavy isotope of hydrogen ($H^3$) that is used in most modern nuclear weapons to "boost" the efficiency of the chain reaction in the "primary" fission explosive.* Like plutonium, it is produced in reactors.†

Fresh tritium will have to be produced for weapons unless the stockpiles are reduced more rapidly than its 5.6-percent annual radioactive decay rate. Tritium's halflife is 12.3 years.

If the amount of tritium in the US nuclear-weapon arsenal is about 100 kilograms, the tritium production required to maintain this stockpile at a constant level would be about 5.6 kilograms per year, which could be generated by production reactors operating at an average total thermal power of about 1 gigawatt. Adequate quantities of tritium to maintain the current US and Soviet nuclear arsenals could therefore be produced by a small number of production reactors.[11] It should be relatively easy to establish arrangements to verify that the reactors were not producing plutonium for weapons.[12] If arms reductions proceeded rapidly enough to stay ahead of the tritium-decay curve, it would also be possible to verify that such reactors had been shut down.

---

* When the fissioning material in the fission explosion reaches a temperature of about 100 million K, its heat can ignite the fusion reaction between the tritium (T) and the natural heavy isotope of hydrogen, deuterium (D): $T + D \rightarrow$ helium-4 + neutron. The released neutrons cause a burst of additional fissions, which "boost" the fission yield of the explosive.

† Tritium (T) is produced in a reactor as follows: neutron + lithium-6 $\rightarrow$ T + helium-4.

*Fuel-cycle Facilities for Naval Nuclear Reactors*

In addition to their nuclear-weapon–related and civilian nuclear activities, the US and USSR each operate more than 100 naval propulsion reactors, for which the total fresh fuel requirements are about 5,000 kilograms of uranium-235 a year.[13] For the US, this uranium-235 is provided in very highly enriched uranium (97.3 percent uranium-235). In the case of Soviet naval-propulsion reactors, it has been stated that the fuel enrichment is only about 10 percent.[14]

The IAEA has not been responsible for safeguarding the fuel cycles of naval reactors, although now-abandoned plans by Canada, an NPT signatory, to build nuclear submarines recently raised this possibility.[15] New arrangements would therefore have to be devised to ensure that no significant fraction of the enriched uranium ostensibly produced to fuel naval reactors was being diverted to weapons use. One obvious method would be to require that an appropriate amount of uranium from spent fuel be returned to safeguarded uses for each annual allotment of fresh fuel released into the naval nuclear-fuel cycle.* Although fuel may spend more than a decade in a propulsion reactor, some naval-reactor fuel is discharged every year, and the exchanges could therefore begin immediately.[†]

## Detection of Clandestine Production Facilities

When, in 1969, the US proposed that a fissile-material–production cutoff be verified by IAEA-type safeguards on declared nuclear facilities, it was

---

* In the US, naval-reactor fuel is reprocessed to recover the unfissioned uranium, which is still highly enriched and therefore has considerable economic value. However, physical evidence for the amount of uranium-235 that has fissioned in the fuel could be obtained even after chemical reprocessing on the basis of the amounts of the artificial isotope, uranium-236, present. When a slow neutron is absorbed by uranium-235, fission results only about 80 percent of the time. The remaining 20 percent of the neutron captures result in uranium-236. The presence of each uranium-236 atom therefore indicates that about five uranium-235 atoms have fissioned.

† In the future, naval propulsion reactors could be fueled using the uranium-235 in highly enriched uranium recovered from dismantled weapons. However, if the US continued to use 97.3-percent enriched uranium in its naval reactors, a small amount of extra enrichment work would be required to raise the approximately 94-percent enrichment of weapon-grade uranium to this level.

implicitly stating that it was confident that it could detect any significant clandestine production facility using US national intelligence capabilities.

The clandestine production of over 1,000 kilograms of plutonium in a year would require distinctive facilities with total capital costs of billions of dollars.[16] Large numbers of a violating state's citizens would therefore be involved in the construction of such facilities, and information about their existence would probably leak out. Uranium diffusion-enrichment plants (the type used in the US) capable of producing 5,000 kilograms of weapon-grade uranium would also be large and very distinctive. However, centrifuge-enrichment plants (used in the USSR and Western Europe) are less readily identifiable, and laser-enrichment plants, which are under development, could be installed in quite small buildings.[17] Such facilities are still quite expensive, however, and the violating government would have to hide the diversion of a great deal of natural uranium to feed the clandestine facilities (about one million kilograms of natural uranium for each 5,000 kilograms of weapon-grade uranium or 1,000 kilograms of plutonium produced).*

Using aircraft for verification tasks, as recently discussed in the "open skies" negotiations, could potentially make the concealment of any clandestine production facilities far more difficult, because sources of gamma radiation, radioactive gases, electromagnetic emanations from pulsed high-power lasers, etc., would all become more easily detectable.

## LIMITATIONS ON STOCKPILES OF UNSAFEGUARDED FISSILE MATERIALS

As described in the next chapter, a fissile cutoff could be followed by reductions in the amounts of fissile material in the nuclear arsenals

---

* Natural uranium is 0.7 percent uranium-235, of which typically 0.2 percent remains in the depleted "tails" from uranium enrichment. In a plutonium-production reactor, about one plutonium-239 atom is produced per uranium-235 atom fissioned and therefore, in principle, no more natural uranium would be required to produce plutonium than would be needed to produce the same amount of highly enriched uranium. However, if the clandestine production reactor were fueled with natural uranium, its fuel would be reprocessed after a low (about 0.1 percent) "burnup" of uranium-235 in order to minimize production of plutonium-240 through neutron capture on plutonium-239.

through verified warhead dismantlement. At some point however, concerns could arise about possibly large disparities in the residual amounts of fissile materials in the US and Soviet stockpiles. Alleviating such concerns would require verified declarations of past production and consumption of fissile materials.

Since fissile material can easily be hidden, the primary way to verify declarations of past production and consumption would be through what we might call "nuclear archeology."

The discipline of nuclear history must already be rather highly developed within the US and Soviet national intelligence communities. Each must have been tracking from almost the beginning of the nuclear era the other's fissile-material production and utilization activities, starting with uranium mining.

In the case of plutonium production, each country has certainly had a very good idea of the number of production reactors the other has had in operation and has used the available indicators to estimate the power of these reactors. Also, the US and USSR would have been able to make direct estimates of the total fission in the reprocessed fuel from each other's reactors from measurements of the amounts of the long-lived fission product, krypton-85, released to the global atmosphere as a result of reprocessing by subtraction of its own releases and the reported and estimated releases by other countries.[18]

In the case of the production of highly enriched uranium, both countries must have worked hard to estimate the amounts of natural uranium going into and depleted uranium coming out of the other country's facilities,* along with such indicators as their electrical-power consumption—which

---

\* Of the approximately 20,000 tonnes of natural-uranium "feed" required to produce 100 tonnes of weapon-grade uranium, 99.5 percent would end up in depleted-uranium "tails." Most of this depleted uranium would probably be stored at the production facility because there have been few large-scale uses for depleted uranium. (In recent years, however, the high density of depleted uranium has led to its use in armor-piercing shells and some other specialized uses.)

would be proportional to the amount of separative work going on inside.* The uncertainties on such notional estimates of the other side's past production of weapon-grade fissile material might be as low as 20 percent.[†] However, it would be desirable to improve confidence in and the accuracy of these estimates, if possible.

Perhaps the most valuable contribution to a cooperative approach to the reconstruction of past production activities would be the sharing of production records. On the basis of the published records concerning past US production activities, for example, it is already possible to estimate the amounts of plutonium and highly enriched uranium that the US has produced for weapons to an accuracy of about 10 percent.[‡]

---

* The US (and, in the past, the USSR as well [Mikerin]) has done all of its enrichment with gaseous-diffusion technology. This technology requires huge amounts of energy. US plants would currently require about 6 gigawatt-years of electric power to produce 100,000 kilograms of weapon-grade uranium (based on 2.2 megawatt-hours per kg-SWU of enrichment work in 1982 [US Department of Energy, *1982 Uranium Enrichment, Annual Report*, p.28]). It is therefore necessary to place gaseous-diffusion plants in locations where large amounts of electrical power are available—and, even there, the plants would consume a large fraction of the power on the regional transmission grid. Knowledge of the total amount of power generated by regional power plants and rough estimates of the size of other industrial and urban demands for electrical power could therefore make possible a rough estimate of the electricity consumption of a large gaseous-diffusion uranium-enrichment plant. Conversion of this estimate into an estimate of actual uranium-enrichment work would require an additional estimate of the relative efficiencies of US and Soviet enrichment plants.

† Such accuracy should be possible because past US and Soviet production of fissile material for weapon purposes has been a very significant fraction of demands of fissile material for all purposes. For example, the 500 tonnes of uranium-235 estimated to be in US weapon-uranium would, if fissioned, release about 1,400 gigawatt-years of thermal energy. By comparison, about 1,200 gigawatt-years of thermal energy had been released by US nuclear power reactors through 1985 (*Statistical Abstract of the United States* [Washington DC: US Government Printing Office, annual]). In terms of enrichment work, the relative importance of the weapon-uranium would be even greater, since the production of highly enriched uranium for weapons consumes about twice as much enrichment work per uranium-235 atom as the production of the approximately 3-percent enriched uranium used to fuel typical power reactors.

‡ To first order, these estimates are based on the records that the US government has made public of: i) the annual separative work done by US uranium-enrichment plants before the US ended production of highly enriched uranium in 1964, and ii) the annual heat output of US production reactors. It was then necessary to estimate what fraction of the pre-1964 enrichment work went to the production of reactor

For such records to be useful for verification purposes, however, it must be possible to authenticate them. This will require consistency checks against related records, historical national intelligence data, and direct physical measurements on-site.

Related records would be those of individual operations in the nuclear-materials production complex. They would include records of: shipments of natural uranium from uranium mills and shipments of natural and enriched uranium between conversion, uranium-enrichment, reactor-fuel fabrication facilities, and reactors; power production on the transmission grids supplying uranium-enrichment plants; receipts of highly enriched uranium at naval fuel fabrication facilities; and power generation by civilian power-reactors. Not all of these records might have been preserved but hopefully enough would be available to provide a significant check on the declarations.

Historical satellite photographs could give information about the dates at which production facilities were brought into operation and later shut down, the growth in the sizes of uranium-mill tailings piles, the capacities of high-level waste storage tanks at reprocessing plants, and the accumulation of canisters of depleted uranium if they had been stored outside uranium-enrichment facilities. Infrared photos would indicate periods when production reactors and enrichment plants were actually in operation and the temperature differences between cooling-water intakes and discharges from production reactors.

Physical measurements could include: the capacities of the production-reactor cooling-water pumps; volumes of tailing piles at uranium mills and their average concentrations of uranium-decay products (which would indicate the original grade of the ore); gas pressures and the areas and

---

fuel and what fraction of the neutrons produced in the production reactors were used to produce isotopes other than plutonium—especially tritium. Finally, estimates had to be made of losses of fissile materials in processing and consumption in nuclear-weapon tests. See Frank von Hippel, David H. Albright, and Barbara G. Levi, *Quantities of Fissile Materials in US and Soviet Nuclear Weapons Arsenals* (Princeton University, PU/CEES Report 168, July 1986); and Thomas B. Cochran, William M. Arkin, Robert S. Norris, and Milton M. Hoenig, *Nuclear Weapons Databook Volume 2: US Nuclear Warhead Production* [New York: Ballinger, 1987] pp.75, 191.

permeabilities of the barriers in gaseous-diffusion plants; inventories of depleted uranium stocks, their enrichment, and their age, as indicated by their accumulation of uranium decay products; the accumulation of activation products in the permanent structural elements of the cores of production reactors (see appendix 4.A); volumes of high-level waste stored at reprocessing plants, their concentrations of long-lived fission products, and their ages; and inventories of unreprocessed spent fuel from civilian reactors and their contents of uranium-235, plutonium, and long-lived fission products.

Of course, with time, the quality of all this evidence will deteriorate. Records will be lost, the scale of civilian nuclear activities will increase and make it more difficult to measure the evidence of older military activities, old production facilities will be dismantled, stored high-level waste will be processed and buried, and depleted uranium will be shipped away from the production sites for various uses. It is therefore important to the long-term hopes for order-of-magnitude reductions of the nuclear arsenals that declarations of past fissile production be made and verified as soon as possible.

## LIMITATIONS ON STOCKPILES OF NUCLEAR WARHEADS

### Relating Fissile-material Declarations to Warhead Inventories

Knowing the quantities of fissile material available to a nuclear-weapon production complex translates into only the roughest estimate of the number of nuclear warheads that could be produced with that fissile material. For example, many more nuclear weapons could be made from the approximately 100,000 kilograms of plutonium and 500,000 kilograms of weapon-grade uranium produced by the US for weapons than the approximately 25,000 nuclear warheads that the US nuclear arsenal is believed to contain.[19,20] Estimates of the weapon-grade fissile material available to the other side will therefore not permit an accurate estimate of the size of its weapon stockpile without actual measurements of the amounts of fissile material in its weapons.

Such measurements could be carried out if the US and USSR have each declared their weapon stockpiles and set up, on at least a pilot-plant basis, the arrangements required to verify nuclear-warhead dismantlement. The dismantlement process could be used to provide a nonintrusive assay of the fissile material recovered from a sample of each type of weapon and then the total fissile material in the declared stockpile could be estimated by multiplying the amount of fissile material in each warhead by the number of declared warheads of that type. If the quantities of fissile material in individual weapons were considered too sensitive to reveal, averages for a number of representative mixes of samples from different classes of weapons could be used instead.

Of course, not all fissile material is in weapons. There must be at least a working inventory of fissile material at the facilities at which warhead maintenance and permitted manufacturing of replacement warheads continued. Such inventories could be declared, and, if greater than required, the excess fissile material could be stored or disposed of under safeguards.

## Declaring Warhead Locations

It would not raise a grave security risk for either country if the US and USSR were to exchange declarations of the exact numbers and locations of each type of nuclear warhead in their arsenals. They have already agreed in the START negotiations to exchange information on the numbers and locations of the nuclear warheads on their strategic ballistic missiles. Since deployed strategic warheads are the ones whose survivability is of greatest concern, it should not imperil either country's security if these information exchanges were extended to their entire nuclear arsenals, including strategic warheads not on ballistic missiles and tactical nuclear warheads.

How could such declarations be verified? Perhaps the easiest method would be baseline inspections such as those agreed for checking declarations of the numbers and locations of mobile ballistic missiles—both deployed and nondeployed—under the INF and START treaties. At the time of this baseline check, it would also be useful to obtain a "fingerprint" of each type of warhead. Possible approaches to fingerprinting that do not reveal sensitive design information are discussed in chapter 5.

Warheads are, of course, small and could easily be hidden. However, certain arrangements might make a hidden stockpile troublesome in the longer term. These arrangements could include widely distributing lists of all the declared locations of warheads. Then anyone who witnessed a warhead being taken from or delivered to a facility not on the list would know that the treaty was being violated—thereby increasing the probability of this information eventually being brought to the other side's attention.

## THE RESIDUAL POSSIBILITY OF A CLANDESTINE NUCLEAR STOCKPILE

The possible effect of all of these measures would be to divide the weapon and fissile-material stockpiles of each of the superpowers into two parts: a declared arsenal and a clandestine arsenal.

The declared arsenals would hold fingerprinted nuclear warheads containing a known total amount of fissile material plus a limited working inventory of fissile material.

Unfortunately, because of the inevitable uncertainties over the amounts of weapon-grade materials originally produced and their disposition, material-accounting techniques could place only upper bounds on the sizes of any clandestine arsenals, and these upper bounds might, after order-of-magnitude reductions, be comparable to the sizes of the residual declared arsenals.

There is, therefore, the possibility of a small group of people hiding a stockpile of warheads in out-of-the-way locations—just as old-time pirates buried their treasure. However, in a world where both the US and USSR still possessed on the order of one thousand survivable warheads, such clandestine stockpiles would not provide either country with any great advantage, and would bring with it an irreducible risk of exposure and a resulting renewal of the arms race with its uncertainties and great costs.[21] And, over the years that the reduction process took up, it would become increasingly difficult to maintain clandestine warheads in useable condition without inadvertently revealing their existence.

It is important to note here that the sooner a warhead declaration, fingerprinting, and verification system is implemented, the less likely it will be that elaborate evasion schemes could precede it; the longer any clandestine stockpile would have to reveal itself; and the sooner it would begin to age.

## FIRST STEPS TOWARDS ESTABLISHING A CONTROL SYSTEM

What has been outlined here is an ambitious proposal which would include large-scale US–Soviet cooperative projects of nuclear archeology to verify their declarations of the past production and disposition of fissile materials. Several smaller-scale efforts are required to lay the foundations for such a program. The cooperative design of a nuclear-warhead–dismantlement facility and of a warhead fingerprinting system should be among the highest priorities of these projects. These efforts will require input from US and Soviet weapon designers as well as verification experts because the systems will have simultaneously to meet stringent verification standards and protect sensitive nuclear-weapon design information.

## NOTES AND REFERENCES

1. A more complete discussion of the verifiability of a fissile-material–production cutoff can be found in Frank von Hippel and Barbara G. Levi, "Controlling Nuclear Weapons at the Source: Verification of a Cutoff in the Production of Plutonium and Highly Enriched Uranium for Nuclear Weapons" in Kosta Tsipis, David W. Hafemeister, and Penny Janeway, eds., *Arms Control Verification: The Technologies That Make It Possible*, (McClean, Virginia: Pergamon-Brassey's, 1986).

2. See for example the "First Report of the United Nations Atomic Energy Commission to the Security Council [Extract], 31 December 1946" in *Documents on Disarmament, 1945–1959* (Washington DC: US Government Printing Office, 1960), pp.50–59.

3. Only two comparisons have been made of the magnitude of the US–Soviet fissile-cutoff safeguards task, with the current safeguards tasks carried out by the IAEA in nonweapon states:

+ A study done by the Swedish government in 1989 (*Costs for a World-wide Safeguards System*, background paper, 15 September 1989). This study concluded that the costs of extending IAEA safeguards to the nuclear complexes of all five nuclear-weapon states and other countries not currently parties to the NPT would approximately equal the 1989 IAEA safeguards budget of $53 million.

+ A more detailed study by Thomas E. Shea, leader of the Systems Studies Group of the IAEA Safeguards Department carried out under the auspices of the US Electric Power Research Institute (EPRI), *Verification of Arms Controls on US and Soviet Fissionable Materials* (Palo Alto, California: EPRI research project 620-650, 1984, unpublished manuscript). Shea estimated that the IAEA would have to add 360–552 additional members to its safeguards staff if it was given the task of safeguarding US and Soviet nuclear facilities under a fissile-cutoff agreement. Because the IAEA safeguards staff numbered about 500 at the time, his estimate is consistent with the Swedish estimate.

4. Frank von Hippel, David H. Albright, and Barbara G. Levi, *Quantities of Fissile Materials in US and Soviet Nuclear Weapons Arsenals* (Princeton University, PU/CEES Report 168, July 1986). Similar estimates of the US stockpiles of highly enriched uranium and plutonium for weapons have been published in Thomas B. Cochran, William M. Arkin, Robert S. Norris, and Milton M. Hoenig, *Nuclear Weapons Databook Volume 2: US Nuclear Warhead Production* (New York: Ballinger, 1987), pp.75, 191. Evgeny Mikerin was shown estimates that the US had produced about 100,000 kilograms of plutonium for weapons and said that the corresponding Soviet quantity was "a little more" (interview, 7 July 1989).

5. See, for example, the "Statement by the Indian Representative (Trivendi) to the Eighteen Nation Disarmament Committee: Nonproliferation of Nuclear Weapons, September 28, 1967" in *Documents on Disarmament, 1967* (Washington DC: US Arms Control and Disarmament Agency), pp.430–440.

6. "Statement by ACDA Deputy Director Fisher to the Eighteen Nation Disarmament Committee, April 8, 1969," *Documents on Disarmament, 1969*, pp.158–164. The Soviet response was that the US interest in a fissile production cutoff was mainly due to the "overproduction in the United States of nuclear materials for military purposes" ("Statement by Soviet Representative Roshchin at the Eighteen Nation Disarmament Committee: Prohibition of the Use of Nuclear Weapons, April 10, 1969" in *Documents on Disarmament, 1969*, pp.164–173). [In fact, there was some truth in this remark. Five years previously, President Johnson had announced unilateral cutbacks of US fissile-material production, asserting that "Even in the absence of agreement we must not stockpile arms beyond our needs...." ("State of the Union Address by President Johnson to the Congress, January 8, 1964" in *Documents on Disarmament, 1964*, p.4).]

7. *IAEA Safeguards Glossary, 1987 Edition* (Vienna: International Atomic Energy Agency, 1987), Table III.

8. A theoretical analysis of a proposed large US uranium centrifuge-enrichment plant estimated an accuracy for material-balance methods of 0.3 percent (D.M. Gordon, J.B. Sanborn, J.M. Younkin, and V.J. DeVito, "An Approach to IAEA Material-Balance Verification at the Portsmouth Gas Centrifuge Enrichment Plant," in *Proceedings of the Fifth Annual Symposium on Safeguards and Nuclear Materials Management, 19–21 April 1983*, pp.39–43). A theoretical analysis of a proposed safeguards system for a medium-sized Japanese reprocessing plant argues that an accuracy of 0.5 percent should be possible (J. Lovett, K. Ikawa, M. Tsutsumi, and T. Sawanata, "An Advanced Safeguards Approach for a Model 200 T/A Reprocessing Facility," (International Atomic Energy Agency Report #STR-140, 1983). However, these are the theoretical performances of advanced systems applied to modern plants designed with safeguards in mind.

9. A 1-gigawatt capacity pressurized-water reactor operating at an average of 70 percent capacity requires 0.11 million kg-SWU of separative work and discharges 160 kilograms of fissile plutonium annually (*Advanced Fuel Cycle and Reactor Concepts* [International Atomic Energy Agency Report #INFCE/PC/2/8, 1980]). It takes approximately 237 kg-SWU to produce one kilogram of 93.5-percent enriched

uranium (assuming 0.2 percent uranium-235 remaining in the depleted uranium). *AEC Gaseous Diffusion Plant Operations* (US Atomic Energy Commission Report #ORO-684, 1972), p.37.

For a civilian nuclear economy approximately the size that the US is expected to have in operation in the year 2000 (100 gigawatts-electric generating capacity), 3 percent of the enrichment work being used would yield annually only 1,500 kilograms of weapon-grade uranium per year—about 0.3 percent of the amount of highly enriched uranium currently available to US nuclear-weapon makers. And even if all of the spent fuel in such a nuclear economy were reprocessed and 3 percent of the recovered fissile plutonium were diverted to weapons, it would only amount annually to about 500 kilograms—about 0.5 percent of the amount of plutonium already in the US or Soviet weapons complex. The Soviet civilian nuclear generating capacity in the year 2000 is likely to be about half that of the US.

10. In 1966, the US described to the Eighteen Nation Disarmament Committee a system for monitoring shut-down production reactors. This system involved sealing metallic tapes containing cobalt-59 into the fueling channels. If the reactor were operated, fission neutrons would convert some of the cobalt-59 into 5-year– halflife cobalt-60 ("Statement by United States Expert [Jensen] to ENDC Delegates: Description of a Monitoring System for Shutdown Nuclear Reactors, August 10, 1966" in *Documents on Disarmament, 1966* [US Government Printing Office, 1967], pp.538–546).

11. The Reagan Administration proposed that two new US production reactors with thermal-power capacities of 1.4 gigawatts and 2.5 gigawatts be built but indicated that the lower-powered reactor could maintain the US tritium supply if necessary (*United States Department of Energy Nuclear Weapons Complex Modernization Report*, Report to the Congress by the President, December 1988, p.15. The powers of the proposed reactors are given in David Albright and Christopher Paine, "The Burdens of Bomb-Building: Arms Control Can Help," FAS Public Interest Report, December 1988, p.1). Evgeny Mikerin stated that it would take 2–3 production reactors to maintain the Soviet stockpile of tritium (interview, July 1989, cited above).

12. As at safeguarded power reactors, it would be necessary to verify that no uranium or thorium exposed to neutrons in the reactor could be diverted to an unsafeguarded reprocessing plant. This could be done by remote monitoring of the reactor refueling and spent-fuel storage areas. No fissile material could leave the site except when an inspector was present to tag and seal its container, and no such container could be opened except at a safeguarded site and after an inspector had checked its tags and seals.

13. About 5,000 kilograms per year for the US naval-propulsion reactors is reported in *US Nuclear Warhead Production*, p.71. The estimated total shaft horsepower of Soviet and US naval power reactors is about the same (see *Jane's Fighting Ships* [Jane's Yearbooks, biennial]).

14. Evgeny Mikerin, July 1989 interview cited above.

15. See, for example, Marie-France Desjardins and Tariq Rauf, *Opening Pandora's Box? Nuclear-Powered Submarines and the Spread of Nuclear Weapons* (Ottawa: The Canadian Centre for Arms Control and Disarmament, 1988) for a discussion of the

issues associated with safeguarding the nuclear fuel cycle of naval-propulsion reactors in non–nuclear-weapon states.

16.  See Von Hippel and Levi, pp.370–371. It would require a large (3,000 thermal megawatt) production reactor operating at two-thirds average capacity and producing one gram of plutonium per megawatt-day of fission-heat released to produce about 700 kilograms of plutonium per year. Because of the low exposures of uranium targets used to produce weapon-grade plutonium, a large reprocessing plant (about 1,000 tonnes of uranium in fuel per year) would also be required. Production of tritium or plutonium with neutrons produced by high-current proton accelerators has been proposed [*Accelerator Production of Tritium*, Executive Report (Brookhaven and Los Alamos National Laboratories; BNL/NPB-88-143, March 1989)]. However, such accelerators, including their power sources, would cost an amount comparable to the equivalent production reactors and have not yet been developed.

17.  See Richard Kokoski, "Laser Isotope Separation: Technological Developments and Political Implications," *World Armaments and Disarmament, 1990* (Oxford: Oxford University Press, 1990).

18.  See the discussions of this method of estimating plutonium production in "Controlling Nuclear Weapons at the Source" and in *Quantities of Fissile Materials in the US and Soviet Nuclear Weapons Arsenals*. Some krypton-85 has been recovered at reprocessing plants but probably not a large enough percentage to significantly affect these estimates.

19.  "U.S., Soviet Nuclear Weapons Stockpiles, 1945–1989: Number of Weapons," *Bulletin of the Atomic Scientists*, November 1989, p.53.

20.  The Nagasaki bomb contained approximately 6 kilograms of plutonium (Leslie R. Groves, "Memorandum for the Secretary of War" 18 July 1945, reprinted in Martin J. Sherwin, *A World Destroyed* [Alfred A. Knopf, 1975], appendix P). This is just under one critical mass for a solid sphere of plutonium inside a thick neutron reflector (*Critical Dimensions of Systems Containing $^{235}U$, $^{239}Pu$, and $^{233}U$* [Los Alamos Laboratory Report #LA-10860-MS, 1986], p.100). Subsequent design improvements made it possible to make a fission explosive with a fraction of a critical mass (Hans A. Bethe, "Comments on the History of the H-Bomb" *Los Alamos Science*, Fall 1982, pp.44–45). Critical masses for weapon-grade uranium are about four times larger than for plutonium.

21.  See, for example, the discussion in chapter 2.

# Estimating Plutonium Production from Long-lived Radionuclides in Permanent Structural Components of Production Reactor Cores

The United States and the Soviet Union face critical decisions about the future of plutonium production for nuclear weapons. The reactors at Hanford and Savannah River, with which the US has produced plutonium and tritium over the last 45 years, are now shut down because of age or safety concerns. (In the case of three of the five Savannah River production reactors, the shutdown is expected to be only temporary.) The Soviet plutonium production complex is also reported to have serious safety and environmental problems.

Both countries could eliminate the economic burden of rebuilding their production complexes by agreeing to ban the production of plutonium for weapons. Such an agreement could also provide important national-security benefits by reinforcing the Non-Proliferation Treaty and by diminishing the ability of both nations to break out of nuclear-arms reduction agreements at a later time—especially if the plutonium in the warheads eliminated by the arms-reduction agreements is put under safeguards (see chapter 5).

A production cutoff would be verifiable. Current plutonium production reactors could be decommissioned under inspection; periodic inspections, entombment, or dismantlement could ensure that the reactors could not be restarted without the knowledge of the other side. IAEA-type safeguards on power and research reactors and reprocessing facilities (and similar controls on naval-propulsion and tritium-production reactors) could verify the lack of weapon-grade plutonium production at other declared sites. Finally, national intelligence could verify the absence of significant clandestine production activities (see chapter 4).

One objection to a ban on the production of plutonium for weapons is that the amounts of plutonium that have already been produced by the US and the USSR are unknown. Production records could be disclosed, but some people would be reluctant to accept these at face value. Reactor-design details (which could be verified by on-site inspections) could be provided and then used to estimate the maximum amount of plutonium that could have been produced, but this upper limit might be as much as a factor of two higher than actual production. As is explained in this appendix, one could, however, accurately authenticate production records even long after the reactors have been decommissioned, by measuring the concentrations of long-lived radionuclides produced by neutron absorption in permanent components in the reactor core.

Under the associated verification arrangements, the monitored party would provide details of production-reactor design (lattice spacing, fuel size, enrichment, etc.) and of its operation from startup to decommissioning (for example, megawatt-days of thermal power produced every day or week). Using this information, the monitoring party would calculate, using a computer model of the reactor and the declared operating history, how much plutonium could have been produced and what the associated concentrations of long-lived radionuclides in permanent components should be. Samples of these components would then be taken from several dozen locations in the reactor core, and the concentrations of radionuclides measured with a mass spectrometer; if the measured and estimated radionuclide concentrations were in good agreement, then the declared operation history could be taken as legitimate.

The buildup of radionuclide $y$ from neutron reactions with a stable isotope $x$ can be described by the following equations:

$$\frac{dN_y}{dt} = N_x\sigma_{xy}\phi - N_y\sigma_y\phi - N_y\lambda \qquad (4.A.1a)$$

$$\frac{dN_x}{dt} = -N_x\sigma_x\phi \qquad (4.A.1b)$$

where $N_x$ and $N_y$ are the atomic concentrations of the stable isotope $x$ and the radionuclide $y$, $\sigma_x$ and $\sigma_y$ are the neutron cross sections in cm$^2$ for transmuting $x$ or $y$ into some other nucleus and $\sigma_{xy}$ is the cross section for transmuting $x$ into $y$, $\phi$ is the neutron flux in neutrons cm$^{-2}$s$^{-1}$, and $\lambda$ is the decay constant of the radionuclide (s$^{-1}$). For the cases considered here, we can ignore other reactions that may produce or deplete $x$ and $y$, and we can therefore set $\sigma_{xy} = \sigma_x$. Solving equation 4.A.1b for $N_x$ and substituting its value into equation 4.A.1a, we have

$$\frac{dN_y}{dt} = N_x^0\sigma_x\phi\exp(-\sigma_x\Phi) - N_y\sigma_y\phi - N_y\lambda \qquad (4.A.2)$$

where $N_x^0$ is the initial concentration of isotope $x$, and $\Phi$, the integrated neutron fluence in neutrons cm$^{-2}$, is given by

$$\Phi = \int_0^t \phi(\tau)\,d\tau \qquad (4.A.3)$$

If $\lambda \ll \phi\sigma_y$, or if $\phi$ is roughly constant, equation 4.A.2 can be solved to give

$$N_y \approx \frac{N_x^0\sigma_x}{\sigma_y - \sigma_x + (\lambda T/\Phi)}\left[\exp(-\sigma_x\Phi) - \exp(-\sigma_y\Phi - \lambda T)\right] \qquad (4.A.4)$$

where $T$ is the total time that the material is exposed to the neutron flux. In reality $\phi$ is not constant, but the dependence of equation 4.A.4 on the flux history will be minimal if the radionuclide has a halflife that is large compared to $T$ (i.e., $\lambda T \ll 1$). Since reactors have been operated for up to 40 years, for the neutron fluences and reactor operation times of interest here, this equation will give reasonably accurate results if the radionuclide has a halflife of at least 100, and preferably 500 years. Note that if $\sigma_x \Phi \ll 1$ and $\sigma_y \Phi \ll 1$, then $N_y \approx N_x^0 \sigma_x \Phi$.

Thus, by measuring the concentrations of various radionuclides in a permanent component of the reactor core one can estimate the total neutron fluence in the reactor at that point; with measurements at several points in the reactor, the total amount of energy (megawatt-days) produced by the reactor can be estimated. Under reasonable assumptions about the design and operation of the reactor, one can then estimate the amount of plutonium that could be produced in the reactor per unit of thermal energy produced (usually about 1 gram of plutonium per megawatt-day).

The neutron-produced radionuclides chosen for measurement should:

♦   have halflives of at least 100 years

♦   not exist in significant concentrations in nature (this eliminates potassium-40 and rubidium-87, for example)

♦   not be noble gases (they might escape the materials)

♦   not be fission products, uranium or transuranic isotopes created by the neutron absorption in uranium, or decay products of transuranics, since reactor components are likely to be contaminated with these substances

♦   be produced in thermal neutron reactions with naturally occurring isotopes.

These conditions are very restrictive. Only 67 radionuclides that do not exist in nature have halflives greater than 100 years; of these, two are noble gases, 23 are in the range of atomic weights produced during fission (79 to 166), and 23 are transuranics or decay products of transuranics. Nine of the remaining 19 radionuclides are not produced, either directly or as decay products, from thermal-neutron reactions with naturally occurring isotopes. The 10 remaining radionuclides that meet the five criteria given above are listed in table 4.A.1.

To be useful, of course, a radionuclide must meet an additional criterion: the element from which it is produced must be present in significant concentrations in a permanent component of the reactor core. The most common production reactor core materials are graphite and some type of metal—steel, aluminum, zircalloy, etc. The concentrations of elements listed in table 4.A.1 in samples of the graphite used in the construction of the original Hanford reactors is given in table 4.A.2. Also given in table 4.A.2 are the concentrations of these elements in certain reactor-grade steels.

The neutron flux in graphite reactors is typically $10^{13}$ to $10^{14}$ neutrons cm$^{-2}$s$^{-1}$; the flux is predominately thermal. If such a reactor has operated for 40 years at an average capacity factor of 0.5, the integrated thermal-neutron fluence would be roughly $10^{22}$ to $10^{23}$ neutrons cm$^{-2}$. Using this range of fluences, and the data in tables 1 and 2, the resulting range of concentrations of long-lived radionuclides in graphite and steel are given in table 4.A.3. (Actual concentrations will be higher in some cases because of the resonance capture of epithermal neutrons. If the concentrations of two or more radionuclides are measured, and if one or more of these has a non-zero resonance

**Table 4.A.1.** Radionuclides that: (1) have halflife ≥ 100 years; (2) do not exist in nature; (3) are not noble gases; (4) are not fission products, uranium isotopes, transuranics, or their decay products; and (5) are produced from thermal-neutron reactions with naturally occurring isotopes. Also given are the thermal-neutron reactions that produce the radionuclide, their cross sections, the abundance of the naturally occurring isotope, and the halflife and thermal-neutron-absorption cross section of the produced radionuclide. Data for tritium and plutonium are also listed.

| Radionuclide | Halflife[*] | Thermal-neutron absorption cross section $b = barns$ | Production reaction | | Thermal-neutron production cross section | Abundance[†] percent |
|---|---|---|---|---|---|---|
| Beryllium-10[†] | 1.6 My | < 1 mb | beryllium-9 | $(n,\gamma)$ | 7.6 ± 0.8 mb | 100.0 |
| Carbon-14[†] | 5.7 ky | < 1 μb | carbon-13 | $(n,\gamma)$ | 1.37 ± 0.04 mb | 1.10 |
| | | | nitrogen-14 | $(n,n)$ | 1.83 ± 0.03 b | 99.634 |
| | | | oxygen-17 | $(n,\alpha)$ | 0.24 ± 0.01 b | 0.038 |
| Chlorine-36[†] | 301 ky | < 10 b | chlorine-35 | $(n,\gamma)$ | 43.6 ± 0.4 b | 75.77 |
| | | | potassium-39 | $(n,\alpha)$ | 4.3 ± 0.5 mb | 93.258 |
| Calcium-41 | 103 ky | 4. b | calcium-40 | $(n,\gamma)$ | 0.41 ± 0.02 b | 96.941 |
| Nickel-59 | 75 ky | 92 ± 4 b | nickel-58 | $(n,\gamma)$ | 4.6 ± 0.3 b | 68.27 |
| Nickel-63 | 100 y | 24.4 ± 3.0 b | nickel-62 | $(n,\gamma)$ | 14.5 ± 0.3 b | 3.59 |
| | | | zinc-66 | $(n,\alpha)$ | < 0.020 mb | 27.9 |
| Rhenium-186m | 200 ky | § | rhenium-185 | $(n,\gamma)$ | 16 ± 3 b | 37.40 |
| Iridium-192m$_2$ | 241 y | § | iridium-191 | $(n,\gamma)$ | 0.16 ± 0.07 b | 37.3 |
| Lead-205 | 19 My | § | lead-204 | $(n,\gamma)$ | 0.66 ± 0.07 b | 1.4 |
| Bismuth-210m | 3.0 My | 54 ± 5 mb | bismuth-209 | $(n,\gamma)$ | 9.6 ± 0.8 mb | 100.0 |
| Tritium | 12.3 y | < 0.006 mb | lithium-6 | $(n,\alpha)$ | 940. ± 4. b | 7.5 |
| Plutonium-239 | 24.1 ky | 1,017.3 ± 2.9 b | U-238 → $(n,\gamma)$ → U-239 ⟶ Np-239 ⟶ Pu-239 | | 2.680 ± 0.019 b | 99.274 |

Source: S.F. Mughabghab, M. Divadeenam, and N.E. Holden, *Neutron Cross Sections Volume 1: Neutron Resonance Parameters and Thermal Cross Sections* (New York: Academic Press, 1981). One barn $b = 10^{-24}$ cm$^2$

\* My = $10^6$ years; ky = $10^3$ years

† Percentage of target isotope in natural element

‡ Present in minute concentrations in nature

§ Unknown; assumed equal to 10 barns for the purpose of estimation

**Table 4.A.2:** Concentrations of the elements listed in table 4.A.1 in six samples of graphite used in the construction of the original Hanford reactors, and in three types of steel

| | Concentration parts per million unless otherwise indicated | |
|---|---|---|
| Element | Graphite* | Steel† |
| Beryllium | < 0.0005 | |
| Carbon | 100 percent | 2,000–5,800 |
| Nitrogen | 10–100 | 40–190 |
| Oxygen | | |
| Chlorine | < 50 | |
| Potassium | < 0.02 | 1–3 |
| Calcium | 0.13–210. | |
| Nickel | 0.02–2.5 | 0.5–16 percent |
| Zinc | 0.06–160. | |
| Rhenium | | |
| Iridium | | |
| Lead | < 0.01 | |
| Bismuth | | 2 |

* William Morgan, personal communication

† Range based on data for SS-316, HT-9, and Fe-1422—kinds of steel—from Steve Fetter, "The Radiological Hazards of Magnetic Fusion Reactors," *Fusion Technology*, 11, 2 (March 1987)

capture integral, then the epithermal as well as the thermal neutron flux could be determined.)

At first glance, carbon-14 might appear to be the best radionuclide to measure in graphite, but the presence of nitrogen in concentrations as low as 1 part per million (ppm) would make an accurate estimate of the neutron fluence impossible, since the concentration of nitrogen would vary during the life of the reactor, and since 8 ppm of nitrogen generates as much carbon-14 as does the 1 percent carbon-13 in the carbon. Calcium-41 and nickel-59 seem better suited for measurement: each is present in very pure graphite, and each has a long halflife, well-known reaction cross section, and is produced in quantity by only one reaction.

In the case of steel, nickel-59 and nickel-63 are the best candidates for measurement, although nickel-63 has a relatively short halflife (100 years). Theoretically, nickel-63 can also be produced from copper-63 with low-energy neutrons (see table 4.A.4), but this should not be a serious problem (at least for metals with small copper concentrations).

One should be able to measure even the lowest radionuclide concentrations in table 4.A.3 without too much trouble. Portable mass spectrometers can accurately measure concentrations smaller than 1 part per billion (ppb); larger machines can go much lower. Concentrations of carbon-14 as small as 0.001 ppt—$10^{-15}$—are routinely measured with high accuracy.* Although radionuclide concentrations and cross sections can be measured with an accuracy of a few percent, uncertainties in the operating record, the rector design, and the computer model of the reactor will probably limit the overall uncertainty of the estimate of the total fission energy released in the production reactor to no less than 10 percent. With accurate records of the design and operation of the reactor, one should be able to estimate the number of excess neutrons per megawatt-day available to produce plutonium with an uncertainty of no more than 10 percent.

Although the total neutron flux or amount of thermal energy produced can be verified in this way, a key problem is that the excess neutrons produced by fission can be used to produce *either* plutonium-239 (from uranium-238) *or* tritium (from lithium-6). This is a problem primarily in deuterium-moderated reactors, which (at least in the US) have been used to produce both tritium and plutonium. One can verify only that the stated production of both materials in a given reactor is consistent with the estimated fission energy released.

Although more research is needed, this preliminary study indicates that measuring the concentrations of long-lived radionuclides in permanent reactor-core components is a feasible and promising way to estimate the integrated neutron fluence during the reactor's operation, and thereby to verify records of past production of weapon-grade plutonium and tritium in a production reactor.

---

* The concentration of carbon-14 in contemporary carbon is 1.3 ppt; carbon-14 dating has been used to establish ages of over 50,000 years, which corresponds to a carbon-14 concentration of less than $1.3 \exp(-50,000/8,270) = 0.003$ ppt.

**Table 4.A.3:** Concentrations of long-lived radionuclides expected in graphite and steel components of reactor cores

| | | Radionuclide concentration | | | |
| | | *in graphite* | | *in steel* | |
| | **Produced** | | | | |
| **Radionuclide** | **from** | low* | high† | low* | high† |
|---|---|---|---|---|---|
| Beryllium-10 | beryllium-9 | | 0.4 ppt | | |
| Carbon-14 | carbon-13 | 0.2 ppm | 2 ppm | 0.3 ppb | 9 ppb |
| | nitrogen-14 | 2 ppm | 20 ppm | 0.7 ppm | 30 ppm |
| | oxygen-17† | | 10 ppt | | 10 ppt |
| Chlorine-36 | chlorine-35 | | 20 ppm | | |
| | potassium-39 | | 5 ppt | 4 ppt | 0.7 ppb |
| Calcium-41 | calcium-40 | 0.5 ppb | 7 ppm | | |
| Nickel-59 | nickel-58 | 0.4 ppb | 60 ppb | 100 ppm | 400 ppm |
| Nickel-63 | nickel-62 | 0.08 ppb | 20 ppb | 20 ppm | 100 ppm |
| | zinc-66 | 0.003 ppt | 30 ppt | | |
| Rhenium-186m | rhenium-185† | | 1 ppb | | 1 ppb |
| Iridium-192m$_2$ | iridium-191† | | 4 ppt | | 4 ppt |
| Lead-205 | lead-204 | | 6 ppt | | |
| Bismuth-210m | bismuth-209 | | | 0.2 ppb | 2 ppb |

* Using the lowest concentrations in table 4.A.2, a thermal-neutron fluence of $10^{22}$ cm$^{-2}$, and ignoring resonance capture

† Using the highest concentrations in table 4.A.2, a thermal-neutron fluence of $10^{23}$ cm$^{-2}$, and ignoring resonance capture

‡ Concentration in graphite or steel not listed in table 4.A.2; set equal to 1 ppm for oxygen and the average crystal abundance for rhenium (5 ppb) and iridium (1 ppb) to give an order-of-magnitude estimate

**Table 4.A.4:** Other neutron reactions that have energy thresholds of less than 6 MeV and produce the radionuclides listed in table 4.A.1[*]

| Radionuclide | Other reactions | | Energy threshold MeV[†] | Cross section[†] |
|---|---|---|---|---|
| Beryllium-10 | boron-10 | (n,p) | 0.0 | ? |
| | carbon-13 | (n,α) | 3.8 | ? |
| Nickel-63 | copper-63 | (n,p) | 0.0 | 0.125 b at 14.7 MeV |
| Rhenium-186m | osmium-186 | (n,p) | 0.4 | < 5.5 mb at 14.7 MeV[§] |
| | osmium-187 | (n,d) | 4.5 | ? |
| Iridium-192m$_2$ | platinum-192 | (n,p) | 0.8 | ? |

[*] Reactions with greater energy thresholds will not occur at significant rates with fission spectrum neutrons

[†] The difference in mass excess (if greater than zero)

[‡] Victoria McLane, Charles L. Dunford, and Philip F. Rose, *Neutron Cross Sections Volume 2: Neutron Cross Section Curves* (Boston: Academic Press, 1988)

[§] To rhenium-186; cross section for rhenium-186m should be considerably smaller

# Chapter 5

*Theodore B. Taylor*

# Dismantlement and Fissile-material Disposal

No nuclear warheads have yet been eliminated by treaty. After warheads have been removed from missiles, the INF treaty allows each country to retain them, without restrictions.[1] According to present expectations similar conditions will apply to the START treaty now under negotiation to reduce numbers of Soviet and US deliverable strategic ballistic-missile warheads by half. Nevertheless, given recent advances in cooperative methods of verification, as well as progress in technical capabilities of detection and monitoring, it is reasonable to hope and expect that dismantlement of nuclear warheads, not just the means for their delivery, will be called for sometime in the future. This possibility has prompted a number of previous studies.[2]

This chapter examines the procedures that would be required to verify that warheads specified by treaty for elimination are, in fact, completely dismantled, their components rendered useless for construction of new warheads, and the contained fissile materials placed under international safeguards or disposed of in such a manner as to make them unusable in weapons.

## THE WORLD'S NUCLEAR WARHEADS

Estimates of present worldwide numbers of nuclear warheads, as well as the numbers of US and Soviet nuclear warheads of various kinds, are shown in table 5.1.[3] Estimates of the total quantities of highly enriched uranium, plutonium, and tritium associated with the warheads are also included. More than 90 percent of these materials are accounted for, roughly equally, by the US and the Soviet Union.[4] These estimates are uncertain, especially the quantities of nuclear materials in Soviet warheads. Nevertheless they are helpful in setting the scales of operations needed to dismantle large fractions of the world's present stockpiles of nuclear warheads.

There are only three fundamentally different types of nuclear warhead. Pure fission warheads derive their explosive energy entirely from rapid fission chain reactions. "Boosted" fission warheads incorporate small quantities of deuterium and tritium that release large numbers of neutrons when they react at the temperatures produced by a fission explosion. These neutrons then speed up the rate at which the fission chain reaction proceeds and increase the overall yield of the explosion considerably above what it would be without boosting. Thermonuclear warheads require pure fission or boosted fission explosions to produce the conditions needed to ignite sufficient quantities of thermonuclear fuels to account for a substantial fraction of the overall yield of the warhead.

Warhead weights range from less than 50 kilograms to more than 4,000 kilograms; diameters range from less than 20 centimeters to more than one meter; yields range from much less than 1 kiloton to about 10 megatons (TNT equivalent).[5]

There are several types of physical coupling between nuclear warheads and their delivery systems. Warheads for land- and sea-based strategic missiles are usually mounted on the missiles, although some warheads may be in storage separately at any given time. Other delivery systems, such as artillery, tactical aircraft, and ships, have associated storage facilities for nuclear and conventional warheads. Although such differences can affect some of the details concerning physical means for identifying and containing warheads at specific deployment sites before they are dismantled, the basic principles explored here apply to all cases.

**Table 5.1:** The world's present nuclear warheads

### The World's Nuclear Warheads

| | |
|---|---|
| United States | 23,000 |
| Soviet Union | 30,000 |
| United Kingdom | 700 |
| France | 500 |
| China | 300 |
| Israel | 50–200 |
| South Africa | ? |
| Pakistan | ? |
| India | ? |
| *World Total* | *about 55,000* |

### US and Soviet Nuclear Warheads

| | US | USSR |
|---|---|---|
| *Strategic* | | |
| Land-based missiles | 2,450 | 6,592 |
| Submarine-launched missiles | 5,312 | 3,256 |
| Bombers | 5,016 | 880 |
| | *12,778* | *10,898* |
| *Nonstrategic* | | |
| Aircraft bombs and missiles | 3,500 | 6,370 |
| Land-based missiles | 1,805 | 4,700 |
| Submarine-launched ballistic missiles | 0 | 50 |
| Submarine-launched cruise missiles | 150 | 400 |
| Antiballistic and surface– air missiles | 385 | 4,200 |
| Artillery | 2,010 | 2,000 |
| Antisubmarine | 1,760 | 1,860 |
| Demolition (ADM) | 300 | ? |
| | *9,910* | *19,580* |

### Nuclear Materials in US and Soviet Nuclear Warheads

| Material | US | USSR | Total |
|---|---|---|---|
| Plutonium | 100 tonnes | 100 tonnes | 200 tonnes |
| Highly enriched uranium | 500 tonnes | 500 tonnes? | 1,000 tonnes |
| Tritium | 100 kg | 100 kg? | 200 kg? |

The effect of nonmilitary disposal of fissile materials from dismantled warheads would be more symbolic than substantive if further production of these materials for weapons were allowed to continue (see chapter 4). But such symbolism may be important politically, contributing to public support for nuclear disarmament. Furthermore, joint development of safe and verifiable procedures for dismantling warheads and transferring recovered materials from military to peaceful use could lay a basis for the verification aspects of future, more stringent disarmament agreements.

## ALTERNATIVE APPROACHES TO THE DISPOSAL OF WARHEADS

There are many different ways in which warheads specified under a treaty can be disposed of. Three are considered here, in order of increasing stringency.

♦   Each nation removes the specified warheads from deployment sites and periodically places negotiated quantities of fissile materials (plutonium and highly enriched uranium) under safeguard. These quantities might be the same for the US and the Soviet Union, and smaller for any other parties to a nuclear-warhead–reduction treaty. Alternatively, they might be proportional to the total numbers of removed warheads. In any case, these quantities should probably be negotiated in the original treaty. The quantities may correspond to significantly more or less of these materials than are actually present in the warheads being eliminated and therefore need not reveal the real amounts of fissile materials in each type of warhead.

♦   Each nation removes and retains all fissile materials and thermonuclear fuel (tritium and deuterium, in compounds or as elements) from the warheads for unrestricted use, and the remaining components are verifiably destroyed.

♦   Each nation recovers the fissile materials and tritium from the warheads. The fissile materials are committed for use under safeguards as fuel for nonmilitary power reactors or for direct disposal in forms that make subsequent recovery for use in weapons impractical. The remaining components are verifiably destroyed. Their material residues, including tritium, may or may not be returned to the owner nation.

A variation of this last option would be to negotiate amounts of fissile materials greater than the quantities to be extracted from the warheads to be committed for nonmilitary use or direct disposal. The excess would be supplied from other sources, such as warheads not yet subject to dismantlement by treaty or stockpiles of unsafeguarded fissile material not in warheads. Once again, the negotiated minimum quantities may differ from country to country, to account for differences in total quantities of weapon materials in national stockpiles. The purpose of such an approach would be to help ensure parity in depletion of fissile materials, considered as fractions of total national stockpiles, as well as parity in giving up specific types of warhead or nuclear-weapon systems.

The first option achieves reductions in the theoretical maximum numbers of nuclear warheads by reducing the accessibility of the key fissile materials that are absolutely required to make nuclear warheads. It is the easiest to implement technically, since it does not require verification of warhead dismantlement operations. But it offers no verifiable guarantee that all the fissile material contained in the warheads is relinquished, or that the other parts of the warheads are destroyed.

The second option ensures that the specified warheads are destroyed, but does not necessarily remove from weapon use the essential components that are most difficult to produce—the plutonium and enriched uranium.

The third option is the most difficult to carry out technically, but is also the only one that ensures that the specified weapons are destroyed and their contained fissile materials are made inaccessible for weapons. It is considered here in some detail, not because it is evidently the most attractive, but because it raises some important technical issues that need to be dealt with in any comparative assessment of these options. Adding fissile materials to those extracted from the warheads to be dismantled, a variation mentioned above, is not analyzed in this paper. Its inclusion would require some minor modifications of the dismantlement process to allow for safeguarded flows of materials from sources other than the specified warheads.

## USE OF WARHEAD FISSILE MATERIALS IN NUCLEAR POWER PLANTS

A worldwide tally of 1988 nuclear power capacity and that projected for the year 2000 is shown in table 5.2.[6,7] Today, more than 95 percent of the fuel for power reactors is uranium of low enrichment (typically about 3 percent uranium-235) or natural uranium. The demand for fissile material for reactor fuel will therefore be overwhelmingly for uranium-235, rather than plutonium, for at least a decade.

The rate of loading of uranium-235 in a 1,000-megawatt-electric light-water reactor fueled with uranium is about 1,000 kilograms per year. A few reactors are beginning to use recycled plutonium to supplement the uranium-235 (but not in the US).[8] In such cases, the fuel is in the form of mixed oxides of plutonium and uranium, with plutonium accounting for a few percent of the mixture. Likely annual loading rates of uranium-235 and plutonium in the mixed-oxide reactors are about 670 kilograms of uranium-235 and 350 kilograms of plutonium per 1,000 megawatt-electric-year. Higher plutonium loadings are possible, but may cause unacceptable reactivity-control problems.

The uranium-235 in the world's stockpiles of nuclear warheads is a

**Table 5.2:** World nuclear power plant capacity

| | Capacity GWe[*] | |
|---|---|---|
| | 1988 | 2000 |
| United States | 100 | 111 |
| France | 49 | 64 |
| USSR | 28 | 85 |
| Japan | 27 | 50 |
| West Germany | 19 | 24 |
| Canada | 12 | 16 |
| United Kingdom | 11 | 11 |
| *Subtotal* | *246* | *361* |
| All other | 51 | 99 |
| Total | 297 | 460 |

[*] 1 gigawatt = $10^9$ watts

potential energy resource worth about $40 billion.[9] Low-enriched uranium contributes about 0.5 cents per kilowatt-hour to the cost of electric power produced by typical nuclear power plants. Most of this cost could be avoided if highly enriched uranium from warheads were used to supply the uranium-235 needed for power reactors.[10]

It is sometimes argued that warhead plutonium should be stored for use in future reactors that will use recycled plutonium or as core material for plutonium breeder reactors. This option is not considered here because the plutonium might again be used for weapons if there were a major breakdown of disarmament treaty restrictions.[11] It is proposed instead that the warhead plutonium be directly disposed of in a way that makes it inaccessible for reuse in nuclear weapons. This proposal, however, is not fundamental to the technical possibilities discussed in this paper. Most would apply equally well if the warhead plutonium were used in reactor fuel.

## PROCESS STEPS FOR ELIMINATING WARHEADS

A system for verifying the elimination of nuclear warheads must ensure that:

♦ All warheads and associated payload hardware identified by the owner country and earmarked for elimination are what they are claimed to be

♦ All items earmarked for elimination are destroyed

♦ None of the nuclear material from the warheads to be dismantled is diverted to unauthorized uses

These guarantees must be provided without the need to disclose sensitive information about the design of the warheads or other associated equipment, such as re-entry vehicles, penetration aids, or shielding against radiation.

Almost all detailed information about the design of specific nuclear warheads is now classified. This includes yields and total weights; quantities of contained materials, including but not restricted to tritium, highly enriched uranium, and plutonium; and dimensions, configurations, and weights of fabricated components. Such information as is now in the public domain is not necessarily reliable. It is therefore assumed here that countries will not be willing to reveal this information in the warhead dismantlement process.

Two key assumptions about secrecy are inherent in the process descriptions that follow.

The first is that the aggregate quantities of uranium-235, uranium-238, and plutonium of any isotopic composition that are contained in a mix of different types of warhead can be declassified in the course of future treaty negotiations. This would allow accurate accounting systems for fissile materials to be set up without revealing information about the fissile-material content of any particular kind of warhead.

The second assumption is that *upper limits* to some of the material quantities, component weights, and dimensions associated with warheads and other payload items can be declassified without national security concerns, provided that the upper limits are sufficiently large compared with their *actual* values. Then each owner nation could mask the true value of quantities it wished to keep secret by adding appropriate materials, in unrevealed amounts, to the objects to be dismantled. An example would be the addition of a certain amount of sand to each of the containers for some type of warhead, without ever revealing what that amount was.

Having made these assumptions, we can describe a verification system which ensures that all fissile materials in the warheads are accounted for and made available for inclusion in reactor fuel or direct, permanent disposal without revealing sensitive design information about specific warheads.

The main steps in the warhead elimination process are shown schematically in figure 5.1. Broadly speaking, the process provides the following assurances:

♦ All materials in the warheads are contained within well-defined boundaries from the time they are placed in shipping containers at the deployment sites until they have been dismantled

♦ Any attempts to divert any of the warhead components to unauthorized purposes will be detected

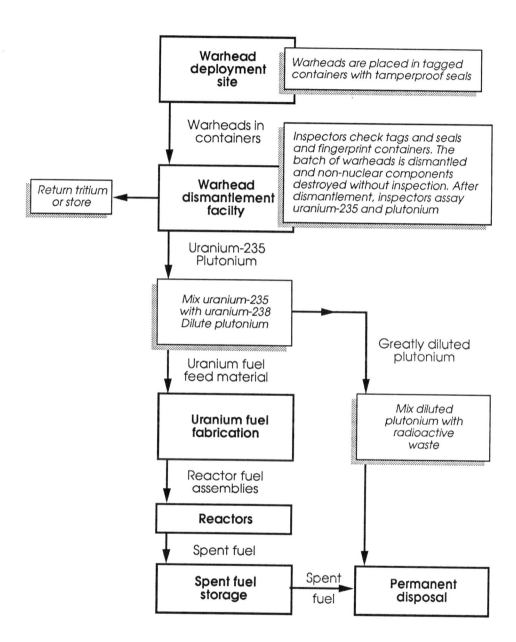

**Figure 5.1:** The main steps in the warhead elimination process

♦ All major components of the warheads or other payload items are destroyed, in the sense that they would require refabrication to be used in other warheads

♦ All uranium-235, uranium-238, and plutonium in the warheads is accounted for in the measured output of these materials from the dismantlement facility

♦ Substitution of fake warheads for real ones at the deployment sites, before dismantlement operations begin, is likely to be detected.

### Fingerprinting

A key concept related to detection of substitutions is "fingerprinting." It denotes any observable indicators that the contents of all the warhead containers claimed to be of the same type are, in fact, the same. Since these indicators must not reveal sensitive information about the warhead designs, it may be necessary to encrypt some of them in such ways that they can be compared accurately enough to reveal significant *differences* between the contents of containers without disclosing restricted data.

The process steps and ways to achieve the above assurances are described briefly in the following sections.

### Tagging, Sealing, and Shipment of Warheads to a Dismantlement Facility

When the dismantlement operations start at a deployment site, all nuclear warheads to be eliminated—possibly along with attached payload components such as re-entry vehicles and guidance packages—are placed inside shipping containers. The containers are provided by the owning country, which is also responsible for removal of the payloads/warheads from delivery vehicles or storage facilities at the site. The containers are not subject to internal inspection on arrival at the site, since they may contain materials that have been added off-site to mask actual weights of warheads or some of their components (see below). Transfer of payloads/warheads from delivery vehicles or storage to the shipping containers is observed by inspectors. The units may be temporarily covered while being transferred to the shipping containers, to avoid revealing sensitive information about their external appearance.

The inspectors then tag and seal the containers. The tags are for unique external identification of each container. The seals are designed to reveal any unauthorized opening of the containers.

Possible methods for tagging the containers include microscopic photography of parts of the outside surfaces or use of spray paint to produce photographed "signatures" that are almost impossible to change or reproduce without detection.

One method of sealing the containers would be to wrap them with bundles of optical fibers. Illumination of selected groups of fibers one end of such a bundle produces a unique and complex pattern at the other end. Before-and-after photographs of these patterns will reveal attempts to cut the bundles of fibers. Such techniques have been used routinely by the International Atomic Energy Agency for safeguarding purposes.[12]

Another sealing option would be to spot-weld any removable access covers to the containers, using the welds themselves as seals. Such seals have unique patterns that can be photographed before and after to reveal unauthorized opening of the containers.[13]

The tagged and sealed warhead containers, which might be temporarily stored at the deployment site, are then shipped to a warhead dismantling and destruction facility in the owner country. At this facility all tagged containers are examined by inspectors to ensure that they have not been tampered with. Inspectors would not need to accompany the shipments in transit, as long as careful accounting for each container is maintained at the deployment sites and the dismantlement facility. After shipment and inspection of the tags, significant numbers of unopened containers would typically be kept temporarily in storage at the dismantlement facility.

## Dismantlement of Warheads and Other Parts of Payloads

The announced nuclear-weapon states all have facilities for dismantling old warheads to recover nuclear materials or other components to be used in new warheads. It is possible that these facilities could be modified to meet the conditions needed for verified dismantlement under a disarmament treaty, especially the need to preserve secrecy concerning some of the warhead design details. Such modification might be difficult, however, in dismantlement facilities that are used both for handling warheads that are to be dismantled by treaty and ones that are not.

Decisions whether to modify existing warhead-dismantlement facilities or build new ones for the purpose of treaty-mandated warhead dismantlement should follow intensive unilateral and bilateral assessments of the alternatives. Lacking access to descriptions of existing facilities, a hypothetical one is described here.

A schematic illustration of such a facility is shown in figure 5.2. Enclosures within which inspectors would not be allowed during dismantlement operations are indicated by double lines. These areas could be inspected between dismantlement operations, to ensure that all fissile material had been removed and had been placed under safeguard and all warheads had been dismantled.

A well-defined boundary surrounds the entire dismantlement area. Portals with access through this boundary are all monitored visually and with appropriate equipment to ensure no passage of unauthorized objects, materials, or people. The main function of the portal-monitoring equipment is to detect unauthorized removals of fissile materials from the facility or the introduction of unauthorized items into the facility. The portal for incoming shipping containers with warheads inside is the only one authorized for incoming fissile materials. The only portal authorized for outgoing fissile materials is the one used for removal of fissile materials after extraction from the warheads, for transfer to an adjoining facility for isotopic dilution of the uranium-235 (if needed) and chemical dilution of the plutonium.[14]

The principal inputs to the facility are the tagged and sealed containers with warheads and other payload hardware. All other inputs, such as process materials or new equipment needed for dismantlement operations, are kept to a minimum.

The principal outputs are the following:

♦ Accurately measured quantities of uranium-235 and -238 mixtures and plutonium (both probably in metallic forms), for secure transfer to an immediately adjoining site for dilution

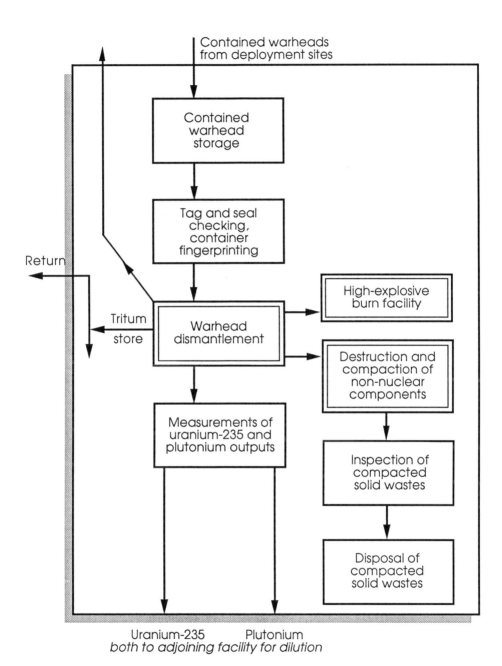

**Figure 5.2**: Schematic diagram of a nuclear-warhead dismantlement facility

- Tritium, in amounts not to be revealed to inspectors, to be returned to the owner nation or disposed of in a safeguarded manner

- Small containers of radioactive materials used for warhead chain-reaction initiators or functions other than directly releasing explosive energy

- Residues of compaction or incineration of all other components of the warheads or other payload items.

Warhead containers intended for re-use may also be considered as facility outputs. After their contents have been removed for dismantlement, the containers are weighed and inspected, to ensure they are empty. The owner nation is then allowed to place into the containers an undisclosed weight of some common material, such as water or sand, to mask the overall weights of warheads. The containers are then sealed by inspectors (but not weighed), and scanned by an external radiation source to ensure that there is no uranium or plutonium inside.[15] The containers remain sealed until their seals are externally inspected at a warhead-deployment site before they are opened to receive more warheads.

The solid and liquid waste outputs from the site are kept to a minimum and subjected to detailed visual and instrumental inspection before they are removed from the site. A radiation scan of the residue from each batch of dismantled warheads would ensure that these residues did not contain significant quantities of fissile materials.

If the high explosives in the warheads are burned, the waste product is mostly gas. This can be vented from the site after passage through an appropriate gas cleanup system for removal of unacceptable pollutants.

Numbers of vehicles entering or leaving the outer boundary of the facility are kept to a minimum—zero if possible. This can be done by using special transfer facilities for moving the warhead containers and other authorized materials or equipment into the inside enclosures. Similar facilities can also be used to transfer all output material through the safeguarded perimeter, so that vehicles leaving the site need not be inspected.

The warhead components are dismantled by nationals of the owning nation. The high explosives and other non-nuclear components are destroy-ed in appropriate facilities inside the containment area. Prior to inspection, the plutonium and uranium are converted to forms such as metallic "buttons" that will not reveal warhead design features. Equipment appropriate for melting or dissolving the fissile materials and any low-enriched uranium in the warheads, and then mixing them, are required, along with the associated criticality and other safety procedures used in fissile-material processing plants.

Accurate measurements of the quantities and isotopic compositions of the recovered uranium and plutonium are made by inspectors, to obtain the initial data needed for accurate materials-accounting so that the enriched uranium can be subsequently incorporated into reactor fuel and the plutonium disposed of.

In initial stages of nuclear disarmament, any tritium might be returned to the owner nation, to reduce requirements for production to make up for tritium decay (with a 12.3-year halflife) in warheads not yet subject to a disarmament treaty. In this case, the amounts of tritium removed from the warheads need not be revealed to inspectors, since the overall tritium content of even mixtures of warheads is likely to be especially sensitive information. Alternatively, the tritium might be placed under safeguards, for possible future use in thermonuclear fusion reactors, or simply allowed to decay. Tritium containers leaving the site would be scanned to ensure they contain no fissile materials.

Radioactive materials used in the warheads to generate neutrons for initiating fission chain reactions would be separated from other components and treated as small quantities of high-level radioactive wastes to be disposed of at an appropriate facility. Their relatively small shipping containers would also be scanned before leaving the dismantlement facility, to ensure that they did not contain fissile materials.

## DETERRENTS TO THE SUBSTITUTION OF FAKE WARHEADS FOR REAL WARHEADS

The procedures just described can ensure that objects claimed to be warheads are dismantled, their components destroyed, and all contained fissile materials accounted for. By themselves, however, these procedures cannot completely ensure that fake warheads may not have been substituted for real ones before the dismantlement operation began at the deployment sites.

Objects substituted for the warheads before tagging and sealing might include any of the following:

♦ Objects similar in all respects to the real warheads, except that natural or depleted uranium has been substituted for some or all of the plutonium and highly enriched uranium that would have been in the real warheads. The purpose of this substitution would be to withhold significant amounts of fissile materials from the dismantlement process. The fake warheads might or might not be capable of producing a nuclear explosion, depending on the amounts of fissile materials withheld.

♦ Objects that might or might not closely resemble the real warheads. Such objects would be much easier to fabricate to less demanding tolerances than real warheads. They might include some fissile materials, but substantially less than in the real warheads. The fake warheads might or might not be capable of producing a nuclear explosion. Their function would be to allow unauthorized withholding of complete real warheads.

♦ Complete warheads that are being retired from stockpile and that have much less fissile material than the warheads that are supposed to be eliminated.

Several measures can be used to help verify that such substitutions have not occurred, without revealing any secret design information:

♦ Verification techniques can be used that will reveal significant differences between the contents, especially in amounts of fissile materials, of any containers for warheads that are claimed to be of the same type. Thus, if illegal substitutions were made, they would have to be made for all warheads of the same type, rather than for some selected fraction. (Specific ways to carry out this type of verification are discussed below.)

♦ The warheads specified for elimination should be placed in tagged and sealed containers as early as possible, starting with sample warheads at a few deployment sites randomly chosen at short notice (for example, less than 24 hours) from sites specified in the treaty.

♦ Inspectors could measure accurately the total quantities and isotopic compositions of mixtures of plutonium or uranium extracted from batches of more than one type of warhead in each dismantlement campaign. Use of this procedure will require that the total plutonium, uranium-235, and uranium-238 extracted from combinations of, say, three types of warhead be declassified. But it is difficult to see how this could reveal information that is critical to the national security of any announced nuclear weapon states.

♦ A few sample sealed warhead containers for each type of warhead could be randomly selected for safeguarded storage for an unspecified time. This will preserve evidence of compliance (or non-compliance) with a treaty, in case more effective verification techniques are developed in the future. Present uncertainty about such possibilities could act as a major deterrent to cheating under a current treaty. Furthermore, the selected warheads could be used as standards against which to match very detailed fingerprints.

♦ The possibility of "whistleblowing" (reporting of treaty violations to a verification authority by nationals of a country whose government orders the violations) is a deterrent that cannot be assessed quantitatively. It may become increasingly important as the universal benefits of nuclear disarmament become more generally apparent and publicized worldwide. Ways to ensure that individuals or groups who report violations can remain anonymous, or at least protected, need to be further developed and assessed.

None of these measures would reveal restricted information, especially if each owner nation were allowed to add unrevealed weights of common materials, such as sand or water, to the warhead containers, to disguise the warhead weights or some aspects of their composition. Uranium would not be allowed for this purpose, since introduction of unknown quantities of it into the process would degrade checks of independent estimates of the total quantities of highly enriched uranium that have been produced by each nuclear-weapon state (see chapter 4). The total weights and configurations of any such added materials must be the same for all warheads of a particular type, to ensure that the total contents of all fully loaded containers for each type of warhead are the same.

## Fingerprinting the Contents of Warhead Containers

As previously indicated, measurements that will reveal differences between the contents of containers of warheads that are claimed to be of the same type, without revealing secret information to the inspectors, can help assure that fake warheads have not been substituted for real ones. The term "fingerprint" is used here to mean the totality of such measurements.

There is a wide variety of possibilities for such fingerprints. They fall into two categories: external measurements before the warhead containers are opened, and measurements of residues from the dismantlement process. In either case, allowable differences in the same measurement for individual warheads of the same type would have to be negotiated, since there may be some variations in, for example, weights and configurations of fusing and firing components.

Possibilities for the elements of a fingerprint include the following:

♦ Total weight of the contents of each warhead container before it is opened for dismantlement. This weight is derived from the difference between the weight of the loaded container, before it is opened, and the weight of the empty container after its interior has been inspected, but before unknown amounts of materials have been added by the owner nation. It is specified that the total weight of the contents should always be the same (within negotiated limits) for the same types of warhead. This does not, however, preclude the possibility of substituting fake warheads and changing the weight of added materials to keep the total weight the same.

♦ Precise measurements of the aggregate quantities and isotopic compositions of plutonium and uranium extracted from each batch of dismantled warheads consisting of known numbers of several specified types. Since isotopic composition, especially of plutonium, can vary significantly between different warheads of the same type, allowable variations in average isotopic compositions would have to be negotiated.

The measurements applied to the fissile-material outputs are taken here to be the principal basis for fingerprinting. They would reveal use of fake warheads that contain less plutonium, uranium-235, or uranium-238 than in the real warheads, unless fake warheads are substituted for all warheads before they are sealed at the deployment sites. Although they would not necessarily reveal differences in the non-nuclear components of warheads that are supposed to be of the same type, such violations would risk being detected eventually by fingerprinting techniques that might be developed for probing the randomly selected sealed containers that had been placed in safeguarded storage.

Among the many other fingerprinting techniques that might be developed and used in the future are an entire class that would produce extremely detailed raw data concerning the configurations, compositions, and masses of materials in the warheads. The raw data would be withheld from inspectors, but combined in a sealed data processing system that would produce scrambled output data that would reveal no classified information, but reveal significant differences between objects inside the containers.[16] Examples of such measurements include weight distributions along several axes and high-resolution scanning with external sources of gamma rays, x-rays, or neutrons.

Some preliminary analysis by the author has shown that passive radiation scanning is not likely to produce a reliable fingerprint. The external fluxes of gamma rays or spontaneous fission neutrons from warheads of the same type, but with uranium or plutonium of differing isotopic compositions, show credible variations that might suggest significant differences in the contents of containers even if they contained what they were supposed to.


## DISPOSAL OF WARHEAD URANIUM AND PLUTONIUM

After accurate measurement of their masses and isotopic compositions by inspectors, the uranium and plutonium would be transferred from the dismantlement facility to an adjoining facility for further processing to prepare them for their ultimate disposal. This facility would also be enclosed by a containment perimeter.

The enriched uranium mixtures from the warhead dismantlement facility are further diluted as necessary with depleted or natural uranium, to provide uranium with about 3 percent uranium-235 that could be used for fuel for light-water power reactors. For use in heavy-water or graphite reactors fueled with natural uranium, which account for a small fraction of the world's nuclear power, the uranium-235 could be diluted with depleted uranium (typically 0.2–0.3 percent uranium-235) to a concentration near the natural level of 0.7 percent. In either case the dilution renders the uranium incapable of sustaining a fast-neutron chain reaction, for which the minimum enrichment required is about 6 percent.

Unlike uranium-235, plutonium cannot be "isotopically denatured" to render it unusable in nuclear warheads after chemical separation from diluting materials. All plutonium isotopes are capable of sustaining a fast-neutron chain reaction.[17] The best plutonium isotope for nuclear warheads is plutonium-239. Substantial concentrations of plutonium-240 (greater than about 6 percent of the plutonium-239) are undesirable because that isotope spontaneously fissions and releases neutrons that can cause a premature chain reaction in a weapon before it is optimally assembled in an implosion. Nevertheless, efficient reliable nuclear weapons, including thermonuclear warheads, can be made with plutonium containing concentrations of plutonium-240 much greater than 6 percent.[18]

A number of steps should therefore be taken to ensure that warhead plutonium, after disposal, would not be practically accessible for making nuclear warheads. The plutonium should be heavily diluted with materials (such as depleted or natural uranium oxide) that are at least as difficult to dissolve and then separate as typical constituents of fresh reactor fuel, in preparation for its irretrievable disposal. Before final disposal at a considerable depth in a safeguarded geological formation, the diluted plutonium should also be mixed with high-level radioactive wastes.

## POSSIBLE PARAMETERS FOR A WARHEAD DISMANTLEMENT FACILITY

Possible parameters for a large warhead dismantlement facility in the US are listed in table 5.3. Its capacity for dismantling eight warheads a day is about as large as may be credibly required for implementing future nuclear disarmament treaties. That is, it would be capable of dismantling all US warheads in less than 10 years if operated six days a week. The main characteristics of a corresponding facility in the Soviet Union might be similar.

The average daily outputs of uranium-235, plutonium, and tritium correspond to averages of 20 kilograms, 4 kilograms, and 4 grams, respectively, per warhead (see table 5.1).

The average weight of a warhead now in the US stockpile is about 350 kilograms.[19] This corresponds to an average daily input of 2,800 kilograms of total warhead weight. The weight of other objects in the warhead shipping containers, such as re-entry vehicle structures and guidance packages, is unlikely to exceed that of the warheads. Additional material in

**Table 5.3:** Preliminary parameters for a US nuclear-warhead dismantlement facility

| | |
|---|---|
| Capacity | 8 warheads/day (25,000 in 10 years) |
| Average uranium-235 output | 160 kg/day |
| Average plutonium output | 32 kg/day |
| Average tritium output | 32 g/day |
| *Storage capacities (100 days throughput)* | |
| Undismantled warheads | 800, in containers |
| Uranium-235 | 16,000 kg |
| Plutonium | 3,200 kg |
| Tritium | 3.2 kg |

the warhead containers, added to mask the weight of the warheads, might also be as much as another 2,800 kilograms per day, for a nominal total of about 8,400 kilograms removed from the containers each day.

If half the average warhead weight is assumed to be high explosive, the corresponding high explosive input is 1,400 kilograms per day. Most of the residue from burning this will be gaseous products, vented, after scrubbing, to the atmosphere.

The remaining average of about 7,000 kilograms per day of non-nuclear materials and thermonuclear fuels not containing tritium could be separated into valuable materials (such as deuterium or beryllium) to be returned to the owner country, and waste materials for direct disposal. If all these materials were compacted into slabs with a bulk density in the vicinity of 4 grams per cubic centimeter, their total volume would be 1.8 cubic meters per day. A reasonable actual size for each slab might be 1 square meter, with a thickness of 4 centimeters, corresponding to an average weight of about 160 kilograms. Each of these slabs (about 40 per day), supported horizontally, could then be conveniently scanned with gamma rays and neutrons to ensure they contained no fissile material or uranium-238.

The warhead and nuclear material storage capacities shown in table 5.3 correspond to 100 days of average throughput. This is a rough estimate that allows for process holdups and fluctuations.

At less than 10 tonnes per day, the facility's daily total input of materials to be processed is similar to that of a commercial mixed (uranium and plutonium) oxide reactor fuel fabrication plant, which may cost several billion dollars. Since none of the final products of a warhead dismantlement facility are components fabricated to exacting tolerances, it seems reasonable to expect that the capital cost of a new dismantlement facility would be lower.

Labor costs for operating such a facility are unlikely to be greater than ten or twenty million dollars per year. A full-time work force of 100 direct labor employees, at $100,000 per person-year (including overhead), would amount to $10 million per year.

It is therefore unlikely that the total costs of dismantling the world's nuclear warheads, and providing the contained fissile materials for use as nuclear fuel or for direct disposal would exceed a few billion dollars.

## TIMING

It is possible that detailed design and construction of facilities needed to eliminate large numbers of warheads may be the pacing items that determine when the complete elimination process can actually begin.

Optimism about new treaties calling for elimination of many warheads should carry with it a considerable sense of urgency about establishing the means for eliminating the warheads. If it is determined that modification of existing dismantlement facilities in the two countries is not appropriate, designing and building new facilities may be necessary.

If a treaty calling for elimination of large numbers of warheads comes into force before the needed facilities exist, the warheads could, of course, be tagged and sealed in containers, and placed in storage to await completion of the dismantlement facility.

## NEXT STEPS

The concepts and analyses presented in this paper indicate that elimination of identified nuclear warheads that are specified in a nuclear disarmament treaty can be verified with high confidence, without revealing national secrets about warhead designs. Much remains to be done, however, to specify procedures for accomplishing this objective in sufficient detail to provide the basis for negotiated formal protocols and the means for carrying them out.

Two consecutive next steps are therefore proposed:

♦ Establish an official joint US–Soviet working group to design and assess specific procedures and corresponding facilities for verified elimination of nuclear warheads. Work by this group should be given high priority by both nations and not require negotiation of further treaties.

♦ Carry out joint US-Soviet demonstrations of the techniques identified in the first step. These demonstrations would be expected to include some field testing of parts of a warhead dismantlement and verification system. Initial tests could be performed using unclassified mockups of warheads. These could be followed with complete system tests, using warheads from each nation.

Results of these two steps could then be incorporated into negotiated protocols for verification of treaties requiring the elimination of nuclear warheads.

## NOTES AND REFERENCES

1. INF treaty, protocols for verification, December, 1987.

2. See, for example, J. Taylor, J. Barton, and T. Shea, "Converting Nuclear Weapons to Peaceful Use," *Bulletin of the Atomic Scientists*, February 1985; James de Montmollin, "Some Considerations Involving Verification of New Arms Control Agreements," unpublished report, December 1985, and "Value of Fissile Material from Dismantled Warheads as Reactor Fuel," unpublished report, June 1986; E. Amaldi, U. Farinelli, and C. Silvi, "On the Utilization for Civilian Purposes of the Weapon-grade Nuclear Material that May Become Available as a Result of Nuclear Disarmament," report of the Accademia Nazionale dei Lincei, Rome, 23–25 June 1988; and Warren H. Donnelly (Congressional Research Service) and Lawrence Scheinman (Cornell University) "Verification of a Fissile Material Production Cutoff and Disposition of Material Retired from Nuclear Warheads" (Sussex University, England: Programme for Promoting Nuclear Non-Proliferation) 14 November 1989.

3. See table 3.1 and Robert S. Norris and William M. Arkin, eds., "Nuclear Notebook," *Bulletin of the Atomic Scientists*, June 1988 and July/August 1988. Estimates of the numbers of warheads are much more uncertain for the Soviet Union than for the US. The estimated number of Israeli warheads is based primarily on revelations by Mordechai Vanunu in the London *Sunday Times*, 5 October 1987.

4. Frank von Hippel, David H. Albright, and Barbara G. Levi, *Quantities of Fissile Materials in US and Soviet Nuclear Weapons Arsenals* (Princeton: Princeton University, July 1986), PU/CEES report 168. We assume that the uranium-235 and tritium in the Soviet stockpile are each the same as those in the US. Either or both of these assumptions may be far from correct.

5. Derived from Thomas B. Cochran, William M. Arkin, and Milton M. Hoenig, eds., *Nuclear Weapons Databook Volume 1: U.S. Nuclear Forces and Capabilities* (New York: Ballinger, 1984).

6. For Western capacities: *The Nuclear Power Plant Capacity of the Western World*, (Alzenau, West Germany: NUKEM, April 1987), NUKEM special report.

7. For capacities of USSR, Eastern Europe, and China: "World Survey," *Nuclear Engineering International*, June 1987, pp.28–39.

8. See David Albright, "Civilian Inventories of Plutonium and Highly Enriched Uranium," in Paul Leventhal and Yonah Alexander, eds., *Preventing Nuclear Terrorism* (Lexington, Massachusetts: Lexington Books, 1987), pp.265–297.

9. We base the estimate of uranium values on the assumption that the value of uranium-235 (in uranium enriched to about 3 percent in uranium-235) to be used for fuel in the world's nuclear power plants during the next decade is about $38,000 per kilogram. Of this total, $19,000 is accounted for by an assumed price of $66 per kilogram for unenriched $U_3O_8$; $2,000 per kilogram by conversion to $UF_6$ prior to enrichment; and $17,000 per kilogram for enrichment to 3 percent uranium-235 (with 0.3 percent in the depleted "tails"). Enrichment costs correspond to $150/SWU. All these costs are consistent with average costs presented in *NUKEM Market Report on the Nuclear Cycle* (Alzenau, West Germany: NUKEM, April 1987). Separative work requirements for uranium enrichment as functions of enrichment of product and tails are from *Standard Table 5 of Enriching Services* (Washington DC: US Department of Energy).

10. This presumes that worldwide use of nuclear power will continue for several decades and show at least moderate growth. If, for whatever reasons, this should not be the case, and international markets cannot absorb the uranium-235 from nuclear warheads dismantled in the course of vigorous nuclear disarmament, it could be rendered useless for nuclear explosives by dilution with natural or depleted uranium.

11. The estimated global weapon plutonium inventory is about 20 percent of the weapon uranium-235 inventory. Since plutonium-bearing fuel costs much more to fabricate than all-uranium fuel, the value per kilogram of plutonium as feedstock for reactor fuel is also less than that for uranium-235.

12. See, for example, G.L. Harvey et al., "Development of Seals for Safeguards," in *Proceedings of the 22nd Annual Meeting of the Institute of Nuclear Materials Management X* (1981), pp.197–203; and R. Günzel et al., "VACOSS-S, The Secret Production Model of the Variable Coding Seal," in *Proceedings of the 28th Annual Meeting of the Institute of Nuclear Materials Management XVI* (1987), pp.318–319.

13. Alex De Volpi, Argonne National Laboratory, private communication, July 1988.

14. For a description of the use of portal monitoring in the detection of fissile materials, see chapter 10.

15. For a description of techniques for fissile-material detection, see chapter 11. The best approach in the present case is probably use of an external, pulsed 14-MeV neutron source to stimulate fissions in plutonium or uranium, and observation of any resulting delayed fission neutrons or gamma rays emitted from the object being scanned.

16. Richard L. Garwin, private communication, March 1989.

17. J. Carson Mark, Theodore B. Taylor, Eugene Eyster, William Maraman, and Jacob Wechsler, "Can Terrorists Build Nuclear Weapons?" in Leventhal and Alexander, 1987, pp.55–65.

18. Ibid.

19. Derived from data in Cochran et al. See note 5.

# Chapter 6

*Robert Mozley*

# Verifying the Number of Warheads on Multiple-warhead Missiles

The US and the USSR currently have most of their strategic ballistic-missile nuclear warheads on missiles equipped with multiple independently targeted re-entry vehicles (MIRVs). START and subsequent strategic arms control agreements to reduce the numbers of these warheads might alternatively mandate a reduction in the number of missiles capable of carrying MIRVs or in the number of warheads actually deployed on such missiles (or a mixture of both). This latter alternative will be termed "deMIRVing."

In general, reduction in the number of missiles (or, in the case of ballistic-missile submarines, in the number of launch tubes on each submarine) should be the preferred alternative to achieve reductions in the number of ballistic-missile warheads. The danger posed by deMIRVing is that of breakout. It may be possible to quickly restore the substantial payloads of missiles that have been tested with a large number of warheads, but have since been deMIRVed.

The risk of such a breakout is related to the time and effort it would require and to how well any breakout activity could be monitored by the other parties to the deMIRVing agreement.

The procedures used in a breakout would probably be similar to those required for maintenance of the re-entry vehicles (RVs) and the "bus" system, which places the re-entry vehicles on their final individual trajectories. With some intercontinental ballistic missiles (ICBMs) this may require only that a missile be lifted slightly from the silo to allow access to the nose-cone region; or possibly the maintenance could be done in place. In these cases, the removal of the nose cone and any other protective shrouds and either the addition of the extra warheads or the replacement of the entire bus and its warheads with another containing more warheads might take little more than a day for a well-trained crew that regularly performs similar work during maintenance. Were this so, several crews might be able to re-equip hundreds of ICBMs with MIRVs in a month or two. In these circumstance, deMIRVing should be regarded, at best, as a temporary measure, pending agreement on less easily reversible reductions.

On the other hand, if it were necessary to remove the missiles from their launchers, as would most likely be the case with submarine-launched ballistic missiles (SLBMs) and missiles carried by mobile launchers, and to take them to a special facility for RV insertion, a week or more might be required to re-equip a single missile, and a breakout involving hundreds of missiles might require several months and would likely be observed. Arrangements might be negotiated as part of a reduction agreement to make breakout still more difficult and time-consuming and visible. In these circumstances, deMIRVing might be a useful part of an arms-reduction agreement.

In case deMIRVing is incorporated into arms-control agreements, methods of counting the number of warheads on a deMIRVed missile must be developed to ensure confidence that the treaty is being observed. The following discussion deals with the technical problem of ensuring that the number of warheads installed is the number agreed upon. A difficulty is that many details of the busing systems and of the re-entry vehicles and the warheads that they contain may be considered secret by the owners; verification procedures must somehow be able to count the number of warheads without revealing these secrets.[1]

The inspection would preferably be by visual means. If it is felt that this reveals too many design details, the use of penetrating radiation might be considered. Radiation emitted by the uranium or plutonium in the warhead could be used or external radiation could be applied for radiography or to increase the output of radiation from the warhead through induction of fission.

It will be shown that the most suitable technique using penetrating radiation is radiography.

## THE VERIFICATION CONTEXT

Under a deMIRVing treaty, negotiations would limit the warheads that could be deployed on each of a declared set of missiles to some agreed number. These missiles would be subject to on-site inspections, both to monitor the elimination of excess warheads and to monitor that the deMIRVing was not later reversed. Monitoring the destruction of excess warheads coupled to safeguards on the recovered fissile material and a cutoff in new fissile-material production for weapons would make breakout more difficult (see chapters 4 and 5).

The deMIRVed missiles would be subject to random on-site inspections to ensure that they remained deMIRVed. Since any missile found with more than the allowed number of warheads would indicate noncompliance, inspection of a relatively limited number of randomly selected missiles would be likely to expose cheating involving a large fraction of the missiles. If, for example, a party to the agreement cheated by not taking the warheads off of 25 percent of its missiles, it would have a 25 percent chance of getting caught by one random inspection and a 44 percent chance of getting caught by inspection of two missiles. In general, if a fraction $C$ of the missiles is in noncompliance, the probability of catching the violator by inspecting $S$ missiles is approximately $1 - (1 - C)^S$. If some of the deMIRVed missiles were later subject to dismantlement, statistical sampling could also be obtained by selecting randomly the missiles to be dismantled.

Sampling only a few missiles each year would make it easier to arrange for ICBMs to be removed from their silos or SLBMs from their launchers to a facility at which the special access required for x-ray radiography was available. There would also be much less chance of damage to the missile from the inspection apparatus, and radiation barriers could be readily installed to prevent excessive radiation doses to the inspectors. In a situation involving inspection of a small fraction of the missiles, even destructive levels of radiation or other measures that might make the missile unusable could be used as part of a procedure to check the number of warheads.

To simplify direct checks of larger numbers of missiles in the field, it is worth investigating the possibility that the missile payload as a whole might have detectable characteristics that change as the number of warheads on the missile is reduced. This may be a pattern of spontaneous radiation, a pattern of induced radiation in response to low levels of neutron or gamma irradiation, a radiation transmission pattern or a pattern of sound waves generated by an inserted acoustic source. A very detailed examination of a single deMIRVed missile might establish both the number of warheads it carried and such a characteristic set of "fingerprints." The fingerprints then could be used as a template in examining other missiles of the same type to determine whether they had been modified in accordance with the deMIRVing agreement.

Tagging and sealing could also be very helpful. If each of the missiles was inspected at the time that it was deMIRVed, and if it was then tagged and sealed, there would be no need for further inspection except to see that the missiles were still tagged and sealed. This would be a particularly useful alternative if there was concern that the high levels of radiation used in repeated active examinations might damage the missiles. For such a system to work, however, it would be necessary that access to the interior of the sealed volume for maintenance be required very infrequently. An alternative suggested by Garwin[2] is to put dummy RVs in place of the missing ones inside the deMIRVed missile and then tag and seal them. This would allow regular maintenance of the approved RVs without breaking any seals.

It might also be appropriate in some circumstances to tag missiles before deMIRVing. Each of the missiles could then be moved into a controlled area, entering as tagged with the original number of re-entry vehicles. The re-entry vehicles to be removed would be withdrawn from the controlled area in conformable canisters that would be checked to confirm that each contained a warhead. The deMIRVed missile would then be tagged and sealed and returned to its launcher.

## METHODS OF ON-SITE INSPECTION

The four methods for counting warheads during an on-site inspection considered in this paper are:

♦ Visual inspection after removal of the nose cone

♦ Detection of the penetrating radiation spontaneously emitted by the fissile material[3]

♦ Use of neutron or x-ray radiography to reveal the dense core of absorptive fissile material that each warhead contains

♦ Irradiating the warhead region with high-energy neutron or x-ray beams to excite fission in the fissile cores thus increasing their emission of penetrating radiation.[4]

As indicated, only the first method requires removing the nose cone of the missile.

## Visual Inspection

The best verification system—one with minimal technical complexity—would be to remove the nose cone of the missile and any other protective shrouds, and to count directly the number of RVs by visual observation. In spite of the design information that might be made available by this procedure, it is not an unlikely scenario. Pictures of US RVs and delivery buses already exist in the open literature. If the Soviet Union maintains its present attitude toward verification, it will likely agree to visual inspection. Indeed, the procedures negotiated in START for verifications of declarations of numbers of warheads per multiple-warhead missile reportedly involve this approach.

Even if complete visual inspection is not acceptable, restricted inspection might be. Thus, for example, in order to make clear the absence of RVs that had been removed after the nose cone had been taken off, conformal covers could be placed over the RVs to allow a view of them that was adequate for counting but did not give away detailed information about the design of the RVs or their delivery bus. These methods seem best suited to delivery-bus designs with all the RVs mounted in a single plane.

This sort of restricted visual inspection would, in most cases, require good access to the missile payload region. For silo-based ICBMs it should be possible to use those access methods normally used for ICBM maintenance at the silos or in special facilities to which the missiles are moved for servicing.

Submarine-based ballistic missiles and mobile ICBMs would probably be inspected at their special service facilities.

These visual methods of inspection might not work if any slots of the delivery bus are occupied by heavy decoys designed to look identical to the real RVs. Some detail of the decoy might distinguish it from the RV but knowledge of the distinction might help to distinguish it from the RV when in flight. Such decoys would either have to be counted as re-entry vehicles, or techniques such as those discussed below would have to be used to check whether they contained fissile material.

## Passive Inspection

An attractive method for counting nuclear warheads is the use of the penetrating radiation that is spontaneously emitted by the uranium or plutonium isotopes that they contain. The effectiveness of this method depends critically, however, on the materials from which the weapons are constructed.

A combination of neutron and gamma-ray detectors should be able to detect the radiation from warheads containing kilogram quantities of plutonium and/or uranium-238 at a distance of a few meters. Shielding to prevent detection would be impractical, given the limited space and launch mass capacity of strategic ballistic missiles.

In chapter 11, estimates are made of the radiation from four hypothetical fission warhead designs. With three of the designs, which incorporate either large quantities of gamma-radiation-emitting uranium-238 in the tamper or significant percentages of neutron-emitting plutonium-240 in a plutonium core, passive detection was found to be possible. In the case of a weapon using a core of weapon-grade uranium (WgU) and a tungsten tamper, however, the radiation emitted—a total of 20 neutrons and 30 gamma rays per second—is so small that it could be undetectable in the presence of other signals and background.

The examples given in chapter 11 are not actual weapon designs but conform to estimates of the average amount of fissionable material used in warheads, to known sizes and weights of ballistic-missile warheads, and to a general "public" understanding that in the fission trigger of a strategic nuclear warhead, the fissionable materials are formed into a spherical shell, which is then compressed by explosive charges to start the chain reaction. The numbers derived are probably accurate to better than an order of magnitude. They relate only to the fission trigger of fusion weapons. Actual weapons might produce much more radiation if uranium were used in the fusion component of the warhead or if there were present highly radioactive contaminant isotopes such as the uranium-232 that was detected in a Soviet cruise missile warhead in the "Black Sea Experiment" (see chapters 13 and 14). For the purpose of verification, however, it is best to assume that, if cheating were done, there would be a great effort to use low-radiation-emitting materials. It is therefore important to note that it is possible to reduce significantly the emission of neutrons from plutonium weapons by using highly purified plutonium-239 instead of ordinary weapon-grade plutonium, which is 6 percent plutonium-240. The use of very pure plutonium-239 in combination with a tungsten tamper would make this type of weapon almost undetectable. The additional cost of the processing required might be of the order of $1 million per kilogram (see chapter 11)—a significant cost but not out of the question.

In any case, using the spontaneous radiation emitted by warheads to count the number carried by a missile is much more difficult than merely noting the presence of radiation. The radiation from the approved warheads can produce a very large background making essential a collimated detector and a reasonably large signal from the hidden warheads.

## Radiographic Inspection

The specific objective for radiographic inspection can be defined as counting the number of high-density, heavy-metal concentrations within the missile nose cone. Any high-density concentration averaging 7 g/cm$^3$ or more,[5] distributed over a volume with a radius of roughly 10 centimeters, would be considered to be part of a nuclear warhead. Because of payload weight limitations, there is no plausible nonweapon purpose for such masses on a ballistic missile.

Both high-energy neutrons and gammas can penetrate moderate quantities of normal structural materials, including aluminum, steel, carbon, and plastics, and give good transmission contrast for weapon-like configurations.

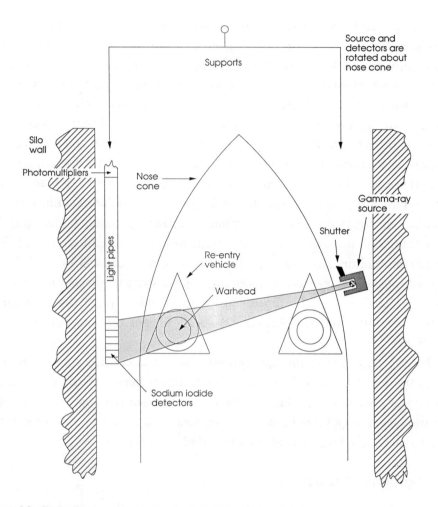

**Figure 6.1:** Radiographic examination of missile warheads in silo

This method of examination would place the source of a neutron, gamma-ray, or x-ray beam[6] on one side of the nose cone and detectors on the other (figure 6.1). Both source and detector would then be rotated around the missile nose cone to produce projected radiographic "images" of the warheads by showing the absorption produced by the fissile material.[7,8] This technique would work best in a configuration with all the RVs located in a circle on a single plane. Some knowledge of the possible positions of RVs on the delivery bus would be needed to make this type of inspection effective.

Although neutron and x-ray accelerator sources can be made small in size and allow some choice of the beam particle energy, isotopic radioactive sources may be more convenient and less expensive.[9] Their use might also be more reassuring to the inspected group because of the inherent limits on the maximum intensity that might be generated "accidentally."

The detection scheme shown in figure 6.1 uses a 1-MeV gamma-ray source. This could be a radioactive isotope enclosed in a well-type container with a remotely operated opening and a collimator. At a greater cost and complexity, an electron accelerator could be used to gain greater penetrating power and control of both intensity and energy.

A very strong cobalt-60 source of about 1 curie emits 300,000 gammas per second into a solid angle of about $10^{-4}$ steradians. On the other side of an approximately 2-meter–diameter missile nose cone, a $20 \times 20$–centimeter array of sodium iodide detectors would be more than adequate to register the transmitted beam. The sodium iodide detectors could be made sensitive only

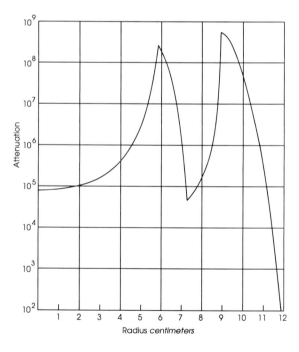

**Figure 6.2:** Attenuation of transmitted gamma-ray intensity from radiographic examination (resolution comparable to diameter of warhead tamper)

to the highest-energy gamma rays so that scattered gamma rays, which always have lower energies, could be ignored. The resolution obtained would be slightly smaller than the size of the individual sodium iodide crystals used.

A diagrammatic cross section of a hypothetical WgU-tungsten warhead can be seen in figure 11.1 while figure 6.2 shows in high resolution the attenuation of 1-MeV gamma rays passing through this warhead as a function of the beam's lateral radial distance from the center.[10]

To prevent the inspection from revealing sensitive design information, the spatial resolution could be limited by controlling the detector size, the length and diameter of the collimator, the number of scan increments, the source strength, and the counting time. Data from the radiation detectors could be delivered at the same time to both host and inspection teams. A preinspection test of resolution could be carried out at the site with a slug of heavy metal (for example, depleted uranium). Radiation-dose monitoring would also be needed to avoid damage to sensitive components.

It would also be possible to produce an algorithm that allowed the

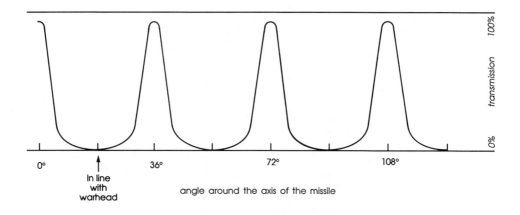

**Figure 6.3:** Transmitted gamma-ray intensity from radiographic examination of a missile with 10 warheads arranged in a circle observed as source and detector are rotated about its nose cone. The resolution (10 centimeters) is comparable to diameter of the warhead tamper. Note that the beam line in figure 6.1 is at an angle so that it intersects only one warhead core at a time.

examination to be performed using a computer interface, which would restrict the detail of the image presented to the verification team of the object scanned. A possible method would have the computer search for large attenuation recorded by a group of sodium iodide detectors and then signal to those persons operating the system that the detector system had shown the presence of an absorbing object at a specified location. The computer program could be jointly produced and monitored by both the verifying and host groups and could be tested on mock-ups.

Figure 6.3 shows the absorption from a circle of these warheads with a resolution just equal to the diameter of the tamper. Only about 10 photons in a million incident would pass through the central region intercepted by the fissile material. The tamper alone causes attenuation by a factor of about 50,000 when the beam passes through the tamper far enough from the center of the warhead that it does not pass through the fissile material. Passage through the thickness of conventional explosive attenuates the beam by less than a factor of 10. Although nuclear warheads may differ from the hypothetical design, they will contain comparable concentrations of heavy metals.

Although a very large cobalt-60 source would be required to allow inspection of a nose-cone region in a reasonable time (an hour), its beam at a distance of a meter would be depositing less than $10^{-5}$ rads per second and at 10 centimeters less than a millirad per second. This is dangerous for human exposure but far too small to damage sensitive weapon components.

Neutrons interact differently from gamma rays, and comparing the transmission of neutron and x-ray beams would allow more information to be obtained about the materials present in the object being examined. This is not necessarily desirable in a verification tool, however, because the missile's owner may wish to keep such information secret. Collimating neutron sources is also more difficult than collimating gamma-ray sources, and the measurement of neutron energy would not seem practical in this type of operation. Gamma rays therefore appear more suitable for the present purpose.

A fundamental difficulty with the use of transmission inspection in the field is that there may not be space to place the radiation source on one side of the missile nose cone and the detectors on the opposite side. In particular in the environment of some launchers, it might be feasible to place equipment only above the nose cone. In this case it would be necessary to raise the missile from its normal position and perform the inspection above the silo or launch tube. It is probable that access to service some silo-based missiles is obtained by raising them in this way. Using radiography to inspect missiles carried by submarines may require removal of the missiles from their launch-tubes and moving them to a nearby facility where radiography could be conveniently accomplished. Similar arrangements may be appropriate for missiles on land-mobile launchers.

## Detection Through Induced Fission

In those situations that do not allow transmission examination, inducing fission with external neutron or gamma beams might be considered as an alternative means of locating masses of fissile material. A possible arrange-ment of equipment for generation of a gamma-ray beam and detection of induced neutrons is shown in figure 6.4.

If "activation inspection" is used to search for extra undeclared warheads on a deMIRVed missile, however, the presence of the declared warheads may cause a severe background. In particular, if the declared warheads have cores containing about 4 kilograms of ordinary weapon-grade plutonium (6 percent plutonium-240), they will be emitting about 400,000 neutrons per second (see chapter 11). A large number of fissions would have to be induced in any undeclared warhead in order for it to become detectable against this very large background.

The most effective particle for producing fission is a neutron, but a neutron beam would be difficult to use in the presence of declared warheads because neutrons in the beam could scatter to these warheads and increase still further the background fission signal from warheads not directly illumi-nated. The low-energy neutrons scattered from the incident beam would be even more effective in causing fission than the higher energy ones present in the original beam.[11] With excellent collimation of the detector, these effects can be reduced, but the use of an energy-sensitive neutron detector would still be required to make possible the rejection of low-energy neutrons scattered into the collimator.[12]

An alternative beam for inducing fission would be an x-ray (bremsstrahl-ung[13]) beam of 10–15 MeV peak energy produced using an electron accelerator. Such an x-ray beam can be collimated to prevent its hitting areas other than that at which it is directed. Its energy would be high enough (over 5 MeV) to stimulate fission and could be varied to give information about backgrounds due to nonfission nuclear reactions. In contrast to neutrons, once a photon has scattered at a large angle it will lose enough energy that it can no longer stimulate fission. Unfortunately, the cross section (a measure of the effective target area presented by a nucleus) for high-energy x-rays producing fission in fissile materials is about a thousand times smaller than that for

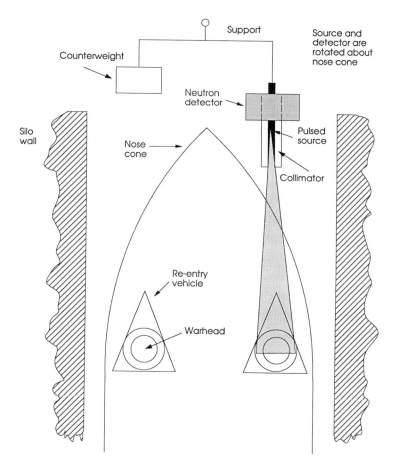

**Figure 6.4**: Search for a hidden warhead by fission excitation

high-energy neutrons. This small cross section can be readily compensated by increasing the intensity of the photon beam from a linear accelerator, but at the cost of more radiation damage.

The particles used to detect induced fission must be different in some characteristic from those of the beam and from those produced by nonfission nuclear reactions caused by the beam. Table 6.1 shows some possible combinations. It would not be possible, for example, to use the detection of prompt gamma rays as evidence of x-ray induced fission because there are too many other sources of such gamma rays. They can come from other nonfission gamma interactions in the material examined and from neutron-induced nonfission interactions (the neutrons coming from spontaneous fission in the declared warheads). Similarly, although the prompt-neutron signal from photofission would be quite strong, the cross section for x-rays to produce neutrons from reactions with nonfissile nuclei could be 10 times larger.

The most distinctive signal from photofission is delayed neutrons. Although there are less than 1 percent as many delayed neutrons from fission as prompt ones, almost no delayed neutrons are produced by a photon beam through nonfission reactions. In contrast, delayed gammas can be produced by nonfissile nuclei excited by a photon beam.

We therefore assume a procedure in which a pulsed photon beam is scanned over the nose cone from above and the delayed fission neutrons are detected by a partly collimated detector (figure 6.4).

It is so easy to produce a well-collimated x-ray beam using a linear accelerator that the distance from the linear accelerator source to the missile nose cone is not a sensitive question. The beam can readily be made so small as to hit only the area subtended by the fissile material in a single warhead from a distance of several meters. In the estimate below of the delayed-neutron signal from a warhead illuminated by such a beam, we can therefore assume that the maximum signal will occur when the x-rays are all incident in a useful region.

An x-ray beam created by the collision of 15-MeV electrons with a dense target material is assumed: such a bremsstrahlung beam contains photons of all energies with approximately equal total photon energy per MeV from 0 to 15 MeV individual photon energy. Only photons with energies above 5 MeV would induce fission. The ratio of the fission cross section to the total cross

**Table 6.1:** Combinations of source beam and resulting radiation useful in detecting fissionable material

| Source beam | Detected radiation |
|---|---|
| Low-energy neutrons | High-energy neutrons |
| X-rays | Neutrons |
| Pulsed neutrons | Delayed neutrons or gamma rays |
| Pulsed x-rays | Delayed neutrons or gamma rays |

section gives the fraction of the incident photons that are usefully absorbed in the fissionable material. In the hypothetical warhead discussed above, the major absorber for the incident x-rays before they reach the fissionable material will be the tungsten tamper. (If a depleted uranium tamper were used, it would be an additional source of induced fission neutrons.) The 3 centimeters of tamper would transmit only about 5 percent of 5–15-MeV x-rays. In the fissile material, an absorption length by electromagnetic interactions is about 1 centimeter and about 1 in 20,000 of the photons reaching the fissile material would cause fission in it. Of those photons with energy greater than 5 MeV incident on the warhead, one in $4 \times 10^5$ would cause fission. As delayed neutrons occur at a rate of somewhat less than 1 percent per fission, the total number of incident photons per delayed neutron would be of the order of 40 million.

If a slab[14] neutron detector with an area of 0.56 square meters and efficiency of 7 percent were located about 3 meters away from a possible hidden warhead, it would detect 1 in 3,000 neutrons emitted by the warhead. The background from the declared warheads could easily be about 400,000 neutrons per second per warhead. If we assume the presence of three of these warheads—also at a distance of about 3 meters—the detector would count about 400 of their emitted neutrons per second. To obtain a five-standard-deviation effect above this background[15] the signal would have to be five times larger than the square root of the integrated background count.

An examination of the time delays for emission of delayed neutrons shows that about half of all neutrons would be observed in about 10 seconds. Unless

there are rate-dependent radiation-damage effects, a good inspection procedure would be to stimulate the fission using a very short beam pulse. The time
could be as short as a microsecond. One would then observe the delayed
neutrons for about 10 seconds.

To express the situation algebraically, $S = 5\sqrt{B}$ where $S$ is the signal
and $B$ is the background,

$$S = \frac{A\varepsilon N_s t}{4\pi r^2}$$

with $A$ the area of the detector, $\varepsilon$ its efficiency, $N_s$ the average number of
delayed neutrons emitted per second from the stimulated material, $t$ the
observation time in seconds and $r$ the distance to the detector,

$$N_s t = IaFDd$$

with $I$ the number of incident photons with energy greater than 5 MeV aimed
at the suspected warhead, $a$ the attenuation factor of the beam before it
reaches the fissile material in the warhead, $F$ the fraction producing fission,
$D$ the fraction of delayed neutrons per fission and $d$ the fraction of the delayed
fission neutrons emitted in the time $t$,

$$B = \frac{A\varepsilon N_B t}{4\pi r^2}$$

where $N_B$ is the number of neutrons emitted by the declared warheads. (It is
assumed that the distance $r$ to the declared warheads is the same as that to
any hidden ones.) The general neutron background is negligible in comparison.

From these relationships we can derive the following expression for $I$, the
required number of photons with energy greater than 5 MeV in the x-ray
beam:

$$I = \frac{5\sqrt{N_B t}\, 4\pi r^2}{aFDd\sqrt{A\varepsilon}}$$

Using the following values of the parameters:

$$N_B = 1.2 \times 10^6$$
$$t = 10 \text{ seconds}$$
$$r = 3 \text{ meters}$$
$$a = 0.05$$
$$F = 5 \times 10^{-5}$$
$$D = 0.01$$
$$d = 0.5$$
$$A = 0.56 \text{ square meters}$$
$$\varepsilon = 0.07$$

we obtain $I = 7.4 \times 10^{13}$ incident photons with energies above 5 MeV to produce a detectable number of delayed neutrons from an undeclared warhead. In the energy region between 5 and 15 MeV a bremsstrahlung spectrum with a peak energy of 15 MeV would contain approximately 1 photon per 15 MeV of energy in the total spectrum ($\ln[15/5] = 1.1$). Therefore $7.4 \times 10^{13} \times 15$ MeV or approximately $10^{15}$ MeV will be deposited by the number of photons needed to make a detectable signal. If the beam irradiates an area equal to the size of the fissile material in the warhead—about 150 square centimeters in the example discussed above—and penetrates to a depth of 20 g/cm$^2$, a mass of about 3,000 grams will be irradiated, and the energy deposited per gram will be about $4 \times 10^{11}$ MeV. Since 1 MeV per gram = $1.6 \times 10^{-8}$ rads this would give a radiation exposure of about 6,000 rads. If the same beam intensity were aimed at a smaller area, as might be appropriate for a plutonium core, the radiation exposure would be correspondingly higher. The inspection procedure could easily double or triple this value. If there was concern about the possible presence of additional shielding, 10–100 times more radiation might be required. Such high exposures would cause radiation damage to sensitive electronics.[16]

## CONCLUSION

In a nuclear-weapon reduction agreement, the achievement of warhead reductions through reductions in the number of warheads carried by existing missiles would be inadvisable if the deMIRVing could be rapidly reversed.

However, if this breakout problem can be reduced to manageable proportions, there are several methods for verifying the deMIRVing. Visual methods of inspection would be the most straightforward but may not always be applicable. The other techniques investigated in this paper employ penetrating radiation. The most suitable of these appears to be the use of radiography. It is possible to design warheads that are very difficult to detect by their spontaneously emitted radiation while the use of neutron-induced fission to increase radiation output is complicated by the likelihood that the radiation background will also be increased through induced fission in the allowed warheads. The use of x-ray induced fission may result in unacceptable large radiation doses to the electronics of the missile.

The choice of system to be used also depends on the available access to the nose-cone area, the structure of the bus system involved (in particular whether it carries more than one layer of warheads), the procedures planned for deMIRVing (whether the missiles will be removed from their launchers during this process), and the confidence placed in tagging and sealing systems. Therefore, in order to determine the methods most appropriate for verification in each specific case, negotiations must provide for some preliminary inspection of the launchers and missiles before the choice of verification system is settled.

## NOTES AND REFERENCES

1. As part of the Strategic Arms Reduction Talks, Soviet and US negotiators have agreed to allow inspection of each other's strategic ballistic missiles to help establish procedures to verify that the missiles hold no more than a declared number of warheads. The Soviets inspected the warhead bus for the US MX missile in a visit to Warren Air Force Base in Cheyenne, Wyoming, on 18 April 1990; inspections of the US Trident II missile and the Soviet SS-N-23 are planned.

2. Richard L. Garwin, "Tags and Seals for Arms Control Verification," 1 August 1988, pp.11,12 (unpublished manuscript).

3. This method was previously discussed in Alex De Volpi, "Expectations from SALT," *Bulletin of the Atomic Scientists*, April 1970, p.6.

4. Ibid.

5. In g/cm³ the density of aluminum is 2.7; iron, 7.9; lead, 11.3; plutonium, 16–18 (depending on crystal structure); and uranium, 19.

6. It is conventional to describe a high-energy photon (energy of the order of 1-MeV or more) as being a gamma ray if it comes from a radioactive nucleus and an x-ray if it is created in a collision of a high-energy electron with a target.

7. Alex De Volpi, *Fast Neutron Radiography with National Security Applications* (Argonne National Laboratory). Summary for submission to ANS Washington DC meeting November 1986.

8. Alex De Volpi, "Applications of Cineradiography to Nuclear Reactor Safety Studies," *RSI*, **55**, 1197 (1984).

9. T. Gozani, *Active Non-destructive Assay of Nuclear Materials, Principles and Applications*, (Washington DC: US Nuclear Regulatory Commission, NUREG/CR-0602) pp.109–120 (gamma-ray sources), pp.83–107 (neutron sources).

10. To a first approximation, a photon or neutron beam passing through a material is attenuated exponentially. Thus, if given thickness of material reduces the beam intensity to one half, twice the thickness will reduce it to a quarter.

11. The fission cross section for a very-low-energy "thermal" neutron can be hundreds of times higher than that for a high-energy neutron.

12. Development work is being done on a double scattering neutron detector capable of obtaining angle and efficiency information concerning incoming fission neutrons. If efficiencies of the order of 1 percent and angular resolution of 10° can be obtained, such a detector could significantly reduce backgrounds and would be very useful in fixed installations. W. Sailor, R. Byrd, Y. Yariv, A. Gavron, R. Hammock, M. Leitch, P. Mc Gaughey, W. Sondheim, and J. Sunier, *A Neutron Source Imaging Detector for Nuclear Arms Treaty Verification*, LA-UR-90-581 (Los Alamos, New Mexico: Medium-Energy Physics Group, Los Alamos National Laboratory).

13. A German word that means "braking radiation." High-energy electrons release x-rays when they are stopped by collisions.

14. The slab detector is described in Gozani, p.218.

15. The natural background radiation would be negligible by comparison.

16. See, for example, G.C. Messenger and M.S. Ash, *The Effects of Radiation on Electronic Systems*, (New York: Van Nostrand Reinhold, 1986).

# Verifying Limits on Nuclear-armed Cruise Missiles

L ong-range ground-launched cruise missiles and their mobile launchers were banned by the 1988 treaty on the elimination of intermediate- and shorter-range missiles (the INF treaty) but thousands of nuclear cruise missiles have been deployed on aircraft and hundreds on ships and submarines. In addition, the US plans to complete the deployment of about 3,000 non-nuclear cruise missiles on 200 US ships and submarines, which are visually indistinguishable from their nuclear counterparts by the mid-1990s. Non-nuclear air-launched cruise missiles (ALCMs) are under development as well. The Soviet Union has deployed hundreds of nuclear and conventional sea-launched cruise missiles (SLCMs), although most of these are short-range.

The SALT II treaty dealt with long-range ALCMs by limiting the types and numbers of aircraft that are permitted to carry them. The START treaty is expected to take a similar approach, including the segregation of nuclear ALCM carriers in designated air bases.

In the case of sea-launched cruise missiles, the US and Soviet Union have agreed to make "politically binding" commitments not to exceed self-imposed limits (presumably on the order of a few hundred) on the numbers of nuclear SLCMs during a five-year period. This approach is a concession to the US position that a treaty limiting nuclear-armed sea-launched cruise missiles could not be effectively verified even with very intrusive on-site inspections. This US view apparently also applies to direct limitations on nuclear air-launched cruise missiles. The START treaty is to limit only the numbers of ALCM-carrying aircraft—stockpiles of non-deployed nuclear ALCMs are not to be restricted.

In a regime involving deeper cuts than those of the START treaty, however, direct limits on nuclear cruise missiles will be needed. Otherwise, secret or overt stockpiles of nuclear cruise missiles could undermine the stability of the regime.

Limits on the stockpiling of nuclear warheads would help, as has been discussed in a previous chapter. But a much stronger system of limits would include mutually reinforcing limits not only on warheads, but also on missiles and launchers. Chapters 7 and 8 consider possible arrangements for verifying direct limits on nuclear cruise missiles and chapter 9 considers such limits on ALCMs. Three basic verification "regimes" are considered, organized into categories requiring "minimal," "intermediate," and "maximal" levels of cooperation.

The minimal level of cooperation would involve very little on-site inspection. This regime would rely on imaging from satellites, interception of communications, and other types of "national technical means." Such intelligence systems could well yield considerable information about loading and testing activities that could make possible reasonable guesses about which air bases and naval vessels have nuclear cruise missiles on them. It would provide little information, however, on how many nuclear-armed cruise missiles were deployed on these air bases and vessels or how rapidly nuclear cruise missiles might be deployed onto other airbases and vessels.

At an intermediate level of cooperation, there would be on-site inspections at the facilities for the production, final assembly, and major maintenance of both nuclear and non-nuclear long-range cruise missiles. At such locations, each side could allow the other to set up permanent portal-perimeter controls such as those that have been set up at two missile-production facilities in the United States and the Soviet Union under the INF treaty. Each missile or missile canister leaving one of these facilities could be checked for the presence of a nuclear warhead, using either passive radiation or radiographic techniques such as those described in chapter 11. A unique tag could then be sealed to each missile or missile canister identified as "nuclear." Then, after a transition period, the discovery of an untagged nuclear cruise missile by any means would be de facto evidence of a violation.

A stronger variant of the intermediate regime would be to require declarations of the locations of cruise-missile storage sites as well as production sites and allow periodic short-notice inspections at these sites to verify the absence of untagged nuclear cruise missiles. The detection of any nuclear cruise missile entering or leaving a nondeclared facility could then be taken as evidence of clandestine production, conversion, or storage activities. A limited number of challenge inspections at suspect sites might also be allowed, with negotiated arrangements to protect national secrets. It might also be agreed to require notifications when SLCMs were being loaded onto or off ships, with provision for the presence of inspectors at shipside to check the SLCMs for nuclear warheads and tags.

At the maximal level of cooperation, inspections would be allowed of cruise missiles and their launchers on ships, submarines, and aircraft. In the case of aircraft, such a regime seems almost to have been achieved in the START negotiations, in that each side has agreed to allow the other to inspect closely both the interior "bomb bays" and the wing attachment points on its long-range aircraft to see that they are not capable of carrying cruise missiles.

Even though warships and submarines are much larger than aircraft, there are relatively few places in them where SLCMs can be stored. Therefore, inspections on most vessels could be limited to SLCM-capable launchers and storage areas, to see that they contained no untagged nuclear SLCMs.

Such arrangements would not answer the concern that non-nuclear cruise missiles might quickly be converted to nuclear SLCMs. Although this is not a serious concern for current US or Soviet long-range SLCMs, it could become serious in the future, because, as is explained in chapter 12, a separated warhead would be quite small and relatively easy to shield from radiation detectors within a ship. Various ideas have been put forward for making such conversion physically difficult or easily detectable.

For any verification regime, it would be important to ensure either that nuclear SLCMs and ALCMs are not interchangeable or that agreements limiting the two types of cruise missiles take the interchangeability into account. Also, the cruise-missile verification problem would be simpler if nuclear cruise missiles were limited without any distinction as to range. Simplest of all would be a limit of zero nuclear cruise missiles. Such a limit might be feasible in the case of SLCMs.

# Chapter 7

Alexei A. Vasiliev

Mikhail Gerasev

Sergei Oznobishchev

# SLCMs: Regimes for Control and Verification

M odern technology has made the cruise missile a strategic nuclear weapon in all essential respects—range, yield, and accuracy—and has thereby created a separate class of systems with a substantial destabilizing potential.

Even a 50-percent reduction in the warheads carried by the three major components of the strategic forces (ICBMs, SLBMs, and bombers) of both sides could create incentives to circumvent a reduction agreement by building up SLCM deployments. After strategic reductions, the relative importance of nuclear-armed SLCMs in the strategic balance would grow significantly, and the uncertainty in their deployed numbers could stimulate a continued arms race.

Claims that the deployment of large numbers of nuclear-armed SLCMs would be much less dangerous than current deployments of ballistic-missile systems appear untenable. Such claims are often backed up by arguing that, because the speed of SLCMs is much less than those of ICBMs and SLBMs, they cannot become first-strike weapons, which allegedly gives them a stabilizing role in the overall strategic balance.

However, the possibility of massive concealed deployment of nuclear-armed SLCMs on a large number of ships and submarines that could operate close to the other side's coasts and the lack of means for reliable early warning of SLCM attacks makes it clear that SLCMs can be just as de-stabilizing as any other kind of strategic nuclear weapon. In addition, the presence of nuclear-armed cruise missiles among hundreds of general-purpose ships lowers the nuclear threshold because such ships could be involved in local naval conflicts. Constraints on nuclear-armed SLCMs therefore appear today not simply desirable but essential. If they are not imposed, the effectiveness of the proposed START agreement could be jeopardized.

Unfortunately, however, the verification of limitations on SLCMs is not simple. SLCMs can be launched from a large number of launchers that are standard on surface ships and submarines, including torpedo tubes on attack submarines, special vertical launch systems (VLSs) that can be installed both on submarines and on surface ships, and armored box launchers (ABLs), which can be placed on the decks of ships (see chapter 8). And conventional SLCMs could in principle be converted into nuclear-armed missiles.

Even sophisticated radiation-detection systems are not, by themselves, a completely reliable tool for verifying limits on deployments of nuclear-armed SLCMs (see chapters 10–12). Therefore, we must consider more complex approaches.

The main verification methods that are available are national technical means (especially satellite-borne telescopes), technical systems for the detection of fissile materials, cooperative arrangements, and inspections (permanent monitoring or periodic or short-notice challenge inspections).

These verification methods can be combined into three basic verification regimes, which can be characterized by different levels of cooperation:

♦ Minimal requirements for cooperation (verification primarily by national technical means)

♦ Intermediate levels of cooperation (including portal monitoring of facilities producing SLCMs)

♦ Maximal plausible levels of cooperation (involving inspections covering every phase of a missile's life cycle including deployment).

The INF treaty illustrates the possibility of close to maximal cooperation.

## DIFFERENT POSSIBLE LIMITATIONS

### Complete Elimination of Nuclear-armed SLCMs

Of all the possible limits on nuclear-armed SLCMs, complete elimination of both long-range and short-range nuclear SLCMs would represent the smallest verification problem: if the history of arms-control agreements has produced one important maxim, it is that it is easier to verify the absence of an entire class of weaponry than its partial reduction.

The elimination procedures in the INF treaty could be borrowed for nuclear-armed SLCMs. The only important remaining problem would be to rule out the concealed manufacture of nuclear-armed SLCMs in the future. This problem would be complicated by the continuing production of SLCMs with conventional warheads and the possibility of their being secretly armed with nuclear warheads.

Two possible approaches for dealing with this situation are:

♦ To set up portal monitors at facilities producing conventional SLCMs. This level of cooperation has already been agreed in the Protocol on Inspections of the INF Treaty (Article 9). It could be quite effective for verifying that no nuclear-armed SLCMs were being produced at declared SLCM production facilities. An extra guarantee could be provided by similar monitoring arrangements at docks where SLCMs are loaded onto ships and submarines.

♦ To seal non-nuclear SLCMs in special canisters at portal monitoring posts. Periodic inspection of the seals would make risky the rearming of the missiles with nuclear warheads. A description of such an approach is given below.

### Non-zero Limits

Non-zero limits on nuclear-armed SLCMs could involve an overall limit on all SLCMs and a sublimit on nuclear-armed SLCMs. In addition, there could be limits on the submarines and ships carrying nuclear-armed SLCMs and sublimits on the number of nuclear SLCMs that could be carried by each vessel. Taken separately, each of these limits presents verification difficulties. In combination, however, they could help ensure effective verification.

Verification of the limitations would require portal checks at SLCM production facilities and an initial data exchange on the numbers of nuclear-armed SLCMs already produced.

The initial data exchange should include in addition:

♦ The basic dimensions and weights of the missiles

♦ The locations of production, final assembly, and storage sites; and of sites at which missiles are loaded onto ships and submarines

♦ The types of vessels carrying SLCMs, the types and numbers of launchers each vessel possesses, and numbers and capacities of any in-ship SLCM storage areas.

This initial data exchange might also cover some shorter-range cruise missiles and their launchers not limited by treaty because of the concern that these other cruise missiles might be upgraded in range or converted to nuclear-armed missiles.

## VERIFICATION REGIMES

### Minimal Cooperation

Under the minimal-cooperation regime, verifying restrictions on the numbers of missiles themselves would be extremely difficult, even if each side agreed to give nuclear-armed SLCMs observable differences and to display them in the open at production sites, final assembly sites, and on ships and submarines for the time required for counting. Such arrangements could not prevent covert production or deployments or provide verification that nominally conventional SLCMs did not in fact carry nuclear weapons.

Verifying limits on numbers of potential launchers might be feasible using national technical means. Indeed, the strongest basis for restricting the numbers of cruise missiles under this regime would be constraints on the numbers of their launchers. For example, vertical launchers on submarines might be limited in this way. The same approach could be applied to verify limits on both vertical-launch systems (VLSs) and armored box launchers (ABLs) on surface ships. However, such limits would be made more difficult by the fact that VLSs on US ships are used to carry weapons other than SLCMs. Similarly, limiting the total number of torpedo tubes on submarines to a small number is impossible, since their primary purpose is for other battle functions. The small inner volumes of submarines, however, would make it impossible to exceed a limit of perhaps 10 torpedo-launch SLCMs per submarine by any significant margin without detracting from a submarine's ability to perform its primary missions—hunting enemy submarines and ships.

To exclude the possibility of launchers being reloaded at sea, it might be enough to monitor the absence of special equipment for reloading alongside cruise-missile launchers with a prohibition against moving such equipment onto the ships.

Thus, the minimal-cooperation regime could make verifiable certain aspects of cruise-missile limitations but a number of other aspects would remain unverifiable, including the problem of a concealed stockpiling of nuclear-armed SLCMs at sea and/or at facilities on land.

## Intermediate Cooperation

The verification situation could be somewhat improved if each side agreed to on-site monitoring at SLCM production and final assembly sites. The total number of nuclear-armed SLCMs produced could be verified by continuous monitoring at the portals of these facilities.

At the time of the agreement's implementation, already-deployed nuclear and conventional SLCMs would be moved to shore for tagging or sealing.*

As is explained in detail in chapter 11 and its appendixes, it should be possible to check containerized missiles radiographically for the presence of nuclear warheads. A failure by the radiation to pass through a shipping container or a missile would indicate a hidden radiation shield, which might be screening a warhead from detection.

Radiographing containers and comparing their images with data

---

* Since it would be difficult to verify that no missiles had been secretly held back, confidence in a future limitation agreement would be improved if there were a moratorium on further deployments of nuclear-armed SLCMs until the verification arrangements were in place.

exchanged at the time of the initiation of the agreement could confirm the missile type. However, verification of the ranges of future missiles could become ever more complicated. Therefore, an important simplification might be to count all nuclear-armed cruise missiles of whatever range against the agreed limit.

Clandestine production operations would become very difficult if portal controls were extended to SLCM storage and to the ports and bases where SLCMs are loaded onto ships and submarines.

Such a system could track every nuclear-armed SLCM if each was provided with a unique identifying tag that could be checked during portal inspections and on-site checks. A range of potentially applicable tagging techniques has already been identified, including techniques for registering the unique and irreproducible characteristics of the materials of parts of the SLCM or its canister. The tag could be checked at a portal where a SLCM was being loaded onto a ship or submarine. This would make it possible to establish where the missile had been produced and where it had been supplied with a nuclear warhead. The discovery of any untagged nuclear SLCM would be evidence of a definite treaty violation.

Because all SLCM-related facilities would be declared, it would become easier to identify clandestine activities by national technical means. Standard itineraries for both the missiles and their carriers would make verification simpler. For example, SLCM-armed ships and submarines could be limited only to ports with monitored portals for checking SLCMs for tags and nuclear warheads.

To prevent conversion to a nuclear-armed weapon, a conventional missile's warhead section could be sealed with a device that would make it impossible to detach the warhead from the missile without the seal being cut. Such a seal could, for example, be a fiber-optic net. Checks on the seals could be made during short-notice challenge inspections. There are also proposals for purely technical methods of checking the status of the seals, in which the inspected side would, on request, relay coded information from an electronic seal to other side via satellite.

Verifying the nonconversion of conventional SLCMs would also be simplified if each side agreed not to manufacture SLCMs with demountable front sections. To ensure compliance with this restriction, it would be enough to provide for challenge inspections of missiles, removed from their shipping containers, at any verification post of the inspector's choice.

## Maximal Cooperation

With adequate and effective monitoring arrangements achieved on land, the principal remaining task of verification at the maximal level of cooperation would be to run random spot checks on ships and submarines.

The smaller the number of ships and submarines armed with nuclear-armed SLCMs, the more effective the verification arrangements.

Restrictions on armored box launchers (ABLs) and vertical launch systems (VLSs) could be easily verified by simply counting them. And inspector groups should be able to satisfy themselves that the launchers have no rapid-reload mechanisms.

The most complicated inspection task on ships and submarines would be to detect any clandestinely deployed nuclear SLCMs. Because of the possibility of blocking the radiation from warheads with shielding, this could only be done by on-ship inspection. One could make provision for the display of nuclear weapons on the deck of the ship or on the dock for a visual check, together with an instrumented check of a ship's inside for nuclear weapons. It would be necessary to have agreed procedures by which to distinguish long-range nuclear-armed SLCMs from any other nuclear weapons that might be present. The inspectors could also check the integrity of seals on conventional SLCMs.

All these arrangements are much more intrusive than the portal inspections discussed above and would still not entirely eliminate the possibility of nuclear-armed SLCMs being hidden on ships or submarines.

## CONTROL OVER MODERNIZATION

The upgrading of short-range nuclear-armed SLCMs could dramatically extend their range and thereby considerably increase their ability to hit targets deep inside enemy territory. The absence of any design improvements capable of achieving this could be verified most effectively through the testing of short-range nuclear-armed SLCMs in the presence of an inspector group from the other side.

The inspecting side would have the right to pick out SLCMs for testing, either from ships or submarines or from storage, final assembly, or other sites. This would make certain that the tested systems were typical of those in service. Each missile would be carrying tags identifying its production and final-assembly facilities.

A possible testing procedure would have the missile's warhead removed, weighed, and replaced with weights and a radio transmitter so that the total replacement weight was the same as that of the warhead. An approximate test flight route would be selected—perhaps circular with the missile to land in shallow water.

Before the test, the inspectors would satisfy themselves that the missile's fuel tanks were filled up. During the test, continuous tracking of the missile with the help of the radio transmitter would make it possible to compute the length of the flight. When the missile was retrieved after landing, the inspecting side would check the tanks to make sure that all the fuel has been used up, while the tag would confirm that the missile was the same one that had been checked before launch.

Of course, the necessity for these verification procedures would not exist if all nuclear-armed SLCMs were limited, regardless of range.

# Chapter 8

*Valerie Thomas*

# Verification of Limits on Sea-launched Cruise Missiles

Sea-launched cruise missiles (SLCMs) have become a serious problem for nuclear arms control. To some extent, this is a consequence of their physical attributes: they need not be deployed in special launchers, and there are both nuclear and non-nuclear versions.

The US long-range SLCM—the Tomahawk—comes in non-nuclear land-attack and antiship versions, and in a nuclear land-attack version. It has been deployed on attack submarines for launch from torpedo tubes and from vertical launch systems, and on surface ships in armored box launchers and in vertical launch systems.

The Soviet long-range SLCM, known in the US as the SS-N-21, has been reported to be nuclear only and to be deployed on submarines for launch from torpedo tubes and from special launchers. A supersonic SLCM, the SS-NX-24, is reported to be under development. (Further details of the US and Soviet SLCM programs are given in appendix 8.A.)

In the START talks, Soviet negotiators sought verifiable limitations on SLCMs, and proposed a system of intrusive verification measures. The US proposed only a "declaration of intent" on SLCM deployment, with no verification measures, which, at the time of this writing, the Soviet Union appears to have accepted.[1,2] Nevertheless, in a context of deeper cuts, it is likely that the issue will have to be revisited.

A central factor in the impasse was the asymmetry between the SLCM programs of the two countries: at the time of writing, the US had deployed far more long-range SLCMs than the Soviet Union. The United States had bought about 2,500 Tomahawk SLCMs as of early 1988 and planned to deploy a total of almost 4,000. Of the 4,000, more than 80 percent are to be non-nuclear, though nearly identical in appearance to the nuclear version. Furthermore, Tomahawks are being deployed on both surface ships and submarines; by the mid-1990s about 200 vessels are intended to have them. In contrast, reports suggest deployment of Soviet long-range SLCMs on only a few submarines.

Therefore, at present, limits on long-range SLCMs would constrain the United States more than the Soviet Union, and on-site inspection of deployed SLCMs would involve greater numbers and types of US ships than Soviet ships.

US government statements about the intractability of the verification of SLCM limits seem to derive less from concern about the threat from Soviet SLCMs than from apprehension about restrictive limits on US SLCM deployment and intrusive verification arrangements. There is also concern that intrusive verification would threaten the US Navy's policy to "neither confirm nor deny" the presence of nuclear weapons on any particular ship.

Even intrusive arrangements would not, however, completely eliminate the uncertainties of SLCM verification. Some on-site inspection could be helpful, but increasingly intrusive measures would provide diminishing returns, as has been noted by some members of the US House Select Committee on Intelligence,[3]

> We face a painful dilemma....We must consider frankly whether major cumulative arms control risks are more or less dangerous than an absence of real or theoretical restrictions on Soviet military power....The current refrain is that treaties must and will be "verifiable."...Politicians and the public must realize, however, that there is no way we could afford to develop collection capability providing 100 percent certainty that the Soviets are or are not violating major arms limitations. And even with unlimited funding, such capabilities often would not be achievable....
>
> Since so many key weapons and capabilities will be difficult to monitor, treaties truly focusing only on clauses monitorable with high confidence

often will be virtually irrelevant and almost certainly will not reduce the overall threat, because military buildup easily could be diverted to non-treaty categories.

This chapter will focus on limits on *long-range nuclear* SLCMs, though auxiliary limits—on non-nuclear SLCMs, short-range SLCMs, and ALCMs—will also be considered. The purpose is to provide the technical information necessary to understand the feasibility and implications of various plans for SLCM verification, and to provide the information necessary to judge the risk of clandestine treaty violations.

## SLCM VERIFICATION PROBLEMS

The main SLCM verification tasks are to ascertain the number of SLCMs produced or deployed and to distinguish between nuclear and non-nuclear SLCMs. One must also take into account the possibility of conversion from non-nuclear to nuclear, of transformation of ALCMs or large short-range SLCMs into long-range SLCMs, and of secret production and storage.

### The Counting Problem

Nuclear SLCMs can be launched from standard torpedo tubes and other multipurpose launchers. Thus the number of potential SLCM launchers is much larger than the number of nuclear SLCMs likely to be deployed. The similarity of US nuclear and non-nuclear SLCMs is a further complication.

Similar difficulties arise for the counting of air-launched cruise missiles (ALCMs), in that more ALCMs could be deployed on long-range aircraft than actually will be deployed. At the START negotiations, it has been proposed that verification of ALCM limits be accomplished by designating certain bomber aircraft to be nuclear ALCM carriers and crediting each of these aircraft with a specific number of nuclear ALCMs according to agreed "counting rules." A similar approach might be used for nuclear SLCMs.

The same approach could work fairly well for Soviet SLCMs, assuming that they all can be counted as carrying nuclear warheads and as long as their deployment remains limited to a relatively small number of submarines. The SS-NX-24, currently under development, is so large that its launchers will be unique and recognizable by satellites. The SS-N-21 is smaller and can, in principle, be launched from any standard 533-millimeter torpedo tube but has reportedly been deployed for torpedo launch on only the most modern Soviet attack submarines. It may be that the SS-N-21, like the US Tomahawk, requires a sophisticated fire-control system, available only on new or refitted submarines. Assuming the SS-N-21 deployments are limited to submarines, each type of submarine from which the SLCM has been tested might be counted as carrying some fixed load of SLCMs.

Designing counting rules for US SLCM deployments would be more complicated, because of the variety of US SLCM launchers and because most US SLCMs are non-nuclear.

In addition to torpedo tubes on attack submarines, the US Navy has developed three other types of SLCM-capable launchers (see figure 8.1): the armored box launcher (ABL2) and vertical launch system (VLS) for surface ships; and the capsule launch system (CLS), which is being installed on new *Los Angeles*-class submarines. The ABL and CLS hold only Tomahawk SLCMs, and, in each case, the number of launchers is externally visible and could be counted from satellites (see figures 8.2 and 8.3). The number of VLS launchers can also be counted, but, because VLSs also carry the Standard surface-to-air missile (see figure 8.4), the US might be reluctant to have all VLSs counted as carrying SLCMs.

The most serious complication with the counting-rule approach to US SLCMs, however, is that only one in five US Tomahawks is planned to be nuclear. Counting rules for SLCMs could therefore significantly overestimate the number of US nuclear SLCMs.

Another approach to the counting problem would be to implement a system of on-site inspections. Monitoring of SLCM production facilities and maintenance and storage sites would allow direct verification of the number of nuclear SLCMs. Spot-checks of SLCM launchers on ships and submarines could directly verify limits on nuclear SLCM deployments.

A ban on nuclear SLCMs would essentially eliminate the counting problem. While verification measures might include some inspection of SLCM launchers, the primary compliance information could be provided by monitoring of the destruction of all nuclear SLCMs, and observation of the elimination of all nuclear SLCM support facilities and related activities.

## The Warhead Switching Problem

It is possible that a non-nuclear SLCM could be transformed into a nuclear SLCM by replacing the warhead. Through such a procedure, a missile that had been designated as non-nuclear could later become nuclear. At present only the US has deployed non-nuclear long-range SLCMs, and conversion between conventional and nuclear versions of the US Tomahawk would be a complex operation. According to Admiral Stephen J. Hostettler, who was then director of the Joint Cruise Missile Project in the US Department of Defense,[4]

> The current cruise missile is a highly complex vehicle which was not designed for field maintenance. Each missile is thoroughly tested before it leaves the factory and remains intact until it is fired or returned for recertification in 30–36 months. During the period the missile is in the fleet, electrical continuity is maintained. To change a variant from conventional to nuclear or vice versa would require replacement of the entire front one-third of the missile. Nuclear surety requirements would dictate a complete retest of the missile requiring each ship be outfitted with highly sophisticated test equipment and highly trained technicians to interpret the results. Clearly, this is beyond the scope of normal Navy maintenance concepts, and will be performed only at shore-based depots. The capability to modify variants in the fleet is not planned for the Tomahawk.

In the future, transformation of non-nuclear weapons into nuclear weapons might become easier. The United States' Lawrence Livermore National Laboratory (LLNL) has developed an "insertable" nuclear warhead for other missile systems. If an insertable nuclear warhead were developed for the cruise missile, conventional cruise missiles could quickly be converted to nuclear. An insertable nuclear warhead was considered for the short-range Harpoon cruise missile; the idea was rejected primarily for arms-control reasons.[5] Agreement not to deploy an insertable nuclear warhead system would be useful.

If warhead switching is taken to be a serious problem for SLCMs, occasional inspections of SLCMs designated as non-nuclear could ensure that nuclear warheads had not been installed. However, this would not remove the potential for warhead transfer, since nuclear warheads could be installed after breaking out of the treaty.

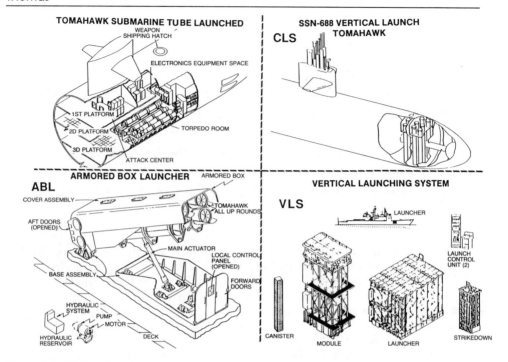

**Figure 8.1:** US Tomahawk launch systems

Source: Cruise Missile Project, US Navy

**Figure 8.2:** Open Tomahawk CLS hatches on a *Los Angeles*-class submarine

Source: US Navy

**Figure 8.3:** Armored box launcher

Source: Cruise Missile Project, US Navy

**Figure 8.4:** Standard missile being loaded into a VLS

Source: Martin Marietta

## The ALCM Problem

In principle, ALCMs and SLCMs differ only in their launch platforms; it therefore might be possible for ALCMs to be launched from SLCM launchers with only minor modifications. If so, a cruise missile designated as an "ALCM" could be used as a SLCM.

The US ALCM, however, is not designed to be launched from Tomahawk launchers. Its diameter is 69.3 centimeters, compared to 53 centimeters for the SLCM. Therefore the US ALCM is too wide to fit in the standard 533-millimeter torpedo tube or other currently deployed SLCM launchers.

However, the Soviet ALCM and SLCM may be more similar,[6] and it is possible that the Soviet ALCM could be launched from SLCM launchers with only minor modifications.[7] A compatible system of ALCM and SLCM limits would remove any advantage of switching from one category to another.[8]

## The Short-Range SLCM Problem

During the Senate hearings on SALT II, it was pointed out that the range of Soviet short-range SLCMs could be upgraded by the substitution of better guidance systems and more efficient propulsion systems.[9] The SS-N-12 and SS-N-19 have a range of 550 kilometers, just short of the usual 600-kilometer definition of "long-range" and are, in fact, larger than the SS-N-21 long-range SLCM (see appendix 8.A). In contrast, the US short-range cruise missile, the Harpoon, has a range of only about 100 kilometers and is only 34 centimeters in diameter and 4.6 meters long.[10] Although some increase in range is possible, it is too small to have the 600-kilometer-plus range of "long-range" SLCMs.

The range of a missile depends on the efficiency of its engine, the amount of fuel carried, and the weight of all missile parts. An older missile could be upgraded by substitution of a more efficient engine, or by use of a lighter warhead. Non-nuclear warheads can be considerably larger and heavier than nuclear warheads. Replacing a non-nuclear warhead with a nuclear one may not only reduce the payload weight, but may also increase space available for carrying fuel. This is why the range of the US nuclear

Tomahawk (2,500 kilometers) is so much longer than the range of the non-nuclear land-attack Tomahawk (1,300 kilometers). Without knowing more about Soviet short-range SLCMs, however, it is difficult to judge the plausibility of substantial increases in range. But short-range cruise missiles could be distinguished from long-range SLCMs on the basis of a combination of size, engine, and other design features. The technology used for large but short-range missiles is probably old, so little risk would be taken in demonstrating the inefficiency of the engine. Small short-range missiles, such as the Harpoon, may rely on sensitive technology, but their size in itself demonstrates that they are short-ranged.

In any case, it is likely that testing would be necessary for a significant upgrade; these tests might be picked up by national intelligence sources.[11]

The problem of distinguishing short-range SLCMs from long-range SLCMs could be eliminated by limiting shorter-range SLCMs as well.

## The Problem of Secret Production or Stockpiles

It is possible that SLCMs could be produced and stored secretly. In principle this is a problem for limits on any weapon system, but it is more relevant for SLCMs because, while ballistic-missile production facilities tend to be distinctive and identifiable by satellite reconnaissance, cruise-missile production facilities do not have distinctive visual characteristics. SLCM storage sites could be even less conspicuous than production facilities.

The possibility of excess production at declared production facilities could be addressed by on-site monitoring at these facilities. The possibility of secret production sites could be addressed by provisions for challenge inspections, which could be particularly useful at production sites for similar weapons such as short-range cruise missiles.

Although the reliability of challenge inspections of suspected storage sites is limited by the difficulty of identifying likely "clandestine storage sites," challenge inspections do provide a mechanism for checking a suspected treaty violation that has not been clearly established by intelligence information.

## SLCM VERIFICATION APPROACHES

Three approaches to verification of limits on long-range nuclear SLCMs will be considered below, in order of increasing intrusiveness.

### Minimal Inspection

A minimal-inspection approach would rely primarily on declarations and on satellite reconnaissance and other national intelligence to monitor treaty compliance. Declarations could detail the number of SLCMs produced and deployed, and, if nuclear SLCMs were not banned, the number of nuclear SLCMs on each ship and submarine. The approximate number of deployed long-range SLCMs might be estimated with reasonable confidence without on-site inspections, but confidence in the number of these that were non-nuclear would be more difficult to achieve. Because the USSR has not deployed a non-nuclear long-range SLCM, current Soviet deployments are relatively easy to monitor but verifying declarations of the number of US nuclear SLCMs in the much larger population of externally identical non-nuclear Tomahawks would be much more difficult.

In the context of a treaty, minimal inspection might be most acceptable if nuclear SLCMs were banned.

### Intermediate Levels of Inspection

A significant improvement in information on the SLCM arsenals could be achieved by monitoring declared production facilities and maintenance facilities. Figure 8.5 shows schematically how an inspection scheme might be designed.

Any SLCMs to be destroyed would be destroyed in the presence of inspectors. For any remaining missiles, inspections would occur at the final production facilities for new missiles or at the maintenance sites for old missiles. At the first inspection, each missile could be tagged for future identification, and possibly sealed. On subsequent inspections—the end of every maintenance cycle—SLCMs would be inspected to check that only tagged missiles passed through the maintenance depots, and that the type of warhead, either nuclear or non-nuclear, matched the tag. If short-range SLCMs were being produced, inspectors could verify that only the desig-

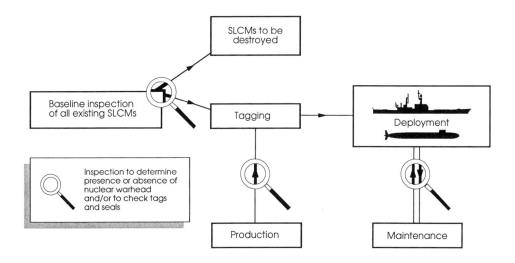

**Figure 8.5**: Schematic diagram of a SLCM monitoring system with an intermediate level of intrusiveness

nated engines, fuel tanks, and warheads were installed.

To reduce concern over warhead exchange at naval bases or on submarine tenders, they might be opened to inspection. Alternatively, non-nuclear SLCMs could be kept in sealed containers that were to be opened only at the designated depot maintenance sites with portal-perimeter controls.

In the United States, SLCMs are brought back for maintenance about once every three years.[12] On this schedule, after three years all SLCMs would have been inspected and tagged, and they would be reinspected at three-year intervals. Since depot maintenance in the US is done at the production facilities, monitoring need occur only at the two production sites.

Tomahawks are produced by General Dynamics Convair Division in San

Diego, California, and by McDonnell Douglas in Titusville, Florida. Figure 8.6 shows Tomahawk production at General Dynamics. The warheads of the non-nuclear Tomahawks are installed at these facilities, but the nuclear Tomahawks are shipped from the factory without warheads. Nuclear warheads are manufactured by the Department of Energy and are installed at naval weapon facilities, of which there are about 16.

Fissile-material detectors would therefore not be able to distinguish non-nuclear SLCMs from prenuclear SLCMs at US production facilities, even though there are internal structural differences. However, each non-nuclear SLCM could be subjected to inspection for fissile material (just to make sure) and its canister tagged and sealed, so that in subsequent inspections it could be identified as non-nuclear. Canisters containing the prenuclear SLCMs could be tagged as identified nuclear SLCMs. These canisters would be opened later, at a naval weapon facility, for warhead installation. Portal monitors could be established at these facilities, though

**Figure 8.6:** Tomahawk final assembly
Source: General Dynamics, Kearney Mesa, California

this might not be considered necessary since these missiles would already be tagged as "nuclear."

There are a number of variations on this approach: for example, all SLCMs might be brought in for an initial inspection and tagging during a period of months after the treaty came into force. But the essential feature that distinguishes this from a maximal-inspection approach is that there would not be regular inspections of ships, submarines, or ports. The plan could be strengthened by data exchanges on the number of SLCMs deployed on each vessel—possibly not distinguishing nuclear from non-nuclear. As with the minimal-inspection regime, this database would support national technical means of verification, since any missile spotted out of place would be evidence of a violation.

A limitation of this method is the absence of a guarantee that all SLCMs will be brought to the designated maintenance site. But if the treaty were to be violated in secret, a separate set of untagged SLCMs and associated production and maintenance facilities would need to be maintained. Such a complex would involve several facilities and many people and would be difficult to conceal.

Sixteen naval ordnance shore facilities have been selected to support the US Tomahawk: six supporting Atlantic Fleet Units and 10 supporting Pacific Fleet Units.[13] Requirements include waterfront facilities, magazines suitable for the storage of high explosives, security facilities, and facilities for intermediate maintenance. The Soviet Union may have fewer support facilities for its long-range SLCMs, because it has both fewer ports and fewer long-range SLCMs.

Monitoring of such storage sites could be continuous portal-perimeter monitoring or it could be limited to periodic inspections. Continuous portal monitoring at these facilities would be a major undertaking, which might be complicated by other naval operations at the same facilities.

The Soviet Union has proposed that facilities suspected of storing clandestine cruise missiles be subject to challenge inspections. Certain highly sensitive facilities, such as military command posts or intelligence centers would be exempted from inspection.[14]

The usefulness of challenge inspections would depend on the likelihood of receiving information on possible clandestine storage sites. Storage sites for nuclear weapons might, for example, be distinguished by extra security and extra fences, but it would be difficult to distinguish storage sites for nuclear SLCMs from storage areas for other naval nuclear weapons. Cruise missile production facilities do not have obvious external features to aid their discovery by reconnaissance satellites. Nevertheless, SLCM manufacture would probably be easiest to hide in facilities that turn out similar products, such as ALCMs, drones, SRAMs (short-range attack missiles), and the Harpoon in the United States, and similar systems in the Soviet Union. Shipment of missile parts in and out of such facilities would not be unusual, nor would employment of a specialized work force or use of specialized machinery.

## Maximal Inspection

A maximal-inspection approach would include inspections of ships and submarines, and SLCM monitoring in all locations from manufacture to elimination. All locations at which SLCMs might be present would be declared, so that any sighting of a SLCM at another site would be evidence of a violation.

A version of the maximal-inspection approach has been proposed by the Soviet government as follows: SLCM production and warhead installation facilities would be subject to portal-perimeter monitoring, where missiles could be tagged and sealed. Inspectors would be stationed at ports to monitor loading of SLCMs into ships and submarines. Suspect sites on land would be subject to challenge inspection. SLCM deployment would be restricted to two types of submarine and one type of surface vessel. Limited inspections of other ships and submarines in ports might be allowed to verify the absence of cruise missiles.[15]

But a maximal-inspection regime need not follow this plan. In particular, restriction of SLCMs to only two types of submarine and one type of surface ship is unlikely to be favored by the United States, which has already deployed SLCMs on several types of surface ship. And although the Soviet Union has proposed stationing inspectors at ports to observe all SLCM loading and unloading, it may be simpler, less intrusive, and more effective to have occasional challenge inspections on ships and submarines (see below).

For the United States, a maximal-inspection regime would include inspection of two SLCM production sites, 16 Tomahawk naval ordnance facilities, about 200 ships and submarines, and any sites chosen for challenge inspection.[16] In the Soviet Union there would be fewer sites to be inspected, with fewer than 10 SS-N-21 SLCM submarines reported at the time of this writing (see appendix 8.A).

Allowing access to ships, submarines, and naval facilities raises the possibility of revealing military secrets. It also causes problems (for the United States at least) for of maintaining a policy of "neither confirming nor denying" the presence of nuclear weapons on particular ships.

The least intrusive inspection arrangement would be to restrict inspections to ships in port, and to restrict on-ship inspections to the vicinity of designated SLCM launchers (or to weapons removed from these launchers). For US SLCMs at least, there are few possibilities for storing SLCMs other than in the launchers.[17] These inspections would not preclude the possibility of transfer of possibly nuclear-armed SLCMs to a ship during a subsequent undisclosed visit to another port.

Each launcher configuration would require customized inspection procedures. The US ABL sits above deck, carries four SLCMs, and does not carry other weapons. If the doors of the ABL were opened, inspectors could check tags on the missile canisters (see figure 8.3). Nuclear versions could be identified either by a special tag or by passive fissile-material detection.

The CLS is a vertical launch system that will be built into US attack submarines and will house 12 Tomahawks. Since it is outside the inner pressure hull of the submarine, it cannot be reloaded from inside the submarine. The CLS will be dedicated to SLCMs and could be inspected fairly easily by opening its hatches from the exterior of the submarine (see figure 8.2).

The US VLS presents more of a problem, since it also stores other weapons. It is built into the ship so that the tops of the launchers are flush with the deck. Much of the VLS capacity is to be taken up by the non-nuclear Standard antiaircraft missile.[18] The VLS system is designed for reloading at sea, though the accompanying crane is certified as only strong enough to transfer the Standard missile, not the heavier Tomahawk. The VLS hatches could be opened for inspection, fissile-material detection, or unloading for a more detailed inspection (see figure 8.4, showing loading of the VLS system).[19]

US (and Soviet) SLCMs are also deployed in attack-submarine torpedo rooms for launch from torpedo tubes. In a US attack-submarine torpedo room there is room for about two dozen weapons.[20] Not all of this space is likely to be taken up by Tomahawks: other torpedo-room weapons include torpedoes, the short-range Harpoon cruise missile, and decoys for antisubmarine warfare. Determination of the contents of a torpedo room would require access to the interior of the attack submarine. Because of the reluctance of the US Navy (and perhaps the Soviet Navy) to allow inspections in submarines, monitoring the loading and unloading of the SLCMs, as the USSR has proposed,[21] could be a preferable alternative.

Under current US plans, SLCMs will soon be the only *nuclear* weapons deployed in SLCM-capable launchers. The US does not have a nuclear torpedo; the nuclear SUBROC, which can be launched from submarine torpedo tubes, is being phased out. Therefore, if arrangements were made to allow detection of nuclear weapons in SLCM-capable launchers or being loaded into torpedo rooms, any US nuclear weapon detected could be counted as a nuclear SLCM.

Even with so much inspection, however, there are possibilities for undetected violation or for sudden breakout from the treaty. The primary violation scenarios are: 1) Some previously manufactured SLCMs might not be declared; 2) Secret production; 3) Even if non-nuclear SLCMs were sealed to deter conversion to nuclear, the conversion might be quickly carried out should the treaty be broken.

## COMPARING VERIFICATION APPROACHES

For some weapon systems, there is a technical method of verification that is inexpensive and highly effective. For example, limits on submarine-launched ballistic-missile (SLBM) launchers can be verified with high confidence by counting the launch tubes on each submarine while it is under construction using existing reconnaissance satellites.

But this is not the case with the SLCMs themselves. Because it is difficult to count the number of SLCMs on ships and submarines or to distinguish nuclear from non-nuclear SLCMs by national technical means, direct verification requires on-site inspections. The disadvantages of such measures in terms of cost, bureaucratic politics, and collateral risks to national security may be significant.

Under the minimal-inspection option, involving only counting rules, the USSR would have no direct way of verifying the absence of nuclear warheads on the 3,000 or so supposedly non-nuclear US Tomahawks.

If, in the future, the USSR deploys a non-nuclear long-range SLCM, the US might also find the minimal-inspection option problematic. However, at present the US sees the disadvantages of on-site inspection as outweighing the advantages of direct verification of SLCM limits.

The value of the maximal-inspection approach, involving on-site inspection of deployed SLCMs, is that it allows direct verification. On-site examination of three types of US SLCM launcher (ABL, VLS, and CLS) could readily determine how many of the launchers contained nuclear SLCMs, and could be done from the exterior of the ship or submarine. For SLCMs in submarine torpedo rooms, however, inspectors would have to go inside the submarine. The alternative proposed by the Soviet Union, monitoring of loading and unloading at ports, would avoid internal inspections, but would require inspectors to be present at all ports that handle SLCMs and the monitoring of every loading and unloading. Although such monitoring would not preclude the possibility of clandestine SLCM unloading at unmonitored ports, the detection of *any* such clandestine operation would be evidence of a violation.

The advantage of the intermediate-inspection regime, in which the SLCM production and maintenance sites would be monitored, is that it provides a method of significantly reducing the probability of clandestine production and deployment of SLCMs without involving the inspection of ships, submarines, or primary naval facilities. The level of intrusiveness at the production site is comparable to that agreed to in the INF treaty, in that it allows portal-perimeter monitoring of declared production sites.

Tagging of SLCMs would greatly enhance the value of inspections under the intermediate regime.

Although neither the minimal- nor the maximal-inspection approach is inherently unreasonable, the intermediate approach would provide a considerable degree of verification without unduly burdensome inspections.

## METHODS FOR SLCM ON-SITE INSPECTION

For any treaty limiting SLCMs, information from reconnaissance satellite and all other sources of intelligence would be available to monitor treaty compliance. Data exchanges concerning SLCM deployments and practices would also clearly be helpful. In addition, there is a range of options available for on-site inspection. Possibilities include checks on SLCMs at deployment sites, storage sites, and production and maintenance sites. Inspection could be one-time-only, on a continuing basis at specified sites, or on demand at numerous possible sites. Fissile-material detectors could be used to discriminate between nuclear and non-nuclear weapons, and tags and seals could be used to identify and ensure the integrity of previously declared missiles.

### Detection of Nuclear Warheads

Nuclear SLCMs can be distinguished from non-nuclear SLCMs either by passive detection of radiation spontaneously emitted by the warhead fissile material or by exposing the warhead to gamma rays or neutrons and measuring the scattered, transmitted, or induced radiation. Detectability of nuclear warheads depends on the warhead design, the method of detection, the sensitivity of the detectors, and any material between the warhead and the detector that might shield the radiation.

The simplest method is passive detection of neutrons or photons emitted spontaneously by the warhead. Chapter 11 finds that the neutrons from an unshielded warhead with a core of weapon-grade plutonium could be detected with a portable neutron detector at a distance of on the order of 10 meters with a detection time of a few minutes. For warheads with depleted uranium tampers surrounding the fissile core, the gamma-ray emissions from uranium-238 would also be detectable at distances on the order of 10 meters using portable equipment.

However, it is also found that warheads that do not contain either weapon-grade plutonium or depleted uranium may not be detectable by passive means. A "stealthy" warhead might, for example, be made with a core of weapon-grade uranium and a tungsten tamper. Or, since it is plutonium-240, not plutonium-239, that emits almost all of the neutrons from weapon-grade plutonium, a low-radiation warhead could also be made from highly purified plutonium-239, in which the plutonium-240 concentration is reduced to less than 0.01 percent.

Shielding would also reduce the detectability of a nuclear warhead. To prevent detection by a portable detector aimed at the weapon for 1 minute from a distance of 1 meter, neutrons from a weapon-grade plutonium warhead could be shielded with a layer of lithium hydride 20 centimeters thick. Similar reduction of the gamma-ray signal from a depleted uranium tamper would require a layer of about 4 centimeters of tungsten.[22] These figures indicate that nuclear warheads can be hidden within ships, rooms, and even large boxes, but that it could be difficult to put shielding within a missile canister to disguise the presence of a nuclear warhead containing either ordinary weapon-grade plutonium or depleted uranium.

An alternative detection method is radiographic analysis. This involves the measurement of transmission of neutrons or gamma rays and can detect the presence of dense fissile material or of radiation shields.

A third method is to detect any fissile material present by inducing a tiny fraction to fission and measuring the associated increased output of radiation. Fission can be induced using a portable neutron source, creating a detectable flux of delayed neutrons or photons from any of the weapon models considered out to distances of more than 10 meters. However, shielding could be as effective in preventing detection by induced fission as it is in preventing passive detection (see chapter 11).

At a portal monitoring station, such as at a production or maintenance facility, any of these methods or a combination might be used to distinguish nuclear from non-nuclear SLCMs. It would, of course, be crucial to minimize the possibility of shielding. As has been noted already, transmission radiography could indicate the presence of shielding.

If SLCMs on ships and submarines are to be checked for nuclear warheads, it would be most convenient to make all measurements with the SLCMs in their launchers. Passive detection would be the most practical method for use on ships. To minimize shielding, there should be nothing but air between the missile canister and the radiation detector. For the ABL, the surface ship VLS, the submarine CLS, or similar launchers, the hatches of the launcher should be opened (see figures 8.2, 8.3, and 8.4).

Monitoring could be done by inspectors on the deck of the ship or possibly from helicopters hovering over the launchers (see chapters 13 and 14). Tests would be necessary to develop these procedures.

To monitor SLCMs in torpedo rooms, however, it would be necessary, because of shielding effects, to take the detectors into the torpedo room. It might therefore be easier to monitor torpedo-launch SLCMs at port during loading. To check that no warheads specially designed to elude passive detection were deployed, SLCM canisters could occasionally be selected at random for radiographic examination.

## Tags[23]

Tags could simplify monitoring. A tag is any unique identifier that can be permanently affixed to a SLCM; an ideal tag must be incapable of being counterfeited or moved without telltale signs. A tag would identify a missile as one of a permitted number, indicating that it had not come from a secret stockpile or production facility. The tag could also indicate whether the SLCM was nuclear or non-nuclear.

Tags could be installed during a baseline inspection of deployed missiles or at the production facility. During subsequent inspections, each SLCM— or its sealed canister—would be checked for its tag, to ensure that it was a declared missile. Any untagged missile or canister would be evidence of a violation. If SLCMs are to be inspected while in their launchers, the tag should be visible through the opened door of the launcher (see figures 8.2, 8.3, and 8.4).

Many types of tag are possible. Tags could be based on an intrinsic property of the missile or canister, such as details of surface features, or a tag could be an item that is applied to the missile. Sandia National Laboratory has developed a tag made of clear plastic embedded with particles of micaceous hematite. For each angle of illumination there is a different pattern of reflections. The tag can be read with a special reader consisting of a still video camera and a number of lights set for illumination at different angles. Readings can be compared with data stored on a computer floppy disk. Electronic tags are also feasible.

A measure similar in effect to tagging would be to declare the position of all SLCMs and to announce all missile movements, as is required for the missiles being eliminated under the INF treaty. Then inspectors could verify by random short-notice site visits that the announced numbers of missiles were at the designated sites. If, during a challenge inspection, SLCMs were found at an undeclared site, that would be a violation of the treaty. This declaration method would require more comprehensive revelation of SLCM locations than would tagging, but it has the advantage of reduced technical complexity.

## Seals

A seal is a device that would reveal whether or not a missile or missile canister had been opened. SLCMs or their canisters could be sealed and tagged as nuclear or non-nuclear either at production facilities or during initial on-site inspections. During subsequent inspections, a broken seal on the canister of a non-nuclear SLCM might indicate that the missile had been tampered with. That missile could then be inspected for the presence of fissile material.

In the United States, SLCMs are returned for maintenance every three years or so. Seals would have to be removed at this time. Portal-perimeter monitoring could be established at these maintenance facilities and inspectors could check the seals and tags of incoming SLCMs, inspect the outgoing SLCMs designated non-nuclear to verify that they were indeed non-nuclear, and reapply the seals and tags.

Seals made with fiber-optic bundles are routinely used by the International Atomic Energy Agency to safeguard nuclear materials. A variety of other designs for SLCM seals is possible.[24]

## CONCLUSION

On-site inspection measures to monitor either a limit or a ban on nuclear SLCMs could include inspection of: ships and submarines where SLCMs are deployed, production facilities, maintenance facilities, and storage sites. Verification plans that involved either very few inspections or, at the other extreme, frequent inspections of ships and submarines might be acceptable. However, a reasonably effective verification plan with an intermediate level of intrusiveness could focus on the monitoring of the production and maintenance of any permitted non-nuclear or nuclear long-range SLCMs and any nuclear long-range SLCMs. Tagging of permitted missiles to allow identification at subsequent inspections at shore-based maintenance depots would significantly decrease the probability that undetected SLCMs could be deployed or that non-nuclear SLCMs might be covertly converted to nuclear versions.

## NOTES AND REFERENCES

1.  Don Oberdorfer, "Mixed Results Cited in Arms Talks," *Washington Post*, 21 April 1988, p.A33.

2.  R. Jeffrey Smith, "U.S., Soviets at Odds on Arms Pact Details," *Washington Post*, 13 February 1990, p.A17.

3.  *Intelligence Support to Arms Control*, Report by the Permanent Select Committee on Intelligence, House of Representatives, November 1987, dissenting views of Reps. Hyde, Cheney, Livingston, McEwen, Lungren, and Shuster, pp. 33–39.

4.  Admiral Stephen J. Hostettler, director, Joint Cruise Missile Project, 8 March 1985. Senate Armed Services Committee, DoD Authorization Hearings FY 1986, part 7, p.3875.

5.  Fred Hiatt and Rick Atkinson, "Insertable Nuclear Warheads Could Convert Arms," *Washington Post*, 15 June 1986, p.A1; letter from John Engehart, *Proceedings of the US Naval Institute*, February 1987, pp.19–20.

6.  "Naval Report: Extent of Soviet Submarine Power," *Jane's Defense Weekly*, 3 December 1988, p.1409.

7.  Paul Nitze reports Soviet INF negotiator Kvitsinsky as saying that it takes 45 minutes to switch between air-launched, ground-launched, and sea-launched modes. Paul Nitze, "The Walk in the Woods," *International Affairs*, Winter 1989.

8.  See also the testimony of Ambassador Maynard W. Glitman, 26 January 1988, INF Treaty Hearings, Senate Foreign Relations Committee, part 1, p.150, regarding the INF treaty provisions for testing of sea-launched cruise missiles in a ground-launched mode.

9.  Testimony of Harold Brown, Secretary of Defense, 23 July 1979, Senate Armed Services Committee, SALT II Hearings, 1979, part 1, p.48; testimony of Ambassador Ralph Earle, 30 July 1979, ibid., part 2, pp.507–510, 535–536.

10. Thomas B. Cochran, William M. Arkin, and Milton M. Hoenig, *Nuclear Weapons Databook Volume 1: US Nuclear Forces and Capabilities*, (New York: Ballinger, 1984), p.188.

11. Ambassador Glitman, INF Treaty Hearings, Senate Foreign Relations Committee, part 1, 26 January 1988, p.150.

12. Admiral Hostettler, Senate Armed Services Committee, DoD Authorization Hearings FY 1986, part 7, p.3875. Statement of Rear Admiral William C. Bowes, director, Cruise Missile Project, before the Defense Subcommittee of the House Appropriations Committee, on the Tomahawk Weapon System, 21 April 1988.

13. Statement of Rear Admiral William C. Bowes, director, Cruise Missile Project, House Armed Services Committee, DoD Authorization Hearings FY 1989, 21 April 1988. These sites include the Naval Weapon Station in Earle, New Jersey, shore facilities in Yorktown, Virginia, and New London, Connecticut, and the Naval magazine facility in Guam, all of which will support Tomahawk submarine operations. Four submarine tenders have also been upgraded to handle submarine-deployed Tomahawks: these include the USS *Fulton* (AS-11) and the USS *McKee* (AS-41). House Armed Services Committee, DoD Authorization Hearings FY 1987, p.970.

14. R. Jeffrey Smith, "Soviets Seek Cruise Data Verification," *Washington Post*, 23 July 1988, p.A1.

15. Ibid.

16. As of early 1988. Information provided by the Cruise Missile Project, US Navy.

17. George N. Lewis, Sally K. Ride, and John S. Townsend, "Dispelling Myths About Verification of Sea-Launched Cruise Missiles," *Science*, 10 November 1989, p.765.

18. Plans for an antisubmarine ASROC or Sealance rocket have been cancelled. John D. Morrocco, "Navy Avoids Major Program Cuts, Face Modest Reductions," *Aviation Week & Space Technology*, 12 February 1990; Frank C. Carlucci, in *Annual Report to the Congress, FY 1990* (Washington DC: US Government Printing Office, 1990) p.151.

19. Paul W. Stiles, "An Alternative to VLS UnRep," *Proceedings of the US Naval Institute*, December 1987, p.129.

20. Norman Polmar, *The American Submarine* (Annapolis, Maryland: Nautical and Aviation Publishing, 1981), p.151.

21. R. Jeffrey Smith, "Soviets Seek Cruise Data Verification," *Washington Post*, 23 July 1988, p.A1. Inspection teams stationed at key ports would count and inspect each missile before it was loaded onto a ship or submarine. Equipment that could be used to load missiles at any other site, including at sea, would be banned. Despite its advantages, this arrangement could result in a greater number of inspections than on-ship launcher inspections.

22.  The shielding spherical shell of tungsten (density 19.2 g/cm³) with an inner diameter of 22 centimeters and weighing 600 kilograms, as discussed in chapter 11, has a thickness of about 4 centimeters. The warhead need not be entirely surrounded by tungsten, only in those directions accessible to the detector.

23.  For a general discussion of tagging, see Thomas Garwin, *Tagging Systems for Arms Control Verification* (Marina del Rey: Analytical Assessments Corporation, 1980), report # AAC-TR-10401/80.

24.  Dennis L. Mangan, *Hardware for Potential Unattended Surveillance and Monitoring Applications* (Albuquerque, New Mexico: Sandia National Laboratory, January 1988), Report SAND87-2840.

*Valerie Thomas*

# US and Soviet SLCMs

## US SLCMs

US long-range SLCMs are called Tomahawks. They are 5.56 meters long and 53 centimeters in diameter. The internal differences of the Tomahawk variants are shown schematically in figure 8.A.1. Note that the nuclear version, the TLAM-N (Tomahawk Land Attack Nuclear), has a smaller warhead than the non-nuclear versions—the TLAM-C (Tomahawk Land Attack Conventional), TLAM-D (Tomahawk Land Attack Dispensed Submunitions), and TASM (Tomahawk Anti-Ship Missile)— so that the nuclear version has room for an additional fuel tank. The US also has a short-range non-nuclear SLCM, the Harpoon.

Table 8.A.1 shows the characteristics of each type of US SLCM. Note that more than four times as many non-nuclear Tomahawks as nuclear are planned. Tomahawk deployment is planned for about 100 surface ships and 100 submarines by the mid-1990s, as shown in table 8.A.2.

The guidance system of all three land-attack versions uses TERCOM (TERrain COntour Matching), which compares terrain profiles gathered by a radar altimeter with computer-stored contour maps. The nuclear version has an accuracy (CEP) of 75 meters.[1] In the non-nuclear land-attack version, guidance is supplemented near the target with DSMAC (Digital Scene Matching Area Correlator), which compares optically sensed scenes with images stored in the guidance computer. This brings

**Table 8.A.1**: US sea-launched cruise missiles

| Type | Mission | Number planned* | Number bought† | Range km | Warhead | Guidance |
|------|---------|-----------------|----------------|----------|---------|----------|
| TLAM-N | land attack | 758 | 367 | 2,500 | Nuclear ~200kt‡ | TERCOM |
| TLAM-C | land attack | 1,500 | 1,197 | 1,300 | HE unitary | TERCOM/ DSMAC |
| TLAM-D | land attack | 1,100 | 273 | 1,300 | HE *multiple* submunitions | TERCOM/ DSMAC |
| TASM | antiship | 593 | 593 | 460 | HE unitary | Radar |
| Harpoon | antiship | ≈2,000§ | ≈2,000 | 110 | HE | Radar |

\* Approximate numbers as of 1989. Information provided by the Cruise Missile Project, US Navy.

† Through FY 1990. Information provided by the Cruise Missile Project, US Navy.

‡ Thomas B. Cochran, William M. Arkin, and Milton M. Hoenig, *Nuclear Weapons Databook Volume 1: US Nuclear Forces and Capabilities,* (New York: Ballinger, 1984), p.79.

§ Estimate based on information provided by the Harpoon Office, OP-354, US Navy. There are also air-launched Harpoons and Harpoons in allied armed forces. More than 4,000 have been produced.

**Table 8.A.2:** Planned US Tomahawk cruise missile deployment*

| Type of vessel | Number of vessels current† | planned | Launcher type | SLCMs/ vessel |
|---|---|---|---|---|
| Battleship BB-66 *Iowa class* | 4 | 4 | 8 ABLs | 32 |
| Destroyer DD-963 *Spruance class* | 7 | 7 | 2 ABLs | 8 |
| Destroyer DD-963 *Spruance class* | 9 | 24 | 1 VLS | 37 |
| Guided-missile destroyer DDG-51 *Burke* | 0 | 29 | 2 VLSs | 16 |
| Guided-missile cruiser CG-47 *Ticonderoga* | 11 | 22 | 2 VLSs | 19 |
| Guided-missile cruiser *nuclear* | 5 | 5 | 2 ABLs | 8 |
| Submarine SSN-719 and later *Los Angeles* | 6 | 37 | CLS, torpedo tubes | 20 |
| Submarine SSN-637 *Sturgeon* through SSN-718 older *Los Angeles* | 39 | 70 | 4 torpedo tubes | 8 |

* Bernard Blake, ed., *Jane's Weapon Systems 1987–88* (New York: Jane's Publishing, Inc.) p.487. Statement of Rear Admiral William C. Bowes, USN, director, Cruise Missile Project, before the Defense Subcommittee, House Appropriations Committee, 21 April 1988. Statement of Rear Admiral Stephen J. Hostettler, USN, director, Joint Cruise Missile Project, before the House Armed Services Committee, DoD Authorization Hearings for FY 1985, part 2, p.361, 14 March 1984. Tomahawks would also be deployed on the planned Seawolf attack submarines.

† Information provided by the Cruise Missile Project, US Navy, March 1990.

**Table 8.A.3:** Tomahawk launchers

| Type | Launchers/ unit | Location | Other weapons | Vessel type |
|---|---|---|---|---|
| Armored box launcher (ABL) | 4 | Above deck | None | Battleships Cruisers Destroyers |
| Vertical launch system (VLS) | 61 or 29 | Internal, top flush with deck | Standard antiaircraft missile | Destroyers Cruisers |
| Capsule launch system (CLS) | 12 | Outside inner pressure hull | None | *Los Angeles* 719 and later |
| Torpedo tubes | 4 tubes per sub | Torpedo room (sub interior) | Torpedoes Harpoon Decoys | Attack subs (*Los Angeles* etc.) |

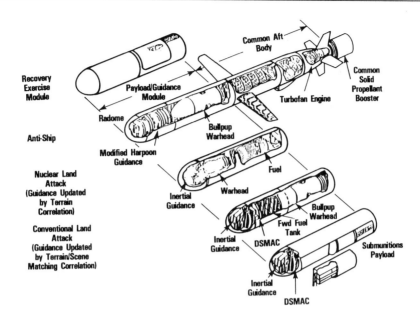

**Figure 8.A.1:** Tomahawk variants

Source: Cruise Missile Project, US Navy

**Figure 8.A.2:** Torpedo launch Tomahawks being loaded into submarine

Source: Cruise Missile Project US Navy

the accuracy to 8 meters, or perhaps 3.[2] DSMAC is not included in the nuclear version, apparently because such accuracy is not needed for nuclear weapons.[3]

Tomahawks are housed in four different types of launcher: the armored box launcher (ABL) and the vertical launch system (VLS) on surface ships, and the capsule launch system (CLS) and torpedo tubes (torpedo rooms) on submarines. These are shown in figures 8.1, 8.2, 8.3, and 8.4 of chapter 8, and described in table 8.A.3.

## SOVIET SLCMS

### SS-N-21

The SS-N-21 is the first long-range Soviet SLCM. It is a development of the AS-15 air-launched cruise missile (deployed in Bear-H bombers), and is essentially the same missile as the SSC-X-4 ground-launched cruise missile, banned by the INF treaty.[4] The SS-N-21 missile is not believed to exist in a conventionally armed version. It is 6.4 meters long and 0.5 meters in diameter, small enough to be launched from a standard 533-millimeter torpedo tube. This makes it the smallest Soviet SLCM; yet its range is the longest, due to an advanced turbofan propulsion system. Earlier types of Soviet cruise missile have less efficient turbojet engines, or solid-fueled rockets for the very short-range antiship missiles (see table 8.A.4).

Testing of the SS-N-21 was first reported in December 1987; the missiles were launched from an Akula-class submarine.[5] A single Victor III submarine with a cylindrical structure forward of the sail has also been reported as an SS-N-21 trials ship.[6]

The first deployment of the SS-N-21 was reported in January 1988, on a refitted Yankee submarine.[7] (The Yankee is an old ballistic-missile submarine from which the launch tubes have been removed and which has been reconfigured to carry cruise missiles.) Norwegian defense officials estimated that the submarine can carry between 20 and 40 cruise missiles. At least 12 Yankee submarines are being refitted, though some of these may be for the larger SS-NX-24. The refitted Yankee submarine, dubbed the "wasp-waisted Yankee" or the Yankee Notch, has been increased in length by 10 meters and has a 3-meter longer and reshaped fin.

Deployment of the SS-N-21 has been reported on the three Akula-class submarines, the two Sierra-class submarines, and the single Mike-class submarine, from which the missiles would be launched from torpedo tubes.[8] These are the most modern attack submarines in the Soviet Navy. It has been reported that the Soviet Union plans to deploy 400 long-range nuclear SLCMs.[9] There has been no indication of plans for deploying the SS-N-21 on surface ships.[10]

### SS-NX-24

The SS-NX-24 is a large (13 meters × 1 meter) supersonic cruise missile. It is too big for a standard torpedo tube, so it will need a dedicated launcher. It has not yet been deployed, though it has been flight tested from a reconfigured Yankee-class submarine.[11]

The Soviet Navy is thought to be building a new cruise-missile submarine specifically designed to carry the SS-NX-24.[12] The SS-NX-24 is not believed to have a conventionally armed version. The SS-NX-24 program has proceeded slowly; no progress has been reported since 1988. It may even have been canceled.[13]

Limits on deployment of the SS-NX-24 could be monitored by counting the number of launchers by satellite reconnaissance; on-site inspection would not be necessary.

**Table 8.A.4:** Soviet sea-launched cruise and short-range antiship missiles

| Missile | Year Introduced | Range km | Launchers* deployed | Length† m | Diameter‡ m | Propulsion | Warhead | Launchers§ |
|---------|-----------------|----------|---------------------|-----------|-------------|------------|---------|------------|
| SS-N-2c | 1959 | 80 | 36 | 6.5 | 0.7 | turbojet | HE | surface ABL |
| SS-N-3 | 1960 | 460 | 312 | 11.7 | 1.0 | turbojet | dual** | surface ABL sub CLS |
| SS-N-7 | 1971 | 60 | 80 | 7.0 | 0.5? | solid fuel | dual | sub CLS |
| SS-N-9 | 1968 | 100 | 236 | 8.8 | ? | solid fuel | dual | surface ABL sub CLS |
| SS-N-12 | 1973 | 550 | 280 | 10–12 | 0.8 | turbojet | dual | surface ABL sub CLS |
| SS-N-19 | 1980 | 550 | 136 | 10–12 | 0.8 | turbojet | dual | surface VLS sub CLS |
| SS-N-21 | 1988 | 2,200 | 50# | 6.4 | 0.5 | turbofan | nuclear | sub CLS? torpedo |
| SS-N-22 | 1981 | 100 | 96 | 9.0? | ? | solid fuel | dual | surface ABL |
| SS-NX-24 | – | 3,000? | – | 13.0 | 1.0 | ? | nuclear | sub CLS |

\*    *The Military Balance 1988–89*, (International Institute for Strategic Studies: Launcher numbers, 6/88); *Soviet Military Power 1987*; US Naval Institute Military Database.

†    *Jane's Weapon Systems, 87–88*. Dimensions of SS-N-12 from Thomas B. Cochran, "Black Sea Experiment Only a Start," *Bulletin of the Atomic Scientists*, November 1989, p.13.

‡    Barton Wright, *World Weapons Database Volume I: Soviet Missiles*, (Lexington, Massachusetts: Lexington Books, 1986). Dimensions of SS-N-19 are assumed to be the same as SS-N-12.

§    Indicates most similar US SLCM launcher. ABL refers to above-deck launcher; CLS refers to dedicated single launchers in submarines; and VLS to launchers below deck of surface ship.

#    Assuming 20 launchers on the converted Yankee, 3 Akulas with 6 tubes each, 2 Sierras with 6 tubes. Tube numbers from Norman Polmar, *Guide to the Soviet Navy*, 4th edition, (Annapolis, Maryland: Naval Institute Press, 1986).

\*\*    Both nuclear and non-nuclear versions are believed to be deployed.

## Shorter-range Missiles ( < 600 kilometers)

The Soviet Union has a variety of short-range sea-launched cruise missiles. The main types are denoted by the US as the SS-N-2, -3, -7, -9, -12, -19, and -22. All of these have a range of less than 600 kilometers; all are considered to be antiship, not land-attack weapons.[14] The SS-N-2 does not have a nuclear warhead; the others are thought to be "dual capable"—to have both nuclear-armed and non-nuclear versions. The SS-N-12 and SS-N-19 have the longest range, estimated at about 550 kilometers.

Many Soviet SLCMs are deployed in launchers that can be identified and counted using satellites. If these weapons were to be limited by arms control, counting the total number deployed should not be a major problem, though distinguishing nuclear from non-nuclear versions would require inspections (see chapters 12–14).

These shorter-range SLCMs could pose a verification problem for an agreement limiting long-range SLCMs, because the range of some may be difficult to verify.

## NOTES AND REFERENCES

1.  Testimony of Commodore Roger F. Bacon, 14 March 1984, House Armed Services Committee, DoD Authorization Hearings FY 1985, part 2, p.392. CEP—circular error probable—is the radius of a circle round the target within which half of the missiles would be expected to fall.

2.  Testimony of Richard Perle, 16 February 1988, Hearings on the INF Treaty of the Senate Foreign Relations Committee, part 3, p.435,

3.  Testimony of Admiral Stephen J. Hostettler, 13 March 1985, House Armed Services Committee, DoD Authorization Hearings, FY 1986, part 2, p.517.

4.  Testimony of Admiral Crowe, USN, chairman, Joint Chiefs of Staff, 4 February 1988, INF Treaty Hearings, Senate Foreign Relations Committee, part 2, p.268.

5.  Edward Neilan, "Soviet Cruise Missile Tested in Sea of Japan," *Washington Times*, 28 December 1987.

6.  Norman Polmar, *Guide to The Soviet Navy* (Annapolis, Maryland: Naval Institute Press, 1986), 4th edition.

7.  Tønne Huitfeldt, "Soviet SS-N-21 Equipped 'Yankee' in Norwegian Sea," *Jane's Defense Weekly*, 16 January 1988, pp.44–45; "Soviets Get Around Arms Treaty by Updating Arms on Older Subs," (Reuters) *Washington Times*, 13 January 1988, p.9.

8.  "Naval Report: Extent of Soviet Submarine Power," *Jane's Defense Weekly*, 3 December 1988, p.1409. The Mike subsequently sank. "42 Die as 'Mike' Submarine Sinks," *Jane's Defense Weekly*, 15 April 1989, p.629.

9.  "U.S.–Soviet to Exchange SLCM Plans," *Defense Daily*, 28 February 1990, p.36.

10. FY 1986 Arms Control Impact Statement, (Washington DC: Government Printing Office, 1985) p.64.

11. Bernard Blake, ed., *Jane's Weapon Systems 1987–88* (New York: Jane's Publishing Inc., 1987), 18th edition, p.115,116; *Soviet Military Power 1987*, p.37.

12. "Soviet Intelligence: Developments in Submarine Forces," *Jane's Defense Weekly*, 12 November 1988, p.1233.

13. Statement of Rear Admiral Thomas A. Brooks, Director of Naval Intelligence, US Navy, before the Seapower, Strategic, and Critical Materials Subcommittee of the House Armed Services Committee, on intelligence issues, 14 March 1990.

14. FY 1986 Arms Control Impact Statement, p.84.

# Chapter 9

*Thomas K. Longstreth*
*Richard A. Scribner*

# Verification of Limits on Air-launched Cruise Missiles

**B**omber-carried weapons, particularly air-launched cruise missiles (ALCMs) are one of the critical areas of both quantitative and qualitative growth in the Soviet–US nuclear arms competition. For, while the United States has had long-range heavy bombers as a major element of its strategic forces since the 1950s, only over the past decade has the USSR expanded and diversified its strategic bomber forces. Both countries now have strategic weapon modernization plans that include substantial roles for improved ALCMs and other air-delivered weapons.

Once concluded, the Strategic Arms Reduction Treaty (START) now under negotiation in Geneva will probably channel each side's strategic modernization further in the direction of bombers. This is because START provides for stricter limits on ballistic-missile warheads than on bomber-carried weapons (see discussion below.) According to General John T. Chain Jr, commander-in-chief of the US Strategic Air Command (SAC), after implementation of a START treaty, if the US were to buy 132 B-2 bombers (as was initially proposed) bombers would carry over half of all US strategic weapons.[1]

The small size and weight of modern ALCMs[2] (as well as modern nuclear bombs), coupled with the payload capacity and mobility of their launch platforms, poses a number of difficult verification questions for strategic arms control. These problems are likely to become more acute in years ahead. The United States has a very ambitious ALCM program encompassing both nuclear and conventional missiles with many ranges, shapes, sizes, and missions. The Soviet ALCM program is behind that of the US but is moving forward rapidly. The importance of these programs to both countries has been demonstrated at the START negotiations, where ALCMs have proved to be a contentious issue; at the time of this writing, however, the two sides have agreed upon many of the elements of the treaty dealing with ALCMs.

Below, we briefly describe current US and Soviet bombers and their nuclear weapons, the status of START limits on bombers and ALCMs, and the potential for enhancing the verification of limits on ALCMs and bombers in a deep nuclear-reductions regime. We conclude that, unless precautions are taken, air-launched cruise missiles and bombers could represent serious obstacles to the achievement of a future deep-reductions agreement.

## US AND SOVIET STRATEGIC BOMBERS AND THEIR NUCLEAR WEAPONS

Current US strategic bombers include the FB-111A, the B-52G and -H models, the B-1B, and the B-2A (which is now in flight testing). Soviet strategic bombers include the Tupolev(Tu)-95 (NATO code name: "Bear"), the Tu-160 "Blackjack" and the Tu-26 "Backfire."[3] All but the US FB-111 and Soviet Backfire are considered "heavy" bombers capable of flying intercontinental missions without refueling. US and Soviet strategic nuclear bombers carry three basic types of nuclear weapon: long-range ALCMs; shorter-range air-to-surface missiles (ASMs); and gravity bombs.

ALCMs are unmanned self-propelled air-breathing vehicles that are powered throughout their atmospheric flight. While early ALCMs were large, inaccurate, and unreliable, current systems are smaller, more reliable and very accurate. Older supersonic ALCMs use turbojet engines while modern US and Soviet subsonic ALCMs use more fuel-efficient turbofan engines to provide for longer range (between 1,500 and 4,000 kilometers).

Current US nuclear ALCMs include: the ALCM-B; which is subsonic, has a nominal range of 2,500 kilometers, is extremely accurate, and carries a variable-yield nuclear warhead of 5 to 150 kilotons; and the advanced cruise missile (ACM), which is also subsonic and uses the same warhead as the ALCM-B but has a longer range (estimated at 4,000 kilometers), is even more accurate, and incorporates "stealth" technology to a greater extent in order to make it even less observable to radar.

Soviet nuclear ALCMs include older, larger, and shorter-range missiles like the AS-3, AS-4, and AS-6, and modern ones, including the subsonic AS-15 "Kent" and the supersonic AS-X-19 "Koala", now in flight testing.

ASMs are shorter-range missiles that are powered by solid-fueled rockets. They are usually supersonic and remain entirely within the atmosphere during flight. Their range varies between 100 and 600 kilometers. They are primarily designed to attack enemy air defense systems and can be programmed to follow unpredictable routes to their targets in

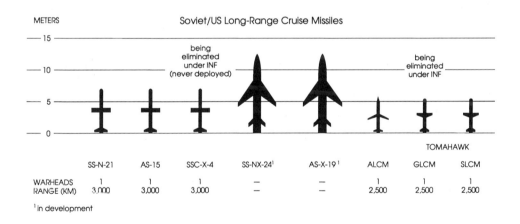

**Figure 9.1:** US and Soviet cruise missiles

Source: *Soviet Military Power 1989*

order to evade enemy interception. Current US nuclear ASMs include the short-range attack missile (SRAM) and its successor (the SRAM II). The USSR has recently begun deploying an advanced ASM, the AS-16 "Kickback," on its strategic bombers.

Gravity bombs are usually carried within an aircraft's bomb bays and are dropped from low altitudes to provide greater accuracy and employ parachutes to allow time for the bomber to escape. Gravity bombs generally have the largest yields of any nuclear weapon carried by strategic bombers; one type of US strategic bomb has a yield of 9 megatons.

The appendixes at the end of this chapter provide more detailed descriptions of US and Soviet bombers, their ordnance, and information on bomber and ALCM production and deployment.

## ALCM VERIFICATION PROBLEMS

The problems of establishing verifiable limits on ALCMs are similar, in many ways, to the corresponding problems for sea-launched cruise missiles (SLCMs), but have some unique aspects as well:

♦ Because ALCMs are relatively small, can be produced in large quantities, and have no production, launch, or support facilities that are easily identifiable by satellite surveillance, sizable hidden stockpiles might be accumulated that could be rapidly deployed in a crisis.

♦ While current long-range ALCMs carry only nuclear warheads—unlike SLCMs—future types may well be dual-capable. A number of deployed Soviet shorter-range (under 600 kilometers) ALCMs are believed to be dual-capable (See appendix 9.B) and the US is developing long-range, conventional ALCMs that could be modified to carry nuclear warheads.

♦ Any range limitations on ALCMs are complicated by the fact that, for a cruise missile of any given size, range can vary greatly, depending on the weight allocation between the warhead and fuel. Thus the strategic nuclear land-attack version (TLAM-N) of the US Tomahawk SLCM has a range almost twice that of the conventional version (TLAM-C) because of the larger size and weight of the conventional high-explosive warhead.

♦ Since individual US and Soviet bombers can carry significant numbers of different types of nuclear bomb, SRAMs and ALCMs, they could be used to quickly move large numbers of nuclear weapons from one location to another, complicating efforts to track weapon inventories.

♦ A certain fraction of former US and Soviet strategic bombers have been converted to either non-nuclear weapon delivery or other missions such as reconnaissance and aerial refueling. While negotiators are developing ways to separate and distinguish between aircraft assigned to these roles and current strategic bombers, there is little that can be done physically to alter a bomber in a manner that is irreversible—i.e., that could prevent the air frame from being modified quickly in a crisis to carry nuclear bombs or ALCMs.

## US–SOVIET NEGOTIATIONS ON ALCM AND BOMBER LIMITS

### Strategic Arms Limitation Talks (SALT)

During the 1969–72 negotiations that resulted in the first Strategic Arms Limitation (SALT) Agreement, the US and USSR did, at one stage, discuss a ban on the testing and deployment of all intercontinental cruise missiles, but this idea was eventually discarded, and the final agreement did not mention cruise missiles. Limits on bombers were also discussed but fell by the wayside when it was determined that a comprehensive strategic offensive limitation agreement would be too difficult to achieve. Instead, the two sides agreed to an "interim" freeze on land-based intercontinental ballistic-missile (ICBM) and submarine-launched ballistic-missile (SLBM) launchers. One of the major obstacles to limiting bombers in SALT I was the Soviet insistence on taking into account US nuclear-capable fighter-bombers based on aircraft carriers and in countries on the periphery of the USSR.[4]

Bombers and ALCMs were a central point of discussion in the 1974–79 negotiations that led to the SALT II treaty. By this time, the US cruise-missile development program that eventually led to deployment of the SLCM, GLCM (ground-launched cruise missile), and ALCM was well under way. Civilian officials were pushing the program along, in part, to provide bargaining leverage at the SALT II negotiations. Unfortunately, rather than facilitating an agreement, the expanding cruise-missile program became a difficult arms-control issue and helped delay the completion of a SALT II treaty until 1979.

In SALT II, each side's heavy bombers were counted against the aggregate launcher ceiling of 2,250, which also included land and sea-based ballistic-missile launchers. There was a complex definition for what constituted a "heavy" bomber: the two sides agreed to include within that category certain existing types—the US B-52 and B-1 bombers and Soviet Bear and Bison—as well as future types that could carry out the same mission. Some former heavy bombers that had been converted to tankers or were in mothballs were excluded from the aggregate ceilings. Also, the US once again managed to exclude its forward-based fighter-bombers from any limits.

The two sides also agreed on both direct and indirect limits on long-range ALCMs. "Long-range" was defined as above 600 kilometers. Observable differences were required to distinguish cruise missiles with a range greater than 600 kilometers from those with a shorter range.

SALT II also prescribed an equal aggregate subceiling of 1,320 on the total number of ICBM and SLBM launchers containing multiple-warhead ballistic missiles and ALCM-equipped heavy bombers. The two sides agreed that existing heavy bombers would carry no more than 20 ALCMs each and that, in the event either side developed a new bomber aircraft to launch ALCMs, the number of ALCMs per bomber could average no more than 28 between all types of ALCM-capable bombers.

The US and USSR also agreed that current bombers converted to carry long-range ALCMs would have to have an externally observable difference (EOD), to distinguish them from non-ALCM carriers of the same bomber type. The identifying feature on US B-52s converted to carry ALCMs was a "strakelet"—a rounded aerodynamic fairing located where the front of the wing meets the fuselage. Future ALCM carriers would have to have a so-called functionally related observable difference (FROD) related to the "ability or inability to carry ALCMs...."[5]

Finally, SALT II banned the flight testing of ALCMs from aircraft other than bombers in order to prevent the conversion of other aircraft, such as commercial transport aircraft, into ALCM carriers. The treaty also provided that, if medium bombers like the US FB-111 or Soviet Backfire were tested with ALCMs, all bombers of that type would be counted as ALCM carriers against the 2,250 aggregate ceiling and 1,320 subceiling.[6]

The relatively loose limits placed on heavy bombers and ALCMs in SALT II were intended to protect all US bomber and ALCM modernization programs. As a result of these programs, the number and capability of US bomber-carried weapons increased substantially through the 1980s.[7]

## START Negotiations on ALCM and Bomber Limits

At the START talks, which began in Geneva in 1982, the question of how to limit ALCMs has been complicated by the US and USSR taking conflicting views on how restrictive the limits should be. This reflects their differing strategic doctrines, operational requirements, and the far greater US emphasis on air-breathing systems as an element of its strategic forces.

The US position has been that "ALCMs should be treated differently from ballistic-missile warheads...because they are slower and do not pose a disarming first-strike threat."[8] The USSR has not completely accepted the US argument that ALCMs are less threatening and has sought to limit them more severely. Soviet negotiators (and some US and Soviet analysts) have argued that, as ALCMs become more "stealthy," approach intercontinental ranges, and achieve supersonic capability, they are becoming a much less stabilizing element in the strategic balance.

Despite these differences, at the time of this writing, most of the limits on bombers and ALCMs were in place. In November 1986, at the Reagan–Gorbachev Reykjavik summit, advisers to the two leaders worked out the basics of the START treaty, including ALCM limits. They agreed to an aggregate ceiling of 6,000 for all ICBM, SLBM, and accountable long-range ALCM warheads plus non-ALCM heavy bombers.[9]

Just before and during the May 1988 Moscow summit, the two sides worked out a number of counting rules to ease the problem of ALCM verification. First, they agreed that heavy bombers equipped to carry long-range ALCMs would be counted as carrying a certain number of ALCMs regardless of the actual number on board. For purposes of the 6,000-

warhead ceiling, a nominal number of ALCMs could then be calculated. The US first proposed six per bomber and later 10; the Soviet Union proposed that each ALCM-capable heavy bomber be counted as having the maximum number for which it was equipped, for example, 20 for the B-52H.

The two sides also agreed to a number of other rules to ease monitoring requirements. For example, just as in SALT II, ALCM-equipped heavy bombers have to be "distinguishable from other heavy bombers." In the Conversion or Elimination (C or E) Protocol to the draft START treaty, the two sides have agreed to some general procedures for making this determination. Each side has proposed detailed procedures for converting ALCM-capable bombers to other nuclear bombers or even conventional bombers, although these procedures have not yet been fully agreed to.

For example, heavy bombers not equipped to carry ALCMs must have their external attachment points or wing adapters removed or modified in a way that makes them incapable of carrying ALCM pylons. Internally, the weapon bay must be modified in a way that makes it likewise incapable of carrying an ALCM launcher. In converting nuclear heavy bombers to a non-nuclear role, weapon-bay attachment devices for nuclear-weapon racks and launchers must be removed and replaced by ones installed for conventional weapons that are distinguishable, although the exact differences have yet to be defined.[10]

Under the proposed agreement, once bombers are converted to another mission, the other party must be informed and thereafter has the right to conduct a short-notice inspection of any bomber that has been converted to confirm that the conversion has taken place according to the agreed procedures. Each side will also be permitted a limited number of "Test Heavy Bombers" that will be kept at designated test centers and will not count against the launcher or warhead ceilings. All other heavy bombers currently in storage are to be destroyed under agreed procedures.[11]

The two sides have also agreed that ALCM-equipped heavy bombers have to be based separately from other types of nuclear or conventional heavy bombers. Finally, to address the emerging problem of conventional ALCMs, the two sides have agreed that all existing long-range ALCMs are assumed to be nuclear and future long-range conventional ALCMs must be distinguishable from nuclear ALCMs.[12] Such observable differences could include length, the presence of nonretractable wing fins, and some type of blade antenna on conventional ALCMs that did not exist on nuclear ones.[13]

At the February 1990 meeting in Moscow between Soviet Foreign Minister Eduard Shevardnadze and US Secretary of State James Baker, the USSR appeared to accept a counting rule of 10 ALCMs per current US ALCM-equipped bomber and the US accepted a counting rule of eight for current Soviet bombers. There was also agreement that the actual number of ALCMs carried by any current US bomber would not exceed 20 and for Soviet bombers would not exceed 12. The maximum number of ALCMs permitted on future US and Soviet ALCM-capable bombers would be 20.[14]

The two sides also made considerable progress on the question of heavy bombers converted to carry only conventional weapons or to reconnaissance or tanker aircraft. The Soviet Union accepted, in principle, a limit on "former heavy bombers" and "conventional heavy bombers." The US accepted intrusive measures, including the right of inspectors to examine the storage areas at bases where conventional bombers are based to verify that no nuclear weapons are stored there. In addition, the US agreed to exhibit any future types of nuclear and conventional long-range ALCM in order to allow Soviet inspectors to examine their external differences.[15]

After the February 1990 meeting, the principal remaining disagreement was over the range of ALCMs to be covered by the treaty. The Soviet Union wanted "long range" defined as more than 600 kilometers—just as in SALT II. The US had previously proposed to include only ALCMs with ranges above 1,500 kilometers in order to exclude several existing and potential conventional ALCM programs. At the February 1990 meeting, the US proposed a range limit of 1,000 kilometers, which the USSR rejected.[16]

At the Moscow Ministerial in May 1990, Baker, Shevardnaze, and Soviet President Gorbachev worked out most of the remaining differences over ALCMs and bombers. They agreed that, for the US, the 10/20 ALCMs-per-bomber counting rule would apply only through the first 150 ALCM-capable bombers, after which each ALCM bomber would be counted with the number of ALCMs "as equipped." For the USSR, the 8/12 counting rule would apply through the first 210 ALCM-carrying bombers.

The two sides also agreed to the Soviet position that ALCMs with a range above 600 kilometers would be considered "long range" and thus treaty accountable. An exception was made for the US Tacit Rainbow conventionally tipped antiradar missile, which although flight tested extensively between 1987 and 1990 and having a range of some 800 kilometers, would be listed as a "future" type of ALCM, making it exempt from the START provision that all existing long-rang ALCMs are assumed to be nuclear. To satisfy the Soviet military's concerns, Secretary Baker provided Shevardnaze with a letter assuring the USSR that Tacit Rainbow would remain a conventionally armed ALCM and listing certain characteristics that distinguished it from nuclear ALCMs.

Finally, the US and USSR agreed to a limit of 107 on former heavy bombers converted to carry conventional weapons or other non-nuclear missions, although differences remained on whether or not to include reconnaissance aircraft. Also, no agreement was reached on how or if to limit the Soviet Backfire bomber.[17] At the Bush–Gorbachev Washington Summit in May 1990, the US and USSR agreed to a 600-kilometer threshold for long-range ALCMs. They also agreed that the USSR would be permitted up to 210 ALCM bombers, while the US would be permitted 150.

*Bomber Display*

In the fall of 1989, US and Soviet negotiators agreed on a series of demonstrations designed to clarify verification and monitoring problems and ease completion of the START protocols on conversion or elimination and inspection.

As part of this package, the two sides held "heavy bomber exhibitions." The first one took place at Urin Air Base in the USSR on 18 April 1990, and the second at Grand Forks AFB, North Dakota, on 11 May 1990. During the exhibitions, each side displayed side-by-side two heavy bombers of the same type, one equipped to carry long-range ALCMs and the other not so equipped. The Soviet Union displayed its Bear-G and Bear-H bombers side by side, and the US displayed two B-1B bombers, one equipped for ALCMs, the other not. The inspecting country was allowed to examine the differences between the two and, for the ALCM bomber, examine it equipped with the maximum number of ALCMs permitted.[18]

Each country's inspectors used "linear measuring devices"—tape measures—to compare the differences between ALCM and non-ALCM bombers. US inspectors were permitted to enter the Bear-H cockpit to examine fire-control devices. However, Soviet inspectors were not allowed such access to the B-1B. Information gathered during these exhibitions is being digested by negotiators and considered in drafting final procedures for inspections under the START treaty.

## IMPLICATIONS OF START VERIFICATION ARRANGEMENTS FOR A DEEP-REDUCTIONS AGREEMENT

Many analyses of the current draft START treaty have explained that, because of the counting rules that exclude thousands of nuclear bombs, ASMs, and ALCMs carried on heavy bombers, each side can legally deploy far more than 6,000 strategic nuclear weapons.

Less commonly recognized is the fact that ALCMs per se—like bombs and SRAMs—are not treaty-limited items (TLIs) within START, and the only indirect limits on ALCMs are those that affect ALCM-capable heavy bombers. Thus, the US and the USSR can legally produce and stockpile thousands of extra ALCMs. Of course, the lack of constraints on missile and warhead production applies to all but mobile ballistic missiles as well. The judgement of the START negotiators is that the more restrictive constraints covering strategic launchers (missiles and bombers), coupled with some limited opportunities for short-notice inspections of deployment and some production locations, will be effective in limiting opportunities for breakout.

For example, the inspection and monitoring provisions that US and Soviet negotiators have agreed to will reduce uncertainties regarding some means of circumvention, such as stockpiling of extra nuclear ALCMs at non-ALCM bomber bases. But direct monitoring of ALCMs will be quite limited, and lack of adequate production monitoring will increase uncertainty regarding the potential for clandestine production and stockpiling of ALCMs in undeclared locations. Such an outcome—thousands of uncounted strategic nuclear weapons—may be acceptable for START because of its relatively high aggregate warhead ceiling, but would present a more difficult problem for a deep reductions agreement.

Another problem pertains to the ALCM carriers. After START is implemented, each side will be permitted to retain two types of heavy bomber that, while not designated as ALCM-carrying platforms, could be converted to carry ALCMs in a crisis. While it normally takes several months to convert a bomber to carry ALCMs, the task could be accomplished in several days if certain procedures were discarded. The actual loading of ALCMs onto modified bombers can be done in just a few hours.[19] START will partially address the potential for either side breaking out of the treaty's limits on ALCM-equipped bombers by requiring the destruction of heavy bombers currently in storage. However, each side may retain hundreds of non-ALCM heavy bombers in its post-START forces that, accompanied by possible stockpiles of unmonitored missiles, could be a source for ALCM breakout.[20]

Increasing the time and difficulty involved in trying to convert bombers to ALCM capability, widening the differences between conventional and nuclear or between ALCM and non-ALCM bombers, and other measures that reduce the potential for utilizing unmonitored ALCMs should all be high priorities of any deep-reductions agreement.

## ALCM/BOMBER VERIFICATION IN A DEEP-REDUCTIONS REGIME

### Verification Objectives

Despite the shortcomings of START, its verification and inspection regime is the most extensive in arms-control history, and will provide a useful framework for establishing verification arrangements under a follow-on deep-reductions agreement. Such a verification regime for bombers and ALCMs should try to achieve the following objectives:

♦ Total numbers of deployed and nondeployed ALCMs should be limited, and the chance of a large hidden or illegal stockpile of weapons should be minimized. This implies a need for random short-notice on-site inspections to supplement satellite surveillance and other intelligence-gathering, as well as additional cooperative measures.

♦ There should be no easy conversion of conventional bombers or other long-range aircraft to nuclear bombers without detection, and there should be as many arrangements as seem useful and workable to provide maximum distinguishability between these different types of aircraft and maximum physical separation between conventional heavy bombers and nuclear-weapon stockpiles.

♦ There should be no easy conversion of conventional ALCMs to nuclear ones.

♦ In general, there should be as many other barriers to easy circumvention as is reasonably practical to introduce into associated military operations. Other means of circumvention (such as increasing the range of shorter-range ALCMs), should be taken into account and made difficult.

The degree to which these objectives can be realized depends upon balancing the necessary demands for compliance assurance, including the confidence that important circumvention routes are closed off, against complexity, intrusiveness, cost, and the disclosure of sensitive information.

One should note particularly here the issue of cost. Already, observers of the START negotiations are noticing that implementation of its vast array of monitoring and verification provisions may prove costly, while experience to date in implementing the INF treaty indicates some of the more elaborate and expensive procedures or equipment may be unnecessary, or cheaper, equally effective alternatives may exist. At very low levels of strategic nuclear warheads, the requirement for high confidence in a verification regime will likely be even greater than for START, but provisions that greatly increase cost while adding only marginally to levels of confidence should be avoided.

## Approaches to Verification

At a minimum, cooperative measures to aid in verifying direct ceilings on long-range nuclear ALCMs (and perhaps all air-delivered nuclear weapons)[21] could include: a detailed initial exchange of data on ALCM production, storage and deployment locations and numbers, and on-site inspections to confirm data exchanges. These data exchanges would be updated periodically, just as is provided for under START's Memorandum of Understanding (MOU) on treaty-limited items.

Just as under the INF and START treaties, one consequence of making ALCMs (or other bomber weapons) treaty-limited items would be the requirement for the destruction of missiles in excess of agreed ceilings, with a provision for inspections to verify the actual destruction of missiles and their components subject to elimination.[22] As has been agreed for heavy bombers under START, a limited number of nondeployed ALCMs for test purposes could be made exempt from the deployment limits, as long as they were located at declared facilities and kept separate from operational ALCMs and bombers.

To provide additional confidence in these direct ALCM ceilings beyond the initial baseline inspections, portal-perimeter monitoring stations—such as the facility constructed by the US at the Soviet SS-20/SS-25 missile production plant at Votkinsk (and the Soviet station at the former US Pershing motor production plant in Magna, Utah) under the INF treaty—could be installed at nuclear ALCM production facilities. In addition, as is apparently under discussion for START in case of mobile missiles, periodic short-notice inspections could be permitted at declared nuclear-weapon storage sites (most of which are located at the bomber bases themselves) to confirm declared inventories.

Portal-perimeter monitoring of conventional ALCM production, coupled with short-notice on-site inspections (OSIs) of conventional-weapon storage facilities, would provide some direct verification that nuclear ALCMs were not in undisclosed locations and that conventional ALCMs did not carry nuclear warheads or could not be easily so converted. Unlike portals at nuclear ALCM production facilities, where all exiting ALCMs would be assumed to be nuclear, portals at conventional ALCM plants would require more sophisticated equipment capable of interrogating exiting containers (see below).[23]

Obviously, each side would continue to rely on its own national intelligence to detect signatures and procedures associated with nuclear handling and storage that would indicate illegal activities or covert stockpiles.

Even with direct limits on ALCMs and certain intrusive measures to ensure compliance with ALCM limits, constraints on potential ALCM-launching platforms would still be needed as an added barrier to guard against breakout. Under a deep-reductions agreement, at the very least,

START's cooperative measures applying to heavy bombers should be retained. These include: strict physical and operational separation of ALCM from non-ALCM heavy bombers; strict physical and operational separation of non-nuclear heavy bombers from nuclear storage areas; and utilization of EODs (and preferably FRODs, if truly functional and observable differences can be found) on different heavy-bomber types to distinguish them from each other.

While these measures would reduce ALCM-ceiling uncertainties and the potential for breakout, they would not eliminate them entirely. There would still exist the possibility of concealed production, assembly, and stockpiling of nuclear ALCMs at undeclared facilities; nuclear loads on bombers that appear to be dedicated to conventional use; and potential loadings of ALCMs above agreed limits.

An additional, more intrusive measure, also under consideration for START, would be to conduct occasional random inspections of bombers and their weapon loadings. This would require at least visual access to the internal bomb bays and direct observation of loading procedures, both of launcher assemblies and of the bombers themselves. Inspectors would have to be satisfied that bomb bays and any external pylons of non-ALCM bombers were incapable of carrying nuclear ALCMs (or, in the case of conventional bombers, any nuclear weapons).

Finally, there is the problem of medium bombers (such as Backfire, F-111, etc.) that already carry certain types of ASM, have significant range and refueling capability, and are therefore potential ALCM carriers. Under both SALT II and START, any type of bomber that conducted a test launch of a long-range ALCM would thereafter be counted as an ALCM-capable bomber. However, range limits on ALCMs are difficult to measure and, furthermore, deployment of advanced ASMs, even with ranges of 600 kilometers or less on medium bombers might represent a strategic threat if forward-deployed, as is permitted under START and planned for US F-111s to be based in England and armed with tactical SRAMs.

Resolution of this breakout threat may require limits on medium-bomber operations from forward areas or direct limits on these bomber types (which may be curtailed, in any case, under the Conventional Forces in Europe agreement being negotiated in Vienna). Of course, even without such limits, direct monitoring of ALCM (or other ASM) production would offer some protection against a breakout in which many medium-range bombers would be equipped with long-range ALCMs.

## Application of Various Technologies to ALCM Verification

Experience to date in implementing existing treaties suggests that high-technology approaches to monitoring and inspection may not always be practical or required. Reportedly, US officials in monitoring the INF treaty have concluded, based on their experience, that some of the high technology developed for verification purposes may be unnecessary for that treaty.[24]

For example, one report said that some sophisticated size and weight monitors didn't work as well as planned under actual weather conditions and "the Soviets agreed to allow US inspectors to open vehicles on the basis of dimensions alone and use tape measures and plumb bobs to do the measuring needed." The report also suggested that, "such sophisticated, and expensive, gear may not be needed for START verification," particularly since direct human inspection of opened containers may be more negotiable than previously thought.

### *Use of Radiation Detectors*

As mentioned above, providing further assurance that conventional ALCMs are not nuclear may prove necessary as more types of conventional ALCM enter operation in the 1990s—particularly since many of these are to be deployed on heavy bombers. Under START, conventional ALCMs must be observably different from nuclear ones, which will facilitate monitoring to some extent. But further confidence could be provided through occasional random sampling of conventional ALCMs by removing them from operational bombers and assuring the inspecting party, through interrogation of the ALCM by a portable radiation detector, that they, indeed, did not contain nuclear warheads.

US laboratories have developed a portable device that is used in INF verification to measure the neutron radiation from the warhead ends of missile canisters to help distinguish between the single warhead SS-25 ICBM and the banned SS-20 three-warhead intermediate-range ballistic missile (IRBM) without opening the missile canister or nose cone.

*Use of Active Radiation Interrogation*

Use of active radiation interrogation devices, such as the US x-ray scanner now installed at the Soviet missile production facility in Votkinsk for INF monitoring purposes,[25] might help assure that the warhead section of a conventional ALCM could not be quickly converted to carry a nuclear warhead. Care would have to be taken that such a technique would not reveal any sensitive aspects of the missile design. Active radiation machines could also be installed at portal-perimeter stations around conventional ALCM production or storage facilities to interrogate exiting missile containers.

Using either passive radiation detectors or active machines to monitor ALCMs may prove somewhat easier than monitoring SLCMs since, unlike SLCMs, ALCMs are not deployed in canisters. This means that direct observation is possible and both passive and active radiation measurements are less easily shielded against.

*Use of Tagging*

Tagging ALCMs to allow for identification during inspections would decrease significantly the probability that "illegal" nuclear ALCMs were being infiltrated into the population of "legal" ALCMs. New nuclear ALCMs could be tagged at the portals of assembly facilities. A tagging system could track every legally produced ALCM if each was provided with a unique identifying tag that could be checked during portal inspections or other random OSIs (see chapter 8).

## CONCLUSION

ALCMs have many characteristics that make them difficult to limit in an arms-reduction agreement. They are relatively small and inexpensive and can be produced, stored, and deployed in great numbers, possibly clandestinely. They are, therefore, a potential breakout concern in any strategic arms agreement.

Given the framework for START that has evolved, the limits are least stringent for bomber-carried weapons. If lower warhead ceilings are achieved, these weapons will have to be limited more stringently.

A verification regime for ALCM limits in a deep-reductions agreement should include a range of measures designed to enhance transparency at all stages of the weapons cycle. Such measures could include: an extensive data exchange on ALCM production, storage, and deployment with inspections to verify baseline data; direct, continuous monitoring of production, storage and deployment through a variety of means (portal-perimeter monitoring, tagging, etc.); random, short-notice inspections of bomber loadings and weapon storage areas; and the use of advanced technological tools (for example, active and passive radiation detectors) to improve the effectiveness of inspections.

US and Soviet negotiators of the START treaty have made considerable progress in reducing some of the verification problems associated with ALCMs and other bomber weapons. However, solutions to some of the most difficult problems have been postponed to a future negotiation. It is in that "deep cuts" negotiation that a more comprehensive implementation of this chapter's ideas could take place.

## NOTES AND REFERENCES

1. See DoD Authorization for Appropriations for FYs 1990 and 1991, Hearings of the Senate Armed Services Committee, part 6, p.331.

2. The US AGM-86 ALCM-B is about 6.4 meters (20.8 feet) long and weighs approximately 1,500 kilograms (3,300 lbs), about 1.7 percent of the weight and less than one third of the length of an MX ICBM.

3. Until recently, the USSR also had a small number of Mi-4 "Bison" bombers. By 1988, all Soviet Bisons had either been dismantled or converted to tankers. All US B-52Ds and some -Gs have also been retired during the past decade. In addition, the US Air Force plans to shift the FB-111 from a strategic to a tactical role over the next several years.

4. These were called forward-based systems (FBS). The USSR continued to press the FBS issue in subsequent negotiations. For a comprehensive account of aborted SALT I proposals on limiting bombers and cruise missiles, see Gerard Smith, *Doubletalk: The Story of SALT I* (Garden City, New York: Doubleday, 1980).

5. See Hearings on the SALT II Treaty, US Senate Foreign Relations Committee, part 1, p.15. While the Air Force claimed that strakelets did have a function—to improve the B-52's aerodynamic efficiency to make up for atmospheric drag caused by external ALCM carriage—they are not really "functionally related" because B-52s do not require them to be capable of launching ALCMs. But each side agreed

that, for existing aircraft like the US B-52 and Soviet Tu-142 (a variant of the Bear bomber used for antisubmarine warfare), putting on a real FROD would be impractical and that they, therefore, would only be required of future bombers.

6. Ibid., pp.32–33.

7. In 1980, the total US heavy-bomber force (including the FB-111) could carry about 2,400 warheads, all of which were either gravity bombs or SRAMs. Today, the number is approximately 4,500, including over 1,600 ALCM-Bs. (Numbers derived from Department of Defense Annual Report for FY 1981, pp.71–77; and "U.S. Strategic Nuclear Forces, End of 1989," in "Nuclear Notebook," *Bulletin of the Atomic Scientists*, January–February 1990.) In fact, primarily because of deployment of ALCMs, the number of weapons carried by US strategic bombers on peacetime alert has tripled since the mid-1970s. (See Committee on Appropriations subcommittee on defense, US House of Representatives, Hearings on DoD Appropriations for 1990, part 1, p.264.)

8. From "Fact Sheet on ALCMs" released by the US Arms Control and Disarmament Agency, June 1988.

9. Under the agreed provisions of START, each heavy bomber not equipped to carry long-range ALCMs would count as one warhead against the 6,000 ceiling (as well as one launcher against the separate aggregate launcher ceiling of 1,600) no matter how many nuclear bombs and SRAMs it carries. This so-called "discounting rule" was a Soviet concession to the US negotiating position that "slow flying" bombers are less threatening than "fast flying" ballistic missiles.

10. TKL conversation with US Government official, 6 February 1990.

11. Ibid.

12. "Fact Sheet on ALCMs," op. cit.

13. TKL conversation with US Government official, February 1990.

14. Ibid.

15. Ibid.

16. See Warren Strobel, "Baker Offers Arms Reductions," *The Washington Times*, 8 February 1990, p.1.

17. Robert Pear, "In Arms Talks, Devil is in the Details," *New York Times*, 21 May 1990; Don Oberdorfer, "U.S., Soviets Agree on Controls on Cruise Missiles," *The Washington Post*, 20 May 1990, p.A1; William Safire, "Taking Baker to the Cleaners," *New York Times*, 21 May 1990, p.21; "U.S. ALCM Work Seen Prompting USSR Tacit Rainbow Concern," *Aerospace Daily*, 23 May 1990, p.314.

18. Under the agreed procedures, 30 days prior to the exhibition, the exhibiting country was to provide the inspecting country with information on the location of the aircraft, the characteristics distinguishing the ALCM carrier from the non-ALCM bomber, and other pertinent information. Upon arrival at the bomber base, the inspection team was given a tour, including locations of various types of heavy bomber, and photos showing the locations of distinguishing features on the bombers. In the case of the Soviet Bear-G and -H bombers, these differences included length of aircraft, length of bomb bay, defensive armament, and fire control devices. (TKL conversations with US Government officials, January–April 1990. See also "Heavy Bomber Inspections" undated 1989 US ACDA factsheet.)

19. TKL conversation with US Air Force official, February 1990. See appendix 9.A for a more detailed description of bomber conversion and loading processes.

20. This would appear to be more of a problem for the USSR than the US, since the planned US post-START bomber force will include over 200 penetrating bombers (B-1Bs and B-2As) as well as some 50 former heavy bombers (conventional B-52s), while the Soviet bomber force will be mostly Bear-Hs and Blackjacks, both of which the USSR has designated as ALCM carriers.

21. While this chapter focuses on limiting nuclear ALCMs, some of the intrusive verification means discussed provide for direct access to storage and deployment areas where ALCMs, bombs, and SRAMs are collocated. Thus, direct limits on these other bomber weapons would not necessarily add to the complexity of the verification regime.

22. A useful description of the destruction of US GLCMs is provided in Theresa M. Foley, "U.S., Soviet Missile Experts Begin INF Treaty Inspections," *Aviation Week & Space Technology*, 11 July 1988, p.25.

23. In July 1988, the USSR tabled as part of its draft START treaty a detailed proposal to monitor ALCM and SLCM limits that, although rejected by the US as too intrusive for START, could provide the basis for a package of additional verification measures under a deep-reductions agreement. The Soviet proposal provided for: nondeployed ALCMs not to exceed a certain percentage of all deployed ALCMs; all ALCMs to be restricted to declared storage and conversion or elimination facilities; portal-perimeter monitoring stations at all heavy-bomber and ALCM production factories; and short-notice inspections permitted at all heavy-bomber bases. (Information based on TKL conversation with Defense Department official, February 1990. See also R. Jeffrey Smith, "Soviets Seek Cruise Data Verification," *The Washington Post*, 23 July 1988, p.A20.)

24. See "INF Shows High-Tech On-Site Monitoring Not Essential, Expert Says," *Aerospace Daily*, 27 December 1989, p.480.

25. The x-ray cargo-screening system installed at Votkinsk is described in some detail in "U.S. Companies Team to Support INF Treaty," news release from Bechtel National, Inc., 5 April 1989. See also chapter 11.

*Thomas K. Longstreth*
*Richard A. Scribner*

# US Strategic Bombers and their Weapons

## STRATEGIC BOMBERS

### B-52 Bomber

The B-52 "Stratofortress" is a subsonic four-engine long-range "heavy" bomber, designed for both nuclear and conventional bombing, that has been in the US inventory since the 1950s. Originally designed as an intercontinental high-altitude nuclear bomber, modifications have given the B-52 an ability for low-level flight and conventional bombing.

Over 744 B-52s in eight different mods (A through H) were built between 1955 and 1962.[1] Although the two models that remain in service, the G and H versions, have the same external appearance as earlier versions, over the years they have been almost completely rebuilt and improved in many ways. Modifications have included: TF-22 turbofan engines and other improvements for longer range; new offensive and defensive avionics; and electrical and carriage modifications for air-launched cruise-missile (ALCM) and short-range attack-missile (SRAM) capability.[2]

Currently, about 200 B-52Gs and -Hs are deployed at 10 Strategic Air Command (SAC) bases in the continental US for nuclear missions (see table 9.A.1). Several others are located at Edwards Air Force Base (AFB), California, and are used for testing. The B-52H has a longer unrefueled range (15,000 kilometers at high altitude) than the B-52G (12,000 kilometers).[3]

Each B-52 can carry about 36,000 kilograms of weapons in its bomb bay and on pylons under the wings. The weapon loads for each bomber vary according to its mission. Only a few B-52Hs remain that carry only SRAMs and gravity bombs. Each one can carry up to eight SRAMs plus four gravity bombs internally and another 12 SRAMs externally on pylons.[4] All remaining B-52Hs are moving from their role as "penetrating" bombers to "shoot and penetrate" bombers, which will carry ALCMs to be launched well before the aircraft reached Soviet territory. The B-52Hs would launch SRAMs later when nearer their targets, and finally, overfly some targets to drop gravity bombs.

All remaining B-52Gs with a nuclear role are assuming a "stand-off" mission, meaning that they will be primarily cruise-missile carriers.[5] Each B-52G can carry up to 12 ALCMs on two external pylons, one under each wing, and can also carry an internal bomb and SRAM load. The B-52H can carry up to 20 ALCMs—12 externally and eight internally on a special rotary launcher, in addition to four gravity bombs. The Air Force plans to deploy advanced cruise missiles (ACMs) only on B-52Hs, each of which can carry 12 ACMs externally and four internally on its rotary launcher.

In 1988, 69 other B-52Gs based at Andersen AFB, Guam, Barksdale AFB, Louisiana, Loring AFB, Maine, and Mather AFB, California, were designated as conventional bombers and removed from their nuclear mission. These converted B-

**Table 9.A.1:** Distribution of US strategic bomber fleet 1990

**B-52Gs**

| | | |
|---|---|---|
| Barksdale Air Force Base (AFB) | Louisiana | 38 |
| Wurtsmuth AFB | Michigan | 19 |
| Griffiss AFB | New York | 19 |
| Castle AFB | California | 25 |
| Eaker (formerly Blytheville) AFB | Arkansas | 17 |
| Loring AFB | Maine | 23 *non-nuclear* |
| Andersen AFB | Guam | 10 *non-nuclear, being retired* |

*Three other B-52Gs used for testing*

**B-52Hs**

| | | |
|---|---|---|
| Carswell AFB | Texas | 28 |
| Fairchild AFB | Washington | 23 |
| K.I. Sawyer AFB | Michigan | 19 |
| Minot AFB | North Dakota | 23 |

*Two other B-52Hs used for testing*

**FB-111s**

| | |
|---|---|
| Pease AFB | New Hampshire |
| Plattsburgh AFB | New York |

*All to be converted to tactical bombers*

**B-1Bs**

| | | |
|---|---|---|
| Dyess AFB | Texas | 33 |
| Ellsworth AFB | South Dakota | 29 |
| McConnell AFB | Kansas | 16 |
| Grand Forks AFB | North Dakota | 16 |

*Three other B-1s at Edwards AFB, California, used for testing*

Source: SAC Headquarters, Offutt AFB, Omaha, Nebraska, February 1990.

---

52s can carry a variety of conventional bombs and stand-off missiles, including the Tacit Rainbow antiradar missile, the Have Nap (also called "Popeye") air-to-surface missile, sea mines, and other munitions. Thirty of the 69 can launch the Harpoon antiship cruise missile.[6] Of the original 69 converted B-52Gs, only about 55 remain in service, the bombers at Mather AFB having been retired in October 1989. That number is expected to be cut by another 20 in 1990, as additional B-52Gs from Andersen AFB are retired to save money. All are expected to be retired by 1992.[7]

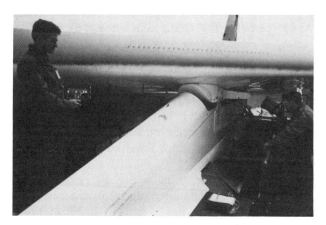

**Figure 9.A.1**: ALCM loading sequence. B-52Gs at Fairchild AFB, Washington, being loaded with ALCM-Bs. Photo sequence shows from top to bottom: hard points under B-52 wing for attaching ALCM-B pylon; fully loaded ALCM-B pylon on hydraulic trailer being maneuvered into position; positioning of ALCM pylon under wing for attachment.

Source: US Department of Defense

*Conversion to ALCM Carriage*

B-52 bombers have been converted to carry ALCMs at several locations. B-52Gs were modified at the Boeing Military Airplane Company, Wichita, Kansas, or the Air Logistics Center (ALC) at Tinker AFB, Oklahoma City, Oklahoma, and at the Air Logistics Center, Kelly AFB, San Antonio, Texas. B-52Hs are all converted at San Antonio ALC.

Ninety-seven B-52Gs have been converted to each carry 12 ALCMs externally. Ninety-five B-52Hs can now carry 12 ALCMs (or ACMs) externally, and another 16 have received the Common Strategic Rotary Launcher (CSRL), allowing them to carry up to 20 ALCMs (or 16 ACMs) internally and externally. Conversion of all B-52Hs to carry the CSRL will be completed in 1993.[8]

To carry ALCMs externally, B-52Gs and -Hs are fitted with jettisonable pylons that fit onto fairings under each wing. In addition to the pylons, each ALCM-equipped B-52 receives the "offensive avionics system" (OAS) to provide the ALCM's internal guidance system with its initial position for greater accuracy. The OAS can be used for other missiles as well. Slightly different electrical interfaces are required between the OAS system and the pylon depending on whether it is carrying ALCMs or other weapons, and the aircraft must be properly configured to accept those interfaces.

To carry ALCMs internally, B-52Hs are fitted with CSRLs during programmed depot-level maintenance, a procedure that involves virtually stripping and reassembling the bomber and that, according to the US Air Force takes between four and five months. Modifying the bomber to carry ALCMs internally and externally usually takes about three months of this time, but the CSRL modification itself could be accomplished in considerably less time if necessary.[9]

To be able to accommodate a CSRL, each B-52H must have its bomb bay and bomb-bay doors modified as well as forward and aft yokes installed inside the bomb bay. During this procedure, B-52s also receive their strakelet modification to distinguish them as ALCM carriers. The CSRL allows the B-52 to carry ALCMs, ACMs, SRAMs, and a variety of other nuclear and conventional gravity bombs.[10]

# FB-111 Bomber

The FB-111 is a medium-sized, two-engine bomber developed in the 1960s for high-speed, low-altitude bomber missions. It can fly supersonically, day or night, and is currently designated for use in strategic nuclear missions. Over 75 FB-111s were deployed between 1969 and 1971, of which fewer than 60 remain operational at two SAC bases (see table 9.A.1).

The FB-111 is a modified version of the tactical F-111 bomber, which is in service in Europe and the US and was originally designed in the early 1960s to be both a Navy and an Air Force bomber. It has a longer wingspan than the F-111 and different avionics, engine inlets, and landing gear because of its different mission and 11,000 kilograms greater weight.

The FB-111 can carry four external fuel tanks and has an unrefueled range of

4,800–5,400 kilometers. Although it would require more refuelings than a B-52 or B-1B on a strategic mission, it is quite capable, with refuelings, of attacking targets well inside the USSR.[11] An FB-111 can carry two bombs or SRAMs internally and four SRAMs on external pylons. It has not been used to test ALCMs although ALCM launch would certainly be feasible.

Before the B-1B was deployed, the FB-111 was the United States' most accurate and effective penetrating bomber. With the B-1 now fully operational and the B-2 expected to begin arriving in 1991, the Air Force plans to change the role of the FB-111 from a strategic to a tactical bomber during 1990 and 1991. Although the Air Force examined the possibility of deploying the FB-111 in Europe as part of NATO's theater nuclear forces, budget cuts have led the service to plan to base the bomber at Cannon AFB, New Mexico, where it will be used primarily for training.[12]

## B-1B Bomber

The B-1B bomber is the most recent US heavy nuclear bomber to enter operational service. It is a modified version of the original B-1A that was developed in the early 1970s and first flew in 1974, but was cancelled in 1977.

The first B-1A converted to B-1B flew in 1983. Introduced into US strategic forces between 1985 and 1988, about 97 B-1s are now operational at four B-1 bases (see table 9.A.1).

Although the B-1A and the B-1B have about 80-percent commonality, a number of changes were made so that the B-1B could carry more fuel and weapons—especially ALCMs. These changes increased the maximum takeoff weight from 180,000 kilograms to about 217,000 kilograms. To cope with improved Soviet air defenses, the B-1B received a new defensive avionics system (which has been plagued by problems) and modifications to engine inlets and other surfaces to reduce its radar cross section.[13] The changes to the engines were also a mandated by the alteration of the bomber's flight profile from a supersonic high-altitude penetrator to a subsonic low-altitude one.

The B-1B, which has an unrefueled range of about 8,000 kilometers, can carry up to 57,000 kilograms of either nuclear or conventional weapons in its three internal weapon bays and on external pylons. It can carry up to 24 B-83 or B-61 nuclear gravity bombs or SRAMs internally and another 14 externally.[14] Because of the smaller size of the SRAM II, the B-1B could carry up to 12 in each bomb bay or 36 total internally.[15]

As with the B-52, SRAM and bomb payloads for nonalert B-1s are stored on fully checked and loaded launchers in storage igloos located near base runways. A large, almost 30,000-kilogram, support vehicle would carry each loaded launcher from storage to the aircraft as needed.[16]

*Conversion to ALCM Carriage*
On all but the first eight B-1Bs produced, the divider wall between the forward two bays is movable, allowing carriage of ALCMs. To accommodate ALCMs internally, the bulkhead between the two forward bomb bays (each of which is only 4.55 meters long)

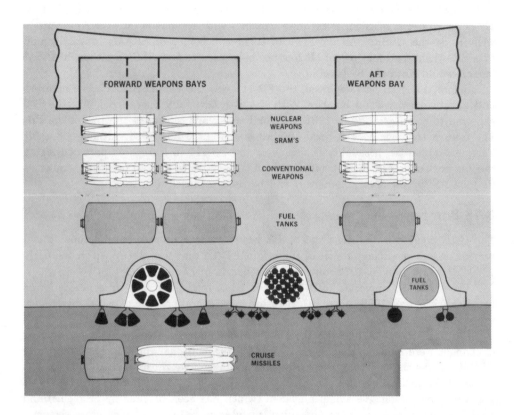

**Figure 9.A.2:** Configuration of the weapon bays in the B-1 fuselage. The dotted line depicts a movable bulkhead between the forward two bays that can be adjusted to accommodate an ALCM launcher.

Source: Rockwell International

must be adjusted to fit the larger CSRL, which is over two meters longer than the regular launcher.[17] This procedure is relatively simple and does not take long. As with the B-52H, changes to the B-1B's electronic systems are also required before it can launch ALCMs. However, the "black boxes" required to communicate with ALCMs located internally already exist on all B-1Bs.

According to the Air Force, ALCM-B testing from B-1Bs converted for the purpose was completed in 1988. The B-1B also completed some initial testing with the ACM, although no test launches of actual ACMs were conducted. However, "final B-1B/ACM integration efforts...were stopped in December 1988 and deferred pending review by the Office of the Secretary of Defense."[18] It is believed that further B-1B/ACM testing was put on hold as a result of both ACM developmental problems and, perhaps, arms-control considerations.[19] Because of the START treaty's counting rules and the premium the treaty places on the reduced weight of warheads carried

**Figure 9.A.3:** Forward dual pylon for ALCMs mounted on a B-1B fuselage. The modified B-1B can carry up to six dual ALCM pylons and two single pylons for a total external carriage of 12–14 ALCMs

Source: Rockwell International

**Figure 9.A.4:** Dual pylon for the B-1B bomber external carriage fitted with ALCM-Bs and loaded on a lift truck. Note the much smaller size of the B-1B pylon compared to the ALCM pylon for B-52s.

Source: Rockwell International

by non-ALCM bombers, the Air Force claims that no B-1s currently have an ALCM capability and that there are no plans to deploy ALCMs or ACMs on B-1s until the B-52H force is retired.[20]

As the B-52Hs are retired, beginning in the late 1990s, B-1Bs will begin receiving the ALCM-capable rotary launcher (CSRL) now being installed in B-52Hs. After conversion, the B-1B could carry eight ALCMs on an internally mounted CSRL, as well as weapons or fuel in the aft bomb bay. Externally, the B-1B can carry two ALCMs or ACMs each on six pylons located along the fuselage, plus one ALCM or ACM each on two single pylons, also located on the fuselage, for a total of 14 externally and 22 total.[21]

However, because of START counting rules and inspection provisions, the Air Force will have to demonstrate to Soviet inspectors that each ALCM-capable B-1B can carry only 20 ALCMs. According to engineers at Rockwell, the B-1B's manufacturer, the Air Force issued an order several years ago for the company to cease production of the single pylons, in order to keep the B-1's external carriage limited to 12, as per START counting rules. However, hard points on the fuselage for these single pylons still exist, thus to enable the B-1B to carry 14 instead of 12 ALCMs externally would simply be a matter of manufacturing the smaller pylons. Unlike the case for internally mounted ALCMs, however, modifying B-1Bs to carry ALCMs externally would require installing additional black boxes, called LRUs, in order to communicate with the ALCMs.[22]

Another aspect of B-1 ALCM carriage relevant to arms control is the fact that installing pylons on B-1Bs is simpler than it is on B-52s. The ALCM pylons for the B-1B are much smaller and lighter, and thus could be installed more quickly using a smaller and more maneuverable lift truck.

As part of the START treaty's cooperative measures on ALCM limits, the US and USSR have been examining ways to distinguish between ALCM-capable B-1Bs and non-ALCM B-1Bs. First, the Air Force has pointed out that each ALCM B-1B would carry a blade antenna to distinguish it from a non-ALCM aircraft—this antenna has no function and is not needed by the B-1B to launch ALCMs. Secondly, Rockwell engineers suggest that the fact that the bulkhead has been moved to accommodate the ALCM launcher can be observed without opening the bomb bay by examining the position of a ring that fits around the bulkhead.[23]

The two sides have examined several ideas for preventing easy installation of ALCM pylons on the B-1B's hard points. One problem is that these hard points are utilized regularly to jack up the aircraft's wings for routine maintenance, which precludes a simple solution like bolting on a plate to cover them permanently. Engineers have looked at simply making some of the hard points along the fuselage too small to permit installation of the ALCM pylon, while leaving intact other hard points used for jacking up the aircraft.[24]

## B-2A Advanced Technology "Stealth" Bomber

The B-2, under development since the late 1970s and built by Northrop, is the latest addition to US strategic bomber forces. It is the most expensive and sophisticated aircraft ever built. Drawing on technological advances in materials and design, it has been designed to reduce its various radar, heat, and other signatures to an absolute minimum in order to drastically reduce the ability of Soviet air defense radars and interceptors to detect and attack it while overflying Soviet territory.

The B-2, a descendant of the YB-49, also built by Northrop in the late 1940s, has a similar "flying wing" appearance. However, the B-2 has a complex trailing edge that, with its saw-toothed shape, helps give the aircraft lift and directional control. All external parts of the aircraft have been designed to reduce radar, infrared, and other signatures. The B-2's four engines are concealed inside the aircraft's structure, and the engine inlets and exhaust ducts have also been designed to reduce signatures. The B-2 is built of carbon-fiber, glass-fiber, and other composite plastics, which absorb radar waves better than metal.

The B-2 is approximately 21 meters long and has a wingspan of 52.4 meters. It flies at subsonic speed and will have an unrefueled range of 9,000–10,500 kilometers, depending on its flight profile. While originally intended to be a high-altitude bomber, the wings of the B-2 were redesigned and strengthened in the mid-1980s to make it capable of sustaining flight and penetrating enemy airspace at low altitude. Its gross takeoff weight, with a full fuel and bomb load, is about 169,000 kilograms.

To minimize drag and radar cross section, the B-2 will carry all weapons on launchers in internal bomb bays located aft of the cockpit. A nominal weapon load would be eight SRAM IIs and eight B-61 or B-83 gravity bombs.[25] Although the B-2 could carry ALCMs and ACMs, because of START counting rules the Air Force has no current plans to deploy cruise missiles on the B-2.

The Air Force had originally explained to Congress that one of the B-2's principal attributes would be its unique ability to seek out and destroy so-called strategic relocatable targets (SRTs) such as Soviet mobile-missile launchers and command centers. Recently, however, the Air Force has conceded that the B-2 would not be effective against SRTs and has instead stressed its versatility as both an advanced nuclear and conventional bomber.

Currently, about 15 B-2s are in various stages of construction, and the Air Force plans to build a total of 75. The first two B-2 squadrons (30 aircraft) are to be based at Whiteman AFB, Missouri. Three other proposed B-2 base locations remain classified. Each B-2 will be kept in its own climate-controlled hangar because of the sensitivity of the bomber's composite structure to temperature extremes.[26]

Because of its high price-tag, the design and production problems that led to schedule delays, and the fact that its details are only now emerging publicly, the B-2 program has become one of the most controversial. Congress may cut back further the Air Force's planned buy (originally 132 bombers) and scrutinize more closely the bomber's cost, schedule, and mission.

One curious aspect of the B-2's weapon capability as it relates to START counting rules is its internal launcher. According to the Air Force, "the B-2 does not have the electrical or hydraulic interfaces to drive the CSRL [Common Strategic Rotary Launcher]" and would require "significant modifications" to accommodate the CSRL (which *can* carry ALCMs) being installed in B-52Hs. Instead, the B-2 is being equipped with the Advanced Applications Rotary Launcher (AARL).

The AARL appears similar in both size, weight, and weapon carriage capability to the CSRL, so it is not clear why a separate launcher was developed for the B-2 (at over $1 billion total cost) or what physical limitation of the AARL, if any, prevents it from carrying cruise missiles. AARLs will be installed on B-2s at intermediate maintenance facilities located at main operating bases.[27]

## US Tanker Aircraft

A major strength of the US strategic bomber force, and one of its advantages over the Soviet strategic bomber force, is its large fleet of tanker aircraft capable of refueling bombers while in flight. This greatly expands the range and flexibility of US strategic bombers and means that they can be loaded with maximum numbers of nuclear weapons for strategic missions.

The US tanker fleet consists of almost 600 KC-135 "Stratotankers" and 60 KC-10 "Extenders." KC-135s are assigned to each SAC bomber base, and, as with the bomber force, a certain percentage is always on ground alert and ready to take off at a moment's notice.

## BOMBER-CARRIED WEAPONS

### Air-launched Cruise Missiles

The United States began developing and deploying ALCMs in the 1950s. Generally speaking, these early ALCMs were large, inaccurate, and unreliable. Later advances in engine design and propulsion, as well as warhead and guidance-system innovation and miniaturization resulted in smaller airframes with increased range, reliability, and accuracy.[28]

There were a variety of technical, military, and diplomatic factors in the late 1960s and early 1970s that helped spur the development of modern cruise missiles. The exclusion of cruise-missile limits from the SALT I agreement provided one motivation. Another was the development of a more advanced turbofan engine that helped extend the range of cruise missiles to well over 1,600 kilometers.

In 1972, the Navy began development of the first of the present generation of long-range cruise missiles, the Tomahawk sea-launched cruise missile (SLCM), which also became the basic airframe for the Air Force's ground-launched cruise missile (GLCM) deployed in Europe in 1983. The Air Force was far less enthusiastic about cruise missiles than the Navy, viewing the long-range ALCM as a threat to the development of the B-1 bomber, so it began development of a shorter-range ALCM, the ALCM-A, which was derived from the subsonic cruise-missile decoy (SCAD). The Boeing ALCM-A was first tested in 1976.

In early 1977, the Office of the Secretary of Defense (OSD) directed the Air Force to begin development of a longer-range ALCM-B. In June 1977, President Carter

**Figure 9.A.5**: 1981 aerial view of Boeing Space Center, Kent, Washington, The assembly building for the ALCM-B (AGM-86B) is at center right of the picture.

Source: Boeing

**Figure 9.A.6**: Final assembly of ALCM-Bs inside Boeing Space Center

Source: Boeing

cancelled the B-1A bomber and directed the Air Force to accelerate development of the ALCM-B for deployment on B-52 bombers.[29] The longer-range ALCM-B was first tested in 1979 and selected by the Air Force in 1980 after a fly-off with a General Dynamics competitor.

## ALCM-B

The AGM-86B ALCM-B is a subsonic long-range (approximately 2,500 kilometers nominal range) nuclear-armed cruise missile that can be launched from well outside Soviet territory to attack important strategic targets. It evades air defenses through a combination of small size, low radar cross section (RCS), large numbers, and a low-altitude flight path. The ALCM-B is launched after being dropped from an internal bomb bay or from pylons mounted underneath a bomber's wing. Its navigation is accomplished by a combination of an inertial navigation system and a terrain contour matching (TERCOM) system.

The ALCM-B is 6.36 meters long and has a diameter of about 0.68 meters. It weighs about 1,432 kilograms at launch. Its W-80-1 Warhead has a variable yield of 5–150 kilotons. The ALCM-B has a median accuracy of less than 0.06 kilometers

Between 1981 and 1986, Boeing Aerospace Company produced 1,715 ALCM-Bs. Pilot production occurred at Boeing's Seattle Developmental Center, but was transferred in 1981 to a 25,000-square-meter production facility in Kent, Washington, located at Boeing's enormous production complex there. The main building housed all subassembly, final assembly, processing, and testing of ALCM-Bs. After final assembly, ALCM-Bs were fueled in a separate, smaller building located nearby before being delivered to the Air Force.[30]

The first ALCM-equipped B-52G went on alert in September 1981, at Griffiss AFB, New York, with the first full squadron following in December 1982. Currently, ALCM-Bs are carried only by B-52G and -H heavy bombers, although they have been tested from B-1B bombers, and other types of tactical combat aircraft as well.

About 200 ALCMs are located at each base. They are preloaded on pylons (for external carriage) or on rotary launchers (for internal carriage) before being mounted on or loaded in the aircraft. Loaded launchers are referred to as "clips" and are attached to the aircraft as needed, using a special heavy-lift trailer. ALCMs, SRAMs, and bombs for nonalert bombers are kept on their launchers and in storage bunkers usually located near the alert bomber area. Bombers on alert (about 30 percent of the force in normal peacetime alert status) are fueled and armed, with their weapons already loaded. ALCMs rotate from alert bombers to bunkers over periods of 30 months, after which the engine is removed for a testing and maintenance period.[31]

## Advanced Cruise Missile

Rapid progress in cruise-missile technology and advances in Soviet air defenses led the Pentagon to begin developing the Advanced Cruise Missile (ACM) or AGM-129A in 1982. The Air Force apparently began limited production of the ACM between 1984 and 1986, with the intention of fielding the first missiles in 1987–88. However, production of the ALCM-B continued through 1986 as a result of problems with the

**Figure 9.A.7:** Close-up view of advanced cruise missiles (AGM-129As) on a B-52 pylon
Source: Jim Burnett, *Omaha World-Herald*

**Figure 9.A.8:** A B-52H armed with AGM-129A ACMs

Source: US Department of Defense

ACM that were first noticed during flight testing.[32]

The ACM provides improved accuracy and range over the ALCM-B, and incorporates advances in stealth technology to reduce the effectiveness of enemy air defense detection. While much about the ACM program remains classified, the ACM is believed to have range of greater than 4,000 kilometers[33] and to be more than twice as accurate as the ALCM-B (which would mean that the ACM would strike within 30 meters of its target).[34] The ACM uses the same W-80-1 nuclear warhead as the ALCM-B.[35]

Because of START counting rules, the Air Force currently plans to deploy ACMs only on B-52H bombers, although ACMs could also be carried by B-1B and B-2 bombers modified for cruise missiles. The B-52H can carry up to 12 ACMs externally, in addition to four internally on the CSRL.[36] Modified B-1Bs could carry four ACMs internally and 12–14 externally.[37]

The ACM is being produced by both General Dynamics' Convair Division in San Diego and the McDonnell-Douglas Missile Systems Company in St. Louis—the same two companies that produce the SLCM. Initial production may have also taken place at an Air Force facility near Seattle.[38] Because of the Air Force's desire to field ACMs as quickly as possible, the first missiles were delivered in 1988, even as developmental flight testing continued.

Because of the problems encountered by the ACM during flight tests, the Congress froze funding in 1989 for ACM production pending better test results. Recent test successes (as of early 1990, 13 of 21 tests have been successful) have fulfilled Congressional requirements, and the ACM is now back in limited production.

A full-scale production decision is now expected by August 1990. As of early 1990, about 30 ACMs had been delivered to the Air Force, although these may be returned to the manufacturer for modifications.[39] Production is planned to continue until final delivery of all 1,461 ACMs by 1996, with the entire program expected to cost about $7 billion.[40] Approximately 1,300 of the 1,461 are to be operational missiles, the rest will be used as spares and for testing.[41]

The ACM's initial deployment is now likely to occur in late 1990 or early 1991. K.I. Sawyer AFB, Michigan, and Carswell AFB, Texas, are the first bases where B-52Hs will carry ACMs.[42]

*Advanced-technology Cruise Missiles*
The Air Force is also working on advanced propulsion, guidance systems, fuels, and materials in order to develop a third-generation of cruise missiles with still greater accuracy, longer ranges, and smaller radar and other signatures than either the ALCM-B or ACM. These advanced-technology cruise missiles could be either conventional or nuclear-armed. Use of recuperative turbine engines, boron slurry or other high density fuels, special composite materials, and other technological advances might allow the next generation of cruise missiles to reach intercontinental ranges with high accuracy. In addition to military service development programs underway, the Defense Advanced Research Projects Agency (DARPA) is also spending considerable effort at developing advanced technology cruise missiles.

**Figure 9.A.9**: Trainer versions of non-ALCM nuclear weapons being loaded onto B-52s. At top, rotary launcher filled with SRAMS on lift vehicle (source: US Department of Defense); at bottom, B-28 gravity bombs

Source: Chuck Hansen

## Other Air-delivered Nuclear Weapons

*SRAM-A*

The AGM-69A short-range attack missile (SRAM), which is propelled by a solid-fuel pulse rocket engine and flies at mach 3.5, has been deployed on B-52, FB-111, and B-1 bombers. About 1,500 SRAMs were produced between 1972 and 1975, of which about 1,100 remain operational.[43] The principal contractor for the SRAM-A was Boeing Aerospace Company.

The SRAM-A is designed to prepare the way for heavy bombers by knocking out air defense systems before the bombers reach enemy territory. It can fly a number of different and erratic trajectories, thus making it very difficult to defend against.

The SRAM-A is 4.27 meters long, weighs 1,010 kilograms, and has a range of 60–220 kilometers.[44] It has an extremely small radar cross section and has a median accuracy of about 30 meters. Each SRAM has a single W-69 nuclear warhead with a variable yield of 5–175 kilotons.

The SRAM has been upgraded over the years with a more powerful and longer-lived motor and a new computer with better targeting software.[45] However, the SRAM-A inventory is declining, and several problems associated with aging of the missile led to an Air Force decision in 1982 to begin developing a new state-of-the-art SRAM—the SRAM II. Recently, several of the nuclear weapon labs directors suggested that the SRAM-A be removed from bombers on alert because of the potential for accidents involving the W-69 warhead.

Each B-52 can carry up to 20 SRAMs, 12 in clusters of three under the wing and eight on a rotary launcher in the aft bomb bay. Each FB-111 can carry up to six, four externally and two internally. The B-1 could carry up to 38 SRAMs, 24 internally and seven externally under each wing.[46] Operational weapons loads on all three aircraft are usually less than the maximum loads. The B-2A will carry only the SRAM II.

*SRAM II*

The AGM-131A SRAM II is a new, more accurate version of the SRAM with longer range, a larger nuclear warhead, and a smaller radar cross section. Engineering development began on the SRAM II, also built by Boeing, in 1986: flight testing from the B-1B will begin in September 1990. The missile is scheduled to be deployed between 1993 and 1998 as SRAM-As are retired. It will carry the 200-kiloton W-89 nuclear warhead and have a maximum range of about 300 kilometers.

The SRAM II is about the same length as the SRAM-A but, at 884 kilograms, weighs slightly less. The Air Force plans to build 1,633 SRAM IIs, of which about 1,400 will be deployed; the rest will be used for testing. The SRAM II is to be carried by B-1B and B-2 bombers.[47]

The Air Force has apparently encountered some difficulties with the SRAM II's solid-fuel rocket motor during development, which, some reports suggest, could delay the system's introduction by at least one year, although the Air Force denies it.[48]

*SRAM-T*

The AGM-131B SRAM-T is a modified version of the SRAM II that is to be deployed as a theater nuclear weapon for NATO forces, beginning in April 1995. The principal difference between the two is that SRAM-T will have a longer range (about 400 kilometers) and a lower-yield warhead (variable yields of 10 and 100 kilotons). The US plans to build about 450 SRAM-Ts initially, but the final number for NATO forces has yet to be determined. The SRAM-T will first be flight tested in late 1992 from an F-15E, and will also be compatible with the F-111, F-16 and Tornado combat aircraft.[49]

*Nuclear Gravity Bombs*

In addition to ALCMs and SRAMS, US strategic heavy bombers carry several types of nuclear bomb. Currently, they include: B-61-0 and B-61-1 bombs, which have an estimated weight of 320–350 kilograms and an estimated yield of 350 kilotons; B-28 bombs, which each have a yield of 70 kilotons to 1.1 megatons and weigh between 900 and 1,100 kilograms; B-53 bombs, which weigh 4,022 kilograms and each have a yield of nine megatons; and B-83 bombs, which weigh 1,100 kilograms and have a yield of up to 1.2 megatons.[50]

The B-83 is the newest strategic bomb, having been first introduced into the stockpile in December 1983. The B-28 is the oldest, having been first deployed in 1958, although it has been upgraded several times since then to improve its safety and security features. The B-53 was first introduced in 1962, and the B-61-1 in 1968. The B-61-1 was also modified in the mid-1980s to improve its safety and security features, and is now called the B-61-7.

Each bomb can be dropped by free fall or by parachute, which allows time for the bomber to escape when delivering weapons from low altitude.

All final assembly of US nuclear warheads occurs at the Pantex facility near Amarillo, Texas. The only warheads currently in production for bombers are the W-80 ALCM-B warhead and the B-61 bomb, although production of the B-83 bomb is expected to resume in the next several years, and the W-89 warhead for the SRAM II will enter production in 1992.

## Bomber Nuclear Weapon Storage

The Air Force has three major depots that provide an intermediate level of storage for its strategic weapons between final assembly and their main operating bases. These storage depots are located at Kirtland AFB, New Mexico, Lake Meade at Nellis AFB, Nevada, and Barksdale AFB, Louisiana.[51] These depots are called Military First Destinations. They are where the Department of Defense (DoD) first takes delivery of warheads from the Department of Energy (DoE). Bomber warheads are then either kept in storage at these depots or transported by air to SAC bases where they are mated with delivery vehicles (such as ALCMs) and signed over to the local custodial units, called munitions maintenance squadrons. At their SAC bases, warheads and delivery vehicles are either kept in bunkers in Weapons Storage Areas (WSA) or loaded on alert aircraft in Victor Areas.

## Conventional ALCMs

The Air Force and Navy are developing and deploying a variety of conventionally armed ALCMs designed to attack both land- and sea-based targets. Table 9.A.2 lists these various conventional ALCMs and what is known of their technical characteristics.

Some of these weapons, all of which are to be carried on B-52 aircraft, are or soon will be deployed. These include: the Harpoon antiship cruise missile, of which thousands are currently deployed; a land-attack derivative of the Harpoon, called the Stand-off Land Attack Missile (SLAM); the Have Nap or Popeye, a TV-guided ASM; and the Tacit Rainbow antiradar missile.

Tacit Rainbow, an Air Force missile that is scheduled to be deployed in the early 1990s, is an 800-kilometer range antiradar missile that can loiter while it waits for enemy radars to be turned on, and then seeks out and destroys those radars. Each B-52G will carry 30 Tacit Rainbows on an internal rotary launcher. Other aircraft such as the F-16 and F-111 may also eventually carry Tacit Rainbow. A ground-launched version is also under development. While the US was planning to buy up to 18,000 Tacit Rainbow missiles, the Navy's withdrawal from the program led to a cut in the planned number down to 5,000.[52]

**Figure 9.A.10:** Tacit Rainbow antiradar missile during a test flight
Source: US Department of Defense

**Table 9.A.2:** US air-launched missiles

*US nuclear air-launched cruise missiles*

| Type | Range<br>*kilometers* | Warhead<br>*kilotons* | IOC* | Number | Carrier |
|---|---|---|---|---|---|
| ALCM-B<br>(AGM-86B) | 2,500 | 5–150 | 1981 | 1,715 | B-52G/H |
| ACM<br>(AGM-129A) | 3,800–<br>4,500 | 5–150 | 1991 | 1,460[†] | B-52H, B-1B, B-2 |

*Other US nuclear air-to-surface missiles*

| Type | Range<br>*kilometers* | Warhead<br>*kilotons* | IOC* | Number | Carrier |
|---|---|---|---|---|---|
| SRAM<br>(AGM-69A) | 60–220 | 170-200 | 1972 | 1,100 | B-52, FB-111, B-1B |
| SRAM II<br>(AGM-131A) | 330+ | 200 | 1993 | 1,633[†] | B-1B, B-2 |
| SRAM-T<br>(AGM-131B) | 420 | 10–100 | 1995 | 565[†] | F-15E, F-111,<br>Tornado, F-16 |

*US conventional air-to-surface missiles*

| Type | Range<br>*kilometers* | Warhead<br>*kilograms* | IOC* | Number | Carrier |
|---|---|---|---|---|---|
| Harpoon<br>(AGM-84A) | 135 | 230 | 1977 | 3,000 | A-6, B-52G, P-3 |
| SLAM<br>(AGM-84) | | | 1988 | 350–<br>2,350[†] | F/A-18, A-6, P-3C<br>B-52? |
| Have Nap<br>(Popeye) | 80 | 340 | 1988 | 100 | B-52G,    F-111F? |
| Tacit Rainbow<br>(AGM-136A) | 800 | 18 | early<br>1990s | 18,000?[†] | B-52G, A-6,<br>F-16?, F-111? |
| LRCSW | 800–<br>3,400 | ? | 2000 | 2,000–<br>10,000[†] | B-52, B-1,<br>others |
| Have Flag | ? | ? | ? | ?[†] | B-52? |
| Have Dark | ? | ? | ? | ?[†] | B-52? |
| Have Slick | 100 | 100 | 1997 | ? | Advanced tactical<br>aircraft (ATA) |
| SLAT | 90 | ? | early<br>1990s | 1,000 | |
| JTACMS | ? | ? | early<br>1990s | | B-52 |

* Initial operational capability
† Planned

Tacit Rainbow has been the subject of controversy during the START talks because it is a long-range ALCM being deployed on heavy bombers, and the Soviets have expressed concern lest it be converted to a nuclear missile.

Other highly classified "black" programs, such as Have Flag, Have Dark, JTACMs, and Have Slick, are believed to have an application to the development of advanced conventional cruise missiles, although they might not be cruise missiles themselves. At least one of these projects may be under development as a follow-on to the SLAM and Have Nap missiles, which are seen as interim ALCMs for the conventional land-attack mission.

For the long term, the Air Force and Navy are jointly developing several variants of advanced conventional cruise missiles under the Long-range Conventional Stand-off Weapon (LRCSW) program. The LRCSW is an ambitious effort to develop very advanced cruise missiles that incorporate state-of-the-art propulsion, stealth and guidance technology. The LRCSW would have an accuracy of several meters over ranges of between 800 and 3,400 kilometers. It would be used for different Air Force and Navy missions and could be launched from surface ships, submarines, and aircraft. It is planned for deployment towards the end of the 1990s.

If the US developed a long-range ALCM such as the LRCSW, under START rules the missile would have to be observably different from current nuclear ALCMs and displayed to Soviet inspectors. In addition, the USSR would be allowed periodic short-notice inspections at bases where such a weapon was deployed to confirm that it did not carry a nuclear warhead.

## NOTES AND REFERENCES

1.  From the Office of the Historian, Headquarters, Strategic Air Command, Offutt AFB, Nebraska, *The Development of the Strategic Air Command 1946–86*, 1 September 1986.

2.  See "The Boeing B-52 Stratofortress," background information paper published by the Boeing Military Airplane Company public relations department, Wichita Kansas, September 1987.

3.  See David L. Baker, B-52 systems manager, "Evolution of the B-52 Weapon System—Past, Present and Future," 1 December 1984.

4.  See Clarence Robinson, "USAF Readies Advanced Cruise Missiles," *Aviation Week & Space Technology*, 10 March 1980, p.12; also "B-52 Stratofortress," USAF factsheet, February 1982.

5.  See DoD Appropriations for 1987, House Appropriations Committee, part 6, p.78.

6.  See Howard Silber, "B-52s being Armed With Navy Missiles," *Omaha World-Herald*, 23 February 1984, p.1.

7. See Annual Report to the Congress for FY 1990, Secretary of Defense, p.190; DoD Appropriations for 1990, House Appropriations Committee, Part 1, pp. 322-323; "B-52's at Anderson to Carry 'Harpoons,'" *Air Force Times*, 20 May 1985, p.31; Richard Halloran, "U.S. Preparing Long-Range Bombers for Nonnuclear Missions," *New York Times*, 25 October 1986, p.6.

8. Ibid. Information is current as of February 1990.

9. Above information provided by Tinker AFB, Oklahoma, media relations office, 13 February 1990, in response to questions by TKL; See also "B-52 catches fire on ground, kills 1," *The Washington Times*, 26 July 1989, p.5.

10. Ibid.

11. DoD Authorization for Appropriations for Fiscal Year 1982, Senate Armed Services Committee, part 7, p.4329; see also David R. Griffiths, "FB-111 Bombers Playing Crucial Role," *Aviation Week & Space Technology*, 16 June 1980, p.145; and Craig Covault, "FB-111's Effectiveness Increased," *Aviation Week & Space Technology*, 10 May 1976, p.103.

12. DoD Authorization for 1990 and 1991, Hearings of the Senate Armed Services Committee, part 6, p.81.

13. Captain Michael B. Perini, "Finally, The B-1B," *Air Force Magazine*, January 1983, p.58.

14. See Tom Diaz, "B-1B is First Bomber to Join U.S. Air Arsenal in 30 Years," *The Washington Times*, 28 June 1985, p.1.

15. See Chuck Hansen, *U.S. Nuclear Weapons: The Secret History*, (Orion 1988), p.109.

16. See "Air Force Nears SAC Upgrade With B-1B Deployment," *Aviation Week & Space Technology*, 2 June 1986, p.46.

17. The CSRL is about 6.7 meters long. The rotary launcher for SRAM or bomb carriage currently on B-1Bs is only 4.55 meters long and can carry up to eight SRAMs, B-61, or B-83 bombs. The launcher for bombs does not require the same wiring and cooling equipment as that for SRAMs. (See *Aviation Week & Space Technology*, 30 May 1983, p.313.)

18. See DoD Authorization Hearings for FY 1990 and 1991, Senate Armed Services Committee, part 6, p.127; and prepared statement of Lawrence W. Woodruff, Deputy Undersecretary of Defense for Strategic and Theater Nuclear Forces, before House Armed Services Committee, subcommittee on research and development, 10 March 1987, p.21.

19. See Department of Defense Appropriations for Fiscal Year 1990, Hearings before the US Senate Committee on Appropriations, part 6, pp.54-55. In the 1990 Defense Appropriations Bill, the Congress directed the Air Force to continue ALCM/ACM testing from the B-1B and earmarked $50 million for the task.

20. The first successful launch of an ALCM-B from a B-1 occurred on 24 November 1987, (See "B-1B/ALCM Successful Launch," DoD news release, 25 November 1987. On ACM capability, see DoD Appropriations for 1990, Committee on Appropriations, US House of Representatives, part 1, p.315. In 1989, in response to a question on the B-1 and ALCMs during Congressional hearings, Secretary of Defense Richard Cheney stated that ALCM testing on the B-1B has been "completed, demonstrating a cruise missile carriage and launch capability. The only test activity remaining before complete cruise missile operational certification is ALCM nuclear certification. This activity is on hold pending agreement on START. Advanced cruise missile testing on the B-1B has been delayed pending further missile development activities." (See Department of Defense Appropriations for FY 1990, Hearings of the US Senate Committee on Appropriations, part 6, pp.54–55.)

21. See *Aviation Week & Space Technology*, 11 June 1984, p.51.

22. TKL conversation with Rockwell International engineer, April 1990.

23. Ibid.

24. Ibid.

25. From USAF factsheet, released December 1989.

26. For detailed descriptions of the B-2 bomber, its characteristics, mission, and production details, see Michael A. Dornheim, "Southern California Facilities Play Major Role in B-2 Production, Testing," *Aviation Week & Space Technology*, 12 December 1988, p.96; "USAF, Northrop Unveil B-2 Next-Generation Bomber," *Aviation Week & Space Technology*, 28 November 1988, p. 20; "B-2 Redesigned for Low Altitude Missions, Aldridge Says," *Aerospace Daily*, 28 November 1988, p.290; Rick Atkinson, "Stealth: From 18-inch Model to $70 Billion Muddle," *The Washington Post*, 8 October 1989, p.A1; "Testing and Operational Requirements for the B-2 Bomber," Hearing before the US Senate Committee on Armed Services, 21 July 1989.

27. When asked whether the AARL was capable of launching ALCMs or ACMs, an Air Force spokesman stated that "since the B-2 is not designated for strategic cruise-missile carriage, no provisions have been made to integrate these weapons into the launcher." TKL telephone conversation with USAF spokesman, Aeronautical Systems Division, Wright Patterson AFB, Ohio, January 1990.

28. Other types of nuclear air-to-surface missiles were also developed and deployed in the 1960s and early 1970s for the purpose of enhancing the effectiveness of strategic bombers. In the early 1960s, B-52s were equipped with GAM-77 Hound Dog supersonic ASMs, which had a maximum range of about 800 kilometers. Also developed during the early 1960s, but eventually terminated because of technical problems, was the GAM-87 Skybolt, which was to be an advanced air-launched ballistic missile with a range of about 1,600 kilometers and a speed of mach 7. Because of the missiles' size and weight, each B-52 could carry only two Hound Dogs and four Skybolts externally, in addition to its internal weapon load. (See DoD Appropriations for FY 1962, US Senate Hearings of the Subcommittee on Defense, Committee on Appropriations, pp.16, 296–297.)

29. An excellent account of the bureaucratic battles involved in the initial development of the ALCM-B is provided in Robert J. Art and Stephen E. Ockenden, "The Domestic Politics of Cruise Missile Development, 1970-1980," in Richard Betts, ed., *Cruise Missiles: Technology, Strategy, Politics*, (Washington DC: Brookings Institute, 1981).

30. See Richard G. A'Lone, "Boeing Gears for Production of ALCM," *Aviation Week & Space Technology*, 12 May 1980, p.43; Jeffrey M. Lenorovitz, "ALCM to Enter Inventory Next Year," *Aviation Week & Space Technology*, 16 June 1980, p.176; and "Last of 1,715 Cruise Missiles Comes Off Assembly Line," news release, Office of Public Affairs, Wright-Patterson AFB, 7 October 1986.

31. "Air Launched Cruise Missile Shows Promise But Problems Could Result in Operational Limitations," report by the Comptroller General of the United States, US General Accounting Office, 26 February 1982, p.6; Robert Ropelewski, "U.S. B-52 Bomber Fleet Being Upgraded," *Aviation Week & Space Technology*, 16 June 1980, p.192; Jeffrey M. Lenorovitz, "ALCM to Enter Inventory Next Year," *Aviation Week & Space Technology*, 16 June 1980, p.176.

32. SAC commander-in-chief General John T. Chain Jr told reporters that the ACM's problems were due to "shoddy workmanship" at the production facility and that the fuel bladders leaked, among other problems. See *Aerospace Daily*, 8 March 1990, p.416.

33. Figure of 2,500-3,000 miles referenced in "Justification of Estimates for FY 1984 Submitted to Congress," Office of the Secretary of Defense, *Research, Development, Test and Evaluation, Defense Agencies*, January 1983, p.121. The figure of 2,300–2,800 nautical miles is referenced in a background paper issued by the Canadian Ministry of External Affairs entitled, "Canada, Security Policy and Cruise Missile Testing," 1 February 1989, p.5; An *Aviation Week & Space Technology* article on 9 July 1984, p.20, states that the ACM has an "80% greater range capability than the ALCM-B."

34. In 1989 testimony, Air Force Secretary James McGovern stated that the ACM "will have almost twice the accuracy of the ALCM-B." (See DoD Appropriations for FY 90, House Appropriations Committee, part 1, p.266.) In 1990 testimony, Air Force Lieutenant General Ronald Yates remarked that the ACM had 2.5 times the accuracy of the ALCM-B (See *Defense Daily*, 16 March 1990, p.417.)

35. Energy and Water Development Hearings for 1989, US House of Representatives, Committee on Appropriations, Subcommittee on Energy and Water Development, part 6, p.838.

36. See *Aerospace Daily*, 19 March 1990, p.473. The Air Force has indicated, however, that the B-52H will not carry ACMs internally (See US Air Force Budget Justification for Program Elements, FY 1990/1991 Biennial RDT&E Descriptive Summary, p.00287.

37. See DoD Authorization Hearings for FY 1985, House Armed Services Committee, part 4, p.428; and DoD Appropriation Hearings for FY 1985, House Appropriations Committee part 6, p.809.

38. February 1990 TKL conversation with knowledgeable source.

39. Jeff Gauger, "Secrecy Lifts Around New Missile," *Omaha World-Herald*, 25 February 1990, p.1.

40. Clarence Robinson Jr, "USAF Planning Stealth Cruise Missile," *Aviation Week & Space Technology*, 8 November 1982, p.18; *Aviation Week & Space Technology*, 1 January 1990, p.34; "Air Force Plans Six Full-Scale Development Flights for ACM," *Aerospace Daily*, 29 January 1990; "ACM Meets Flight Test Criteria to Unfence Procurement Money," *Aerospace Daily*, 13 December 1989, p.409.

41. See DoD Authorization Hearings for FY 1986, House Armed Services Committee part 4, p.1061; Also, "Last of 1,715 Cruise Missiles Comes Off Production Line," USAF news release, 7 October 1986.

42. "Full-Rate Production of ACM to Begin in 1992," *Aviation Week & Space Technology*, 29 January 1990, p.32.

43. See *The Development of the Strategic Air Command 1946–86*.

44. See Thomas B. Cochran, William M. Arkin, and Milton M. Hoenig, *Nuclear Weapons Databook Volume 1* (New York: Ballinger 1983), pp.154–155.

45. See *The World's Missile Systems*, p.292.

46. Ibid.

47. DoD Authorization Hearings for FYs 1990 and 1991, US Senate Committee on Armed Services, part 6, pp.377–380, 391.

48. See David Lynch "SRAM II Being Hampered by Propellent Problems," in *Defense Week*, 22 January 1990, p.1; and John Morrocco, "Problems With Rocket Motor Delay Initial Flight of SRAM 2," in *Aviation Week & Space Technology*, 29 January 1990, p.31.

49. Ibid., pp.426, 429.

50. *Nuclear Weapons Databook Volume 1*, pp.42–43, 58, 66–67; also Tom Diaz, *The Washington Times*, 28 June 1985, op. cit.

51. See William M. Arkin and Richard W. Fieldhouse, *Nuclear Battlefields*, (New York: Ballinger 1985), pp.190, 200.

52. Timothy McCune, "Air Force Thinks Twice about Two Tacit Rainbow Builders," *Defense Week*, 4 June 1990, p.14.

*Thomas K. Longstreth*
*Richard A. Scribner*

# Soviet Strategic Bombers and their Weapons

## STRATEGIC BOMBERS

Although the Soviet Union's long-range bomber force has historically received less investment than its land-based and submarine nuclear-missile forces, Soviet interest in a modern bomber force and air-launched cruise missiles has increased during the past decade. Currently, bomber-carried weapons make up slightly more than 10 percent of the approximately 10,000 nuclear warheads in the USSR's strategic arsenal. However, that number is expected to increase as a result of current programs.

As of early 1990, the Soviet long-range bomber force included 140 of several versions of the Tupolev(Tu)-95 (NATO code name: Bear), 350 Tu-22M/Tu-26 (Backfire) and 10 of their newest aircraft, the Tu-160 (Blackjack).[1] Three types of Soviet bomber are currently in production: Bear-Hs, Backfires, and Blackjacks. Backfire and Blackjack bombers are produced at the S.P. Gorbunov Kazan aircraft production facility about 700 kilometers east of Moscow. Bear-H bombers are built at the Kuybyshev aircraft plant 850 kilometers southeast of Moscow on the Volga River.[2]

## Tu-95 Bear-A–G Bombers

The Bear bomber, first deployed in 1955, is the only turboprop-powered strategic bomber in the world still in operational service. It has an unrefueled range of 13,000 kilometers and a maximum takeoff weight of 154,000 kilograms.[3]

Four models of the Bear bomber remain in operation: the B, C, G and H. There are approximately 15 Bear B/C bombers, which are believed to carry four bombs internally or a single large AS-3 Kangaroo nuclear subsonic air-to-surface missile (ASM) externally. There are over 45 Bear-Gs, which were first introduced in 1984. The Bear-G is essentially a B or C model converted to carry the AS-4 Kitchen dual-capable supersonic ASM. According the US inspectors at the April 1990 Sovie4t Bomber exhibition, each Bear-G bomber carries three AS-4s.[4]

## Tu-142 Bear-H Bomber

The USSR's principal long-range air-launched cruise-missile (ALCM) carrier is the Tu-142 Bear-H, a new production version of the venerable bomber designed specifically to carry the AS-15 Kent long-range nuclear ALCM. It is also expected that the Bear-H will carry the new AS-X-19 ALCM, when it becomes operational.[5]

First operational in 1984, as of early 1990 there were about 80 Bear-Hs, although US intelligence expected production to end soon. Estimates of Bear-H ALCM capabilities vary. In April 1990, US inspectors observed that the Bear-H carries one internal rotary launcher with six AS-15s as its only armament. Soviet military officers also demonstrated the lack of an external ALCM carriage capability on the Bear-H or even adequate fire-control devices to support such a modification. Nevertheless, although Bear-Hs carrying external ALCMs have never been observed, some US intelligence agencies continue to believe that the Bear-H is capable of carrying up to four more AS-15s on external pylons under the wings.[6] Under the pending START treaty's Memorandum of Understanding, the Bear-H will be allowed a maximum ALCM capability of 12, and will be counted for treaty purposes as carrying eight.

Bear-H aircraft are deployed at Dolon, in central Asia, as well as possibly a base in the Far East along the Chinese border (Ukraina air base at Svobodny) and southeast of Moscow (Engels air base).[7]

## Other Bear Models

The USSR also uses the same basic Bear airframe for several other types of military aircraft. There are about 20 Bear-D and -E models, which are maritime reconnaissance aircraft used to monitor US naval activity. The USSR has about 65 Bear-F (Tu-142) antisubmarine-warfare (ASW) aircraft. Each Bear-F can carry torpedoes or nuclear depth charges in its bomb bays. In 1985, the USSR also began deploying the Bear-J, a submarine-communications aircraft. Both the Bear-F and Bear-J are in production and use the same airframe as the Bear-H ALCM carrier. Of the above types, only the Bear-F carries nuclear bombs.

## Tu-160 Blackjack-A Bomber

The newest Soviet strategic bomber is the supersonic Tu-160 Blackjack, the largest and heaviest intercontinental bomber in the world.[8] Both in design and appearance it closely resembles the US B-1B, although it is not believed to be as capable an aircraft and, despite its maximum takeoff weight of 275,000 kilograms, is believed to have a payload of only 16,000 kilograms.[9] Like the B-1, the Blackjack also has a variable-sweep wing, although a much simpler one. It also has four turbofan engines, two mounted under each wing. The Blackjack has an unrefueled range of 15,000 kilometers.

Under development since the late 1970s, the Blackjack began flight testing in December 1981 but only became operational in 1988, and reportedly encountered developmental problems that persist today.[10] It is about 54 meters long.

According to one Soviet Air Force officer, the Tu-160 can achieve top speeds of mach 2. He also claims that "the engine system of the Tu-160 is more powerful than that of the B-1B" and that the Tu-160 is superior to the B-1B in speed, stealth, maneuverability and other ways, as well.[11] According to *Soviet Military Power*, "the Blackjack can cruise subsonically over long ranges, perform high altitude supersonic dash, and attack utilizing low altitude, high subsonic penetration maneuvers."[12]

The Blackjack is produced at the Gorbunov Kazan aircraft production plant. Only 17 Blackjacks are now operational, whereas US intelligence once believed that

**Figure 9.B.1:** A Soviet Bear-H bomber, intercepted off the east coast of Canada by a US F-15 during a 1988 training flight

Source: US Department of Defense

**Figure 9.B.2:** Artist's drawing of AS-16 SRAMs being loaded onto a rotary launcher of a Soviet Tu-160 Blackjack bomber. The Blackjack can also carry AS-15 ALCMs.

Source: *Soviet Military Power 1989*

100 would be produced. That estimate has been revised downward to 50.[13] Western officials were first able to view the Blackjack up close at the Kubinka air base about 65 kilometers west of Moscow, where the first operational regiment was apparently formed. However, public sources indicate that this unit has been relocated to Dolon.[14]

The Blackjack has two large internal weapon bays. The longer bay, some 12.16 meters long, is located ahead of the wing carry-through box, and the shorter bay, about 7.6 meters long, is behind it.[15] Each can accommodate six AS-15 ALCMs or 12 AS-16 short-range attack missiles (SRAMs) on a rotary launcher similar to the one carried in the US B-1B. While a Soviet military description of the Tu-160 states that it "has no external [weapon] store attachments," and existing Blackjacks have not been observed carrying any weapons under the wings, it may nevertheless have that capability.[16]

Despite its developmental problems, the Blackjack is a very capable bomber that appears to be ideally suited for the "shoot and penetrate" mission in which it would first launch ALCMs well outside US territory and air defenses, then penetrate at supersonic speed in order to drop bombs and SRAMs.

## Tu-26 Backfire-B/C Bomber

The Tu-26 (Soviet designation Tu-22M) Backfire bomber is a sweep-wing medium-range bomber initially deployed in 1974. About 190 Backfires are currently deployed with the Soviet Air Armies (SAA) and another 160 are deployed with Soviet Naval Aviation (SNA), their primary mission being to attack US naval vessels.[17]

The Backfire-B was the initial production version. The Backfire-C is the more advanced production version with different air intakes. Each Backfire can carry one AS-4 Kitchen ASM semi-recessed in the underside of the fuselage or two AS-4s mounted on pylons, one under each wing. Alternatively, the Backfire can carry two internally mounted gravity bombs.[18] There are also recent reports of the Backfire being tested with the AS-16 SRAM.

During negotiations and subsequent ratification hearings on the SALT II treaty, the Backfire was a source of controversy because of disputes within the US intelligence community over its range and mission. The Defense Intelligence Agency (DIA) concluded that Backfire had a much longer range than CIA's estimate, supporting claims by opponents of SALT II that the Backfire had an intercontinental capability and should have been counted as a heavy bomber under the treaty. The Soviet Union rejected these assertions but agreed, in a letter attached to the treaty, to keep production at or below 30 Backfires per year and not to give it an intercontinental capability through in-flight refueling.

During the 1980s, after US intelligence had been able to observe the Backfire more closely, estimates of its unrefueled combat radius were revised downward from 5,500 kilometers to 4,000 kilometers. Intelligence analysts concluded that its primary mission was to attack targets in Europe and around the periphery of the USSR. However, some variants of the Backfire can be fitted with a refueling probe

**Figure 9.B.3:** A Soviet Tu-26 Backfire bomber, carrying a single AS-4 Kitchen ASM under its fuselage, is intercepted by a Swedish Draken fighter

**Figure 9.B.4:** 1988 photograph of a Soviet Blackjack bomber at Kubinka Air Base west of Moscow. Notice the open forward bomb-bay door. MiG-29 interceptor in foreground.

Source: *Nuclear Weapons Databook Volume 4*

that, according to US intelligence, would "permit in-flight refueling so that it can be used against the continental US if sufficient tankers are available."[19] Earlier production models of Backfire had a refueling probe but these were removed as part of Soviet assurances upon signing SALT II that the Backfire would not be given an intercontinental capability.

## Soviet Tanker Aircraft

One of the principal shortcomings in the Soviet bomber force, and a reason that the Backfire bomber is viewed by US intelligence as a very limited strategic threat, is the lack of a large Soviet air-refueling capability.

Until recently, the Soviet strategic tanker force consisted solely of about 50 converted Bison bombers. In 1987, the USSR began deploying a new tanker, the Il-78 Midas, which US intelligence believes will replace the Bison tanker over the next few years. The best public estimate is that the Soviet tanker force now consists of about 30 Bisons and 25 Midas. Each Midas can refuel up to three bombers simultaneously.[20]

## BOMBER-CARRIED NUCLEAR WEAPONS

## Long-Range ALCMs

The Soviet Union has deployed nuclear-tipped air-to-surface missiles since the early 1960s, but only began developing modern long-range ALCMs in 1981 and deploying them in 1984.[21]

The Soviet Ministry of Medium Machine Building is responsible for cruise-missile production. The M.I. Kahnin Machine Building Plant in Sverdlovsk was identified during the INF treaty negotiations as the production site for Soviet ground-launched cruise missiles (GLCMs) and may also be involved in ALCM production. Soviet cruise missiles are also produced at plants in Novosibirsk, Omsk, Plesetsk, Podberesye-Kimli, Riga, and Rostov. Dubna Airfield near Moscow also produces ASMs.[22]

### *AS-15 Kent*

The AS-15 Kent ALCM is a long-range, subsonic, turbofan-powered missile which began flight testing in 1981. First deployed in late 1984 on the Bear-H, the AS-15 is also carried on the Blackjack. One report suggested that the AS-15 may have been flight tested from the Backfire, but this has never been confirmed.[23] According to US officials who have seen AS-15s loaded on the internal rotary launcher for the Blackjack, that launcher is very similar to the one used by the B-1B.[24] AS-15s are also mounted on a rotary launcher within the Bear H's bomb bays.

According to *Soviet Military Power*, "Today, the majority of the [USSR's] current strategic air-delivered weapons inventory comprises AS-15s..."[25] which suggests that there are at least 500 of them. *Soviet Military Power* also states that the AS-15 is "similar in design to the [US] Tomahawk" and uses a guidance system similar to TERCOM although, at 7 meters long and with a wing span of 3.2 meters, it is somewhat larger than Tomahawk.[26] The AS-15 has a range of about 3,000 kilometers and is almost identical to the Soviet SS-N-21 sea-launched cruise missile (SLCM) and SS-CX-4 GLCM. Only a nuclear-armed version is believed to exist.[27]

### AS-X-19

The AS-X-19 (NATO: Koala) is a very large supersonic ALCM under development for deployment on the Bear-H and possibly the Blackjack in the early 1990s. It may also be carried on a new cruise-missile carrier aircraft reportedly in development. US intelligence believes that it will have a range of over 3,000 kilometers and will be only nuclear-armed. Its size (almost 12 meters long) suggests external carriage, although it could possibly be carried internally on the Blackjack or Bear-H.[28]

## Other Air-to-Surface Missiles

For many years, Soviet bombers have utilized a number of shorter-range (under 600 kilometers) nuclear-tipped air-to-surface missiles (ASMs) to enhance their effectiveness. While they vary in size, weight, and capability, these ASMs remain operational with Soviet Air Armies.

### AS-3

The AS-3 (NATO: Kangaroo) is the largest known ASM with a length of almost 15 meters, a diameter of 1.9 meters, and a weight of about 11,000 kilograms. First deployed in 1960, the AS-3 uses a turbojet engine and can achieve a speed of mach 2. It was the subject of some attention during SALT II hearings because US intelligence believed it had been tested in the early 1960s to ranges above 600 kilometers. The Soviet government denied this, and the AS-3 was not counted as a long-range ALCM in SALT II.[29] US intelligence believes that the AS-3 can carry both nuclear (800 kilotons) and conventional warheads. The AS-3 has been largely replaced by the AS-4 and AS-6, but may still be in the inventory and deployed on some older Bear-B bombers.[30]

### AS-4

The AS-4 (NATO: Kitchen) is a supersonic turbojet-powered ASM with a range of 250–400 kilometers believed to be primarily an antiship missile, but which could also be used against land targets. The USSR has reconfigured about 45 older Bear aircraft, which were carrying the AS-3, to carry the AS-4. Each aircraft, designated Bear-G, can carry two or three AS-4s. The AS-4, which was first deployed in 1967, can also be carried by the Backfire (up to two on wing pylons) and Blinder bombers, for which the AS-4 was originally designed. It is about 11.3 meters long and weighs 6,500 kilograms. The AS-4 is believed to be dual-capable with a nuclear-warhead yield of about 350 kilotons or a conventional warhead weighing about 500 kilograms.[31]

**Table 9.B.1:** Soviet air-launched cruise missiles

*Soviet long-range (600+ kilometers) nuclear air-launched cruise missiles*

| Type US designation | Range kilometers | Warhead kilotons | IOC* | Number | Carrier |
|---|---|---|---|---|---|
| AS-15 (Kent) | 3,000 | 250 | 1984 | 700 | Bear-H, Blackjack |
| AS-X-19 (Koala) | 3,000 | nuclear | early 1990s | ? | Bear-H Blackjack? |

*Soviet short-range nuclear air-to-surface missiles*

| Type | Range kilometers | Warhead kilotons | IOC* | Number | Carrier |
|---|---|---|---|---|---|
| AS-3 (Kangaroo) | 500–700 | 800 | 1960 | ? | Bear B/C |
| AS-6 (Kingfish) | 250–400 | 250 | 1970 | ? | Backfire, Badger-G |
| AS-16 (Kickback) | 250? | 200? | 1989 | ? | Blackjack, Bear-H, Backfire? |

*Soviet dual-capable and conventional ALCMs/ASMs*

| Type | Range kilometers | Warhead kilograms | IOC* | Number | Carrier |
|---|---|---|---|---|---|
| AS-1 (Kennel) | 100 | ? | 1958 | ? | |
| AS-2 (Kipper) | 165 | 1,000 dual-capable? | 1961 | ? | Badger-C |
| AS-4 (Kitchen) | 250–400 | 500/350 kt[†] | 1967 | ? | Blinder, Backfire, Bear-G |
| AS-5 (Kelt) | 150–300 | 1,000 | 1965 | ? | Badger-C/G |
| AS-11 (Kilter) | 500–800 | 480 | ? | ? | Fencer |

\* Initial operational capability
† dual-capable

### AS-6

The AS-6 (NATO: Kingfish) is a supersonic, turbojet-powered ASM with a range of 250–400 kilometers. It is about 10.7 meters long and weighs 5,000 kilograms. It is believed to be dual-capable with a nuclear warhead of about 200 kilotons and a conventional warhead weighing about 500 kilograms. It is carried by both Badger and Backfire bombers and is designed primarily for antiship missions.[32]

### AS-16

The AS-16 (NATO: Kickback) is a new Soviet short-range attack missile first deployed in 1989 on Blackjack bombers. It is believed to be similar in design and performance to the US SRAM-A. According to the Pentagon, each Blackjack can carry up to 24 AS-16s on its internal rotary launchers, and they "could be used against theater and intercontinental targets."[33]

## Soviet Conventional Air-to-Surface Missiles

While the USSR has not deployed modern, conventionally armed ALCMs comparable to US conventional ALCMs, it has developed and deployed several types of conventional ASM.

### AS-5

The AS-5 (NATO: Kelt) is a conventional-only ASM carried by Badger bombers for use against ships and some land targets. It is about 9.5 meters long, 0.90 meters in diameter and weighs 3,000 kilograms. It can carry a 1,000-kilogram high-explosive warhead to a range of 150–300 kilometers, depending on the flight profile. It was first deployed in 1966.

### AS-9

The AS-9 (NATO: Kyle) is reported to be an antiradar missile with a range of about 90 kilometers It is supersonic and carries a 150–200-kilogram warhead. It is carried by Su-24 Fencer, Tu-16 Badger, and Tu-26 Backfire aircraft.[34]

## NOTES AND REFERENCES

1. This count excludes approximately 300 older Tu-16 "Badger" and Tu-22 "Blinder" bombers which have a combat radius of under 3,000 kilometers and would be used only for attacking theater targets. All Mya-4 "Bison" bombers have either been dismantled under informal SALT procedures or converted to tankers. All Bear-A models are also believed to have been retired by 1989. Bear B/C bombers are also being retired as Bear-Hs and Blackjacks are deployed. In December 1989, the Soviet Defense Ministry released figures showing that the USSR had 162 heavy bombers, of which it indicated 97 (presumably Bear-Hs and Blackjacks) were ALCM-capable.

2. Thomas B. Cochran, William M. Arkin, Robert S. Norris, and Jeffrey I. Sands, *Soviet Nuclear Weapons* (New York: Ballinger 1989), p.77. (The 1984 edition of the

US Defense Department's publication *Soviet Military Power* also mentions the Dimitriev Airframe Plant in Tagangrog as a production location for the Bear-H.)

3. *Soviet Nuclear Weapons*, pp.237–238. Combat radius is new DIA estimate from "DIA Force Structure Summary," February 1990, p.6.

4. TKL conversation with US Government official.

5. See testimony of General John L. Piotrowski, commander-in-chief, US Space Command, before Senate Armed Services Committee, 7 March 1990.

6. Ibid; see also *Air Force Magazine*, March 1988, p.77.

7. *Soviet Nuclear Weapons*, pp.60–61, 77, 237.

8. US Department of Defense, *Soviet Military Power 1988* (Washington DC: US Government Printing Office, 1988), p.50.

9. DoD Authorization Hearings for FY 84, House Armed Services Committee, p.414; also Piotr Butowski, "More on Blackjack," in *Jane's Soviet Intelligence Review*, November 1989, p. 494.

10. One developmental model crashed in 1987. See *Aviation Week & Space Technology* , 25 May 1987, p.19. See also John Morrocco, "Soviet Union Unveils Blackjack Bomber," *Aviation Week & Space Technology*, 8 August 1988, p.14.

11. Colonel Yevgeni Vlasov, "Soviet Tu-160 and U.S. B-1B Bombers: The Price of Competition for Air Superiority," from *Military Bulletin*, Novosti Press Agency, USSR, January 1990.

12. *Soviet Military Power 1988*, p.50.

13. David Lynch, "Soviets to Halt Blackjack Production at Halfway Point," *Defense Week*, 4 June 1990, p.1.

14. See *Aerospace Daily*, 3 August 1988, p.177; *Aviation Week & Space Technology*, 9 May 1988, p.43; *Jane's Soviet Intelligence Review*, November 1989, p.494; *Soviet Military Power 1989*, p.46; and John W.R. Taylor, "Gallery of Soviet Space Weapons 1990," *Air Force Magazine*, March 1990, p.71.

15. Bill Sweetman, "Blackjack: Air Defense Challenge for the 1990s," in *Interavia*, October 1988, p.1012.

16. See Yevgeni Vlasov, op. cit. See also Butowski, op. cit., and David North and John Morrocco, "Blackjack Shares Aspects of U.S. B-1B and XB-70," *Aviation Week & Space Technology*, 15 August 1988, p.16.

17. A DIA estimate from February 1990 indicates that only 160 Backfires are deployed with Soviet Air Armies and 190 with Naval Aviation.

18. "Gallery of Soviet Space Weapons 1990."

19. *Soviet Military Power 1987*, p.37.

20. "Gallery of Soviet Space Weapons 1990."

21. *Soviet Military Power 1987*, p.35; *Soviet Military Power 1988*, p.40.

22. *Soviet Nuclear Weapons* pp.73–77.

23. "Gallery of Soviet Space Weapons 1987," *Air Force Magazine*, March 1987, p.98.

24. North and Morrocco op. cit.

25. *Soviet Military Power 1989*, p.46.

26. *Soviet Military Power 1987*, p.37.

27. Ibid., p.38.

28. *Soviet Military Power 1989*, pp.47–49; *Aviation Week & Space Technology*, 28 March 1988, p.15; ibid., 9 May 1988, p.43; "Gallery of Soviet Space Weapons 1987," p.98.

29. "Military Implications of the SALT II Treaty," Hearings of the Committee on Armed Services, US Senate, part 2, pp.508–509.

30. "Gallery of Soviet Space Weapons 1987," and General Dynamics, *The World's Missile Systems*, August 1988, pp.130–131.

31. See *Soviet Military Power 1987*, p.36; Neil Munro, "Soviet Antiship Missiles," *ICA*, July 1987, p.25; DoD Appropriations for 1988, House Appropriations Committee, part 4, p.1185; *Soviet Military Power 1984*, p.29; "New Soviet Bombers, Fighters Heighten Alaska's Strategic Role," *Aviation Week & Space Technology*, 9 May 1988, p.43; *Soviet Nuclear Weapons*, pp.115, 233, 237, 238.

32. Ibid.

33. *Soviet Military Power 1989*, p.46.

34. See "Gallery of Soviet Space Weapons 1987," p.88.

# The Technical Basis for Warhead Detection

The detection of nuclear warheads or the materials that could be used to make them is a requirement of many of the verification arrangements described in the previous chapters. The purpose of this section and its associated appendixes is therefore to provide a technical understanding of both the potential and the limitations of various detection techniques.

All the techniques discussed here involve the detection of penetrating radiation—either high-energy photons ("gamma rays") or neutrons. Some are "passive," limited simply to detecting radiation emitted spontaneously by fissile material. Others are "active," involving the irradiation of the object being examined with gamma or neutron radiation to obtain an x-ray–type image of its interior density distribution or to induce fission in any fissile material present and thereby increasing its output of radiation, making it more detectable.

The primary limitation of these techniques comes from the possibility of radiation shielding, which can greatly reduce the amount of gamma or neutron radiation escaping from or penetrating an object containing fissile material. Five centimeters of tungsten shielding will cut the intensity of gamma radiation from a fission source by almost a factor of 100. Twenty centimeters of lithium hydride shielding will similarly cut the intensity of the emitted neutrons by a factor of 600. It is therefore relatively simple to prevent the remote detection of radiation from a nuclear warhead hidden in a ship or building.

As the preceding chapters illustrate, however, there are many situations in which it would not be possible to have such massive radiation shielding without the presence of the shielding being detectable. In other cases, the presence of such shielding could be detected by occasionally rolling a random sample of the object in question (for example, a shield inside a missile canister or a space-launch payload) through a portal monitoring system equipped with a high-energy x-ray machine.

Although the techniques discussed in this section will not solve every verification problem, they are likely to be a standard part of the verification equipment associated with any agreement limiting warheads or fissile material.

We start this section with a description of current portal-monitoring techniques (chapter 10). Chapter 11 and its appendixes then analyze, starting from fundamentals, the detectability of warheads made with different combinations of fissile and surrounding materials. Chapter 12 discusses the detectability of nuclear warheads on ships, and finally, chapters 13 and 14 present the results of measurements made in July 1989 on the Black Sea near Yalta of the gamma and neutron radiation from a specific Soviet SLCM warhead on the Soviet missile cruiser *Slava*.

# Chapter 10

*David Albright*

# Portal Monitoring for Detecting Fissile Materials and Chemical Explosives

The "portal monitoring" of pedestrians, packages, equipment, and vehicles entering or leaving areas of high physical security has been common for many years. Many nuclear facilities rely on portal monitoring to prevent the theft or diversion of plutonium and highly enriched uranium. At commercial airports, portals are used to prevent firearms and explosives from being smuggled onto airplanes. An August 1989 Federal Aviation Administration (FAA) regulation requires US airlines to screen luggage on international flights for chemical explosives.

Portal monitoring is now being introduced into arms-control agreements. Under the 1987 Intermediate-Range Nuclear Forces (INF) Treaty, the United States and the Soviet Union have each established a portal-monitoring facility on the other's territory at the perimeter of a missile-production plant to verify that prohibited missiles are not being produced inside. The START treaty is expected to use portal-monitoring facilities in a similar way to help verify limits on the numbers of mobile missiles. As discussed in chapters 7 and 8, a treaty verifiably limiting cruise missiles could require portals that distinguished between nuclear-armed and non-nuclear missiles. And, as discussed in chapters 4 and 5, portal monitoring might also be needed to verify agreements requiring warhead dismantlement or a bilateral cutoff in the production of fissile materials. The small size of the objects that would have to be checked under such agreements means that more stringent perimeter controls will be required than those provided by the INF and START agreements. Equipment requirements could include tunneling-detection equipment, radar able to detect low-flying helicopters, fences with cameras aimed along the perimeter, and fiber-optic seals that would break if the perimeter were crossed.

Because some of the portal-monitoring equipment that would be useful in verifying arms-control agreements is already widely used as part of the physical security systems at nuclear facilities and commercial airports, we review these uses of portal monitoring, as well as its role in verifying the INF treaty. Then we survey the major types of portal-monitoring equipment that would be most useful in detecting nuclear warheads or fissile material.

Methods that detect the relatively large amounts of high explosives contained in nuclear weapons—on order of 100 kilograms—could complement methods that can detect their fissile material. We therefore also briefly survey recent developments in systems aimed at detecting concealed high explosives.

## CURRENT APPLICATIONS OF PORTAL MONITORS

### Fissile-material Safeguards

Portal monitoring at nuclear facilities is a key part of security measures taken to guard against the theft or unauthorized diversion of nuclear weapons, plutonium, or highly enriched uranium. Individuals, packages, and vehicles are exposed to fixed and hand-held radiation-detection equipment and metal detectors. In addition, packages pass through x-ray imaging equipment. Any suspicious object or activity can trigger closer inspection by the guard force stationed at that portal. To increase the probability of detecting unauthorized shielded nuclear materials, guards sometimes conduct random searches at portals. All activities can also be monitored by closed-circuit television to record events. Permitted shipments containing fissile materials are usually placed in containers of authorized design that are sealed and registered. The portal operator's main task is to check that the seals are intact.

To increase the effectiveness of portal monitoring, the number of portals leading to safeguarded areas is minimized as is the passage of large vehicles that might contain shielded fissile materials. In addition, private motor vehicles are excluded from the more sensitive areas.

## Preventing the Spread of Radioactive Contamination

A second important purpose of portal monitoring at nuclear facilities is to prevent persons, objects, or vehicles from leaving a facility inadvertently contaminated with radioactive materials. This type of monitoring is routinely done at nuclear power plants, nuclear reprocessing facilities, research facilities, and hospitals.

In 1984, a serious radiation accident was uncovered by a vehicle portal monitor designed to prevent radioactively contaminated material leaving the proton accelerator at Los Alamos National Laboratory in New Mexico. A truck, which had made a wrong turn off the highway into the lab, triggered a portal alarm. Photos taken automatically when the alarm sounded enabled authorities to track down the truck, which was carrying steel contaminated with cobalt-60 from an abandoned cancer-treatment radiation source.[1] (Cobalt-60 emits gamma rays with energies near 1.2 MeV.) This cancer treatment machine had been stored at a medical clinic warehouse in Juarez, Mexico. Without realizing the danger, clinic employees disassembled the machine and sold the steel container holding thousands of cobalt-60 pellets to a local scrap yard. During the process, the steel cylinder was damaged, resulting in pellets being scattered throughout the junkyard. In December 1983, contaminated scrap was shipped to two Mexican foundries, which melted the material down and made it into reinforcing bar—rebar— and restaurant table legs. Before the radiation alarm sounded at Los Alamos, significant amounts of the contaminated material had been distributed in Mexico and the United States.

## Pre-board Screening at Civilian Airports

At some commercial airports, over 2 million persons are screened annually to prevent firearms, explosives, and other dangerous weapons from being carried onto airplanes. The screening equipment currently consists of walk-through metal detectors and x-ray inspection systems to detect suspect items in carry-on items. Hand-held metal detection devices are used as backup support for the walk-through detectors. In addition, screening personnel may require physical searches of carry-on baggage that appears suspicious when x-rayed.

According to US Federal Aviation Administration data, between 1973 and 1987, over 38,000 firearms were detected and at least 117 hijackings and related crimes may have been prevented by these security measures.[2]

A major weakness in this screening process, however, is its dependence on personnel alertness. According to the US General Accounting Office, there are many personnel-related problems, including a high turnover, low wages, and inadequate training.[3] In tests conducted from September through December 1986, screening personnel detected only 79 percent of the test weapons in x-ray tests, 82 percent in metal detector tests, and 81 percent in physical search tests. Earlier results were similar. Such problems have led to an increased emphasis on automated equipment that requires human inspection only when an alarm sounds.

As a result of increased concern about explosives in checked luggage following the destruction of Pan American World Airways Flight 103 on 21 December 1988 over Lockerbie, Scotland, with a loss of 270 lives, the FAA announced in August 1989 that it will require US airlines to use automated explosive-detection systems to screen checked luggage at about 40 of the busiest international airports in the US and abroad.[4] Although the FAA does not specify a particular type of detection equipment, the only device that currently satisfies the FAA's requirements of being able to operate reliably for long periods of time and to spot explosives in luggage without human intervention is the thermal neutron analysis (TNA) device manufactured by Science Applications International Corporation of Santa Clara, California. This heavily shielded device, which weighs 9,100 kilograms, bathes the luggage with thermal neutrons and detects primarily the 10.8-MeV gamma from neutron absorption on nitrogen-14. A large nitrogen concentration may indicate the presence of high explosives.

Nitrogen is also found in many textiles and certain foods, however, so the machine has a high false-alarm probability of about 4–6 percent per item of baggage. For a jumbo jet carrying 450 passengers, each with two pieces of luggage, this means that about 35–55 pieces of luggage would require further inspection. To reduce the false-alarm rate, the TNA device is coupled to a special x-ray scanner that provides correlated information about the density of the objects in the luggage.[5]

## INF Treaty

Under the terms of the Intermediate-Range Nuclear Forces (INF) Treaty, the United States and the Soviet Union have each established a continuously operating portal-monitoring facility at a missile production site on the other country's territory. Inspectors at these portals examine vehicles leaving the site to help verify that missiles prohibited by the INF treaty are not being produced there.

The US portal-monitoring facility is at the main gate of the Votkinsk missile assembly plant, about 600 miles east of Moscow (see figure 10.1). This plant was formerly used to assemble the now-banned SS-20 intermediate-range missile. It is still involved in the final assembly of the Soviet SS-25 ICBM, which has a first stage similar to the first stage of the SS-20.

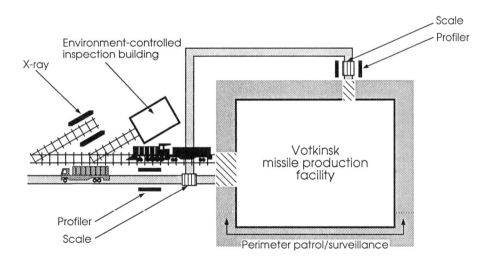

**Figure 10.1:** Votkinsk missile production facility
Based on information from the US On-site Inspection Agency

At the Votkinsk facility, up to 30 personnel of the US On-Site Inspection Agency (OSIA) monitor all items leaving the production site. Any large vehicle leaving the Votkinsk facility that is declared not to be carrying a missile is measured to determine whether it is large enough to carry a missile. If it contains a canister or shrouded object large enough to hold an SS-20, the Soviet Union is responsible for demonstrating that an SS-20 is not inside.

US personnel also measure any launch container in a vehicle declared to contain a missile.[6] Under the treaty, the United States can visually inspect the missiles inside these containers up to eight times a year. Monitoring is made easier by the relatively light traffic leaving the plant—about 45 railcars and 120–150 wheeled vehicles per week (roughly one per hour).[7] So far, SS-25 missiles have been transported only on railcars.[8]

Although the treaty also allows inspectors to weigh a missile container, in practice this is not done.[9] Weighing a missile stage and its container on a railcar is more difficult to accomplish than might be expected, since it requires that the container be lifted off its carrier.

In March 1990, inspectors began using an x-ray system to scan each declared missile. The system, developed by a consortium of Bechtel National Inc., the American Science and Engineering Corporation, and Varian Associates Inc., images portions of the missiles through the canisters and railcars to provide additional assurance that banned SS-20 missile stages are not within the canisters (see appendix 11.D for a description of x-ray radiography).

In early March, when the x-ray system became operational, the Soviet Union refused to allow US inspectors to x-ray missile canisters leaving the plant.[10] In response, US inspectors declared an "ambiguity" concerning Soviet compliance with the INF treaty, which is the most serious charge the United States could make. The dispute apparently centered on differing interpretations of how the equipment was actually to be operated. Following extensive consultations between the US and Soviet governments that resolved this dispute, US inspectors began examining the missile canisters with x-rays on 21 March.

The INF treaty also allows the US inspectors to patrol the four-kilometer perimeter of the Votkinsk facility, which is surrounded by a concrete wall approximately 3 meters high. Twice daily, US inspectors walk around the perimeter.[11] Such infrequent monitoring of the perimeter is acceptable to the United States because the size and weight of the SS-20 make it unlikely that the Soviet Union could secretly transport missiles across the perimeter wall.[12] Specifically, according to officials of the OSIA, the Soviet Union does not have a helicopter large enough to transport an SS-20 missile. Using a crane to lift a missile over the wall also seems unlikely. Such an operation would be time-consuming, and would also leave evidence in the ground and vegetation near the wall, since no roads run directly adjacent to the perimeter wall.

The Soviet portal-monitoring facility is at the Hercules Plant at Magna, Utah, which once produced stages for the banned Pershing II missile and now manufactures solid rocket motor stages for the MX and Trident strategic missiles. The Soviets use similar measurement technology to the US team, but less elaborate; the Soviet team has been satisfied with its procedures, which rely on visual inspection of the containers leaving the plant, and has not felt the need to install an x-ray system.

## MAJOR TYPES OF PORTAL MONITOR

### Detecting Passive Gamma and Neutron Radiation[13]

Automatic passive portal radiation monitors detect nuclear sources by measuring increases in radiation intensity. When the portal is unoccupied, the detectors measure the average background radiation levels from natural and other sources. When the portal is occupied, the detectors integrate the radiation intensity over a short period and compare it to the previously measured background levels. An alarm sounds when the count exceeds the average background count plus a threshold number of counts, called the alarm increment. (See appendix 11.C for a discussion of radiation detectors and backgrounds.)

*Pedestrian Portals*
All US facilities that handle plutonium or highly enriched uranium are required to monitor pedestrians for these materials each time an individual

exits a materials access area or protected zone. Facilities licensed by the Nuclear Regulatory Commission (NRC) are required to have pedestrian portals with detection systems that can detect the passage of 3 grams of unshielded uranium-235 or 0.5 grams of plutonium. In addition, these systems are required to be able to detect the passage of 100 grams of lead (which could be used as gamma shielding around fissile material).

A typical walk-through portal monitor equipped with passive gamma-radiation detectors is illustrated in figure 10.2. It contains an occupancy sensor, power supplies, either sensitive plastic or sodium iodide scintillator detectors, signal-conditioning electronics, signal-processing electronics, devices to communicate the output to the attendants, and usually a metal

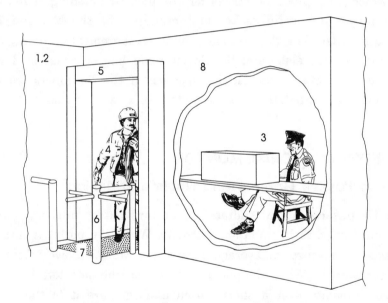

1. Provide a monitoring area with low-intensity, low variability background radiation

2. Provide adequate climactic control for the monitor, either by room heating and airconditioning or by heating and cooling the monitor cabinets

3. Supervise the SNM monitoring area. Station a guard in the area as an observer or let him observe remotely with good television surveillance

4. Adequately test and maintain the monitor

5. Keep the monitor's detectors as close as possible to pedestrians

6. Slow pedestrian passage speed as much as possible

7. Allow only one person at a time near the monitor

8. Provide bypass barriers

**Figure 10.2:** A pedestrian portal monitor
Source: Paul E. Fehlau, *An Applications Guide to Pedestrian SNM Monitors*

detector. The entire monitoring process typically takes about 1 second.

Table 10.1 compares four categories of walk-through pedestrian portal that have been developed in response to increasingly stringent fissile-material detection requirements. Under unfavorable conditions, category-1 monitors, can detect 1 gram of unshielded plutonium, but only about 60 grams of weapon-grade uranium. Category-4 monitors are designed to detect 1 gram of weapon-grade uranium.[14]

One way to improve the detection sensitivity of a portal monitor is to add more radiation detectors or to decrease the width of the portal. For example, the easiest way to go from a category-2 to a category-3 monitor is to install head and foot detectors.[15] All monitors also perform better if the passage of people through them is slowed by installing a turnstile or requiring people to stop briefly in the portal. Background radiation is reduced by placing the monitor in a lead-shielded room. Greater sensitivity is also obtained by using computer software that permits several measurements while a person is moving through the portal, thereby capturing the most intense portion of the source signal.

Hand-held gamma radiation monitors can supplement automatic radiation-detection equipment by pinpointing the exact location and

**Table 10.1:** Detection sensitivities of pedestrian monitors equipped with gamma detectors for unshielded weapon-grade plutonium (WgPu) and weapon-grade uranium (WgU)

| Category | WgU $g$ | WgPu* $g$ |
|---|---|---|
| 1 | 64 | 1 |
| 2 | 10 | 0.3 |
| 3 | 3 | 0.08 |
| 4 | 1 | 0.03 |

The table estimates performance under a set of unfavorable conditions: 25 microroentgens per hour background radiation and a standard compact metallic test source attached below an interior ankle of an individual walking at normal speed, with pace adjusted to swing the source through the monitor. Test results must give a probability of detection of 50 percent or greater when the alarm threshold is set so that there is less than a 5-percent chance that a measurement will produce a false alarm due to a fluctuation in the background.

* Weapon-grade plutonium that is freshly separated from daughter products.

Source: Paul E. Fehlau, *An Applications Guide to Pedestrian SNM Monitors*, LA-10633-MS (Los Alamos, New Mexico: Los Alamos National Laboratory, February 1986), p.9.

sometimes type of radioactive material that caused an alarm.[16] To be effective, hand-held monitors must be within about 5–15 centimeters of any special nuclear material that may be present. Therefore, operators must be properly trained, motivated, and supervised to conduct effective searches, and given adequate time for a thorough search.

*Detection of Shielded Fissile Material.* Table 10.2 lists published results for detecting shielded weapon-grade plutonium and weapon-grade uranium with a category-2–type monitor as a function of shielding thickness under conditions of relatively high background. The minimum detected masses of weapon-grade uranium increase more rapidly because the principal gamma rays emitted by uranium are less energetic and therefore less penetrating than those from plutonium.

One approach that increases the probability of detecting shielding materials in a walk-through monitor is to install highly sensitive metal detectors. Some of the more sensitive detectors are capable of detecting 100 grams of solid lead, which is less than the amount needed to shield 1

**Table 10.2:** Detection sensitivity for gamma rays from shielded compact metallic weapon-grade plutonium and weapon-grade uranium in a prototype walk-through category-2 monitor

| Lead shielding | | |
| Thickness mm | Mass* g | Weapon-grade uranium g |
| --- | --- | --- |
| 1.6 | 60 | 300 |
| 3.2 | 280 | 1,000 |
| | | **Weapon-grade plutonium** g |
| 1.6 | 2.5 | 1 |
| 3.2 | 13 | 3 |
| 6.4 | 70 | 10 |
| 12.7 | 450 | 50 |

* The mass is calculated by assuming that the fissile material is in the form of a metal sphere. The plutonium is assumed to be in delta phase.

Source: W.H. Chambers et al., *Portal Monitor for Diversion Safeguards*, LA-5681 (Los Alamos, New Mexico: Los Alamos National Laboratory, December 1974).

kilogram of weapon-grade uranium against detection in a category-2 monitor (see table 10.2). A detector this sensitive, however, can be set off by even the foil from a gum or cigarette wrapper, and therefore such a detector must be accompanied by changing rooms where employees and visitors put on metal-free clothing. Even so, current metal detectors cannot detect noncontinuous gamma-ray shielding, such as that made from granular lead.

Another method for dealing with the shielding problem in the case of plutonium is to use a portal monitor that detects neutrons. Neutrons are more difficult to shield than gamma rays and the neutron background intensity generally varies less than the gamma background. However, the neutron emission rate from weapon-grade plutonium is quite low, typically about 50 neutrons per second per gram (see table 11.2). Achieving good detection sensitivity for such a small flux of neutrons therefore requires a large-area neutron detector.

Los Alamos National Laboratory has developed a relatively inexpensive, large-area neutron pedestrian portal monitor.[17] The sides of the portal are hollow chambers with polyethylene walls, each containing a helium-3 proportional counter. The hydrogen atoms in the polyethylene slow the neutrons to the low energies at which they can be efficiently captured by the helium-3 and thereby detected in the proportional counters (see chapter 11). The designers of this monitor have determined that the two neutron detectors detect 1.4 percent of the fission spectrum neutrons emitted at the center of the portal.[18] At sea level, such detection efficiency means that this portal can in 1 second detect a radiation source that emits 1,000 neutrons per second 50 percent of the time. This corresponds to roughly 20 grams of weapon-grade plutonium.[19] If the source is in the portal for 2 seconds, the portal counters can detect about 6 grams of weapon-grade plutonium.

*Vehicle Portals*
A simple portal gamma monitor for vehicles is illustrated in figure 10.3. The portal's radiation detectors are typically large plastic scintillation detectors contained in weathertight cabinets next to the roadway. They separately measure the radiation from each meter of a vehicle's length as it passes slowly through the portal.[20] Under conditions where a vehicle

moves at about 2 meters per second, each vehicle requires less than 10 seconds to monitor. This type of portal could also monitor vehicles moving at normal speeds in roadways, albeit with decreased sensitivity. Because most vehicles are large and can contain significant amounts of shielding and the detectors are further away than in pedestrian portals, this type of vehicle portal monitor can detect only relatively large amounts of fissile material.

The simplest way to improve a vehicle portal monitor's detection ability is to require the vehicle to remain stationary during monitoring. This type of portal is illustrated in figure 10.4 and uses plastic scintillation detectors that are located beneath the canopy and in the roadway. This design enables the detectors underneath to be close to the vehicle and its cargo space, and gives the detectors in the canopy an unobstructed view through the vehicle's roof.

Table 10.3 shows the results of monitoring a small recreational van with a hand-held detector, and in portals such as those illustrated in

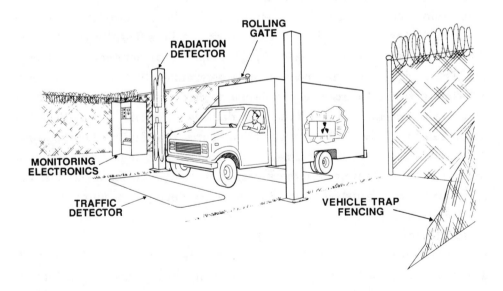

**Figure 10.3:** The basic elements of a passive-radiation–based vehicle portal that monitors vehicles with gamma counters as they move slowly through the detectors. The monitoring electronics are normally located in a guard station.
Source: Paul E. Fehlau, *An Applications Guide to Vehicle SNM Monitors*

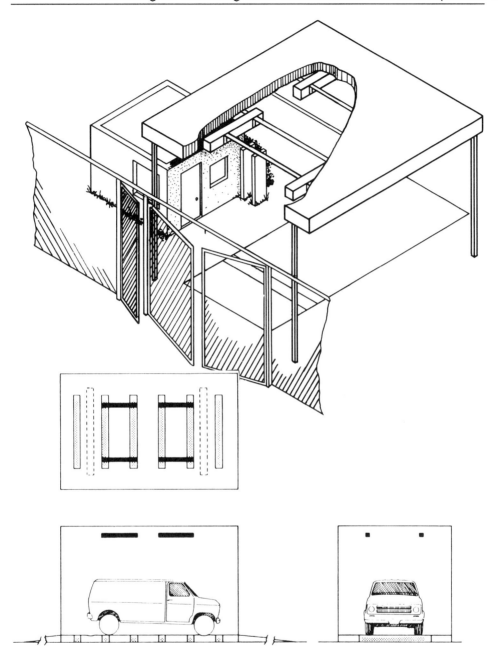

These orthogonal views of the Los Alamos vehicle SNM monitoring station illustrate the detector cabinet positions. Underground detector cabinets are in trenches covered by aluminum diamond plate

**Figure 10.4:** Passive-radiation detection equipment in a vehicle portal that monitors stationary vehicles with gamma counters
Source: Paul E. Fehlau, *An Applications Guide to Vehicle SNM Monitors*

figures 10.3 and 10.4. In all cases, gram quantities of unshielded weapon-grade plutonium were detected. But only stationary monitoring detected small quantities (less than 100 grams) of unshielded weapon-grade uranium.

Los Alamos National Laboratory is developing relatively low-cost neutron detectors to detect plutonium in vehicle portal monitors that do not require vehicles to stop.[21] This type of portal monitor has two large-area neutron chamber detectors positioned beside the roadway.[22] The prototype neutron vehicle portal monitor has a detection probability for a 10,000-neutrons-per-second source (about 200 grams of weapon-grade plutonium metal) of 99.9 percent at an alarm threshold set at four times the standard deviation from the average background. This detection capability is comparable to that of a stationary-vehicle portal monitor to detect plutonium shielded by 5 centimeters of lead, enough to absorb more than 99.8 percent of the principal gamma rays from plutonium.[23,24]

**Table 10.3:** Detection sensitivity for gamma rays from unshielded weapon-grade plutonium (WgPu) and weapon-grade uranium (WgU) in a 1-tonne recreational vehicle

|  | Minimum detected mass* | |
|  | WgPu | WgU |
|  | g | g |
| Hand-held | 3–9 | 100–300 |
| | | |
| *Vehicle portal monitor* | | |
| Moving vehicle† | 10 | 1,000 |
| Stationary *one minute* | 0.3 | 40 |

* Under unfavorable conditions: background intensity is 20 microroentgens per hour, and shielding by vehicle structure is significant. Detection implies a detection probability of 50 percent or greater. In all cases, compact metallic shapes were assumed for the fissile material.

† 2 meters per second.

Source: Paul E. Fehlau, *An Applications Guide to Vehicle SNM Monitors* LA-10912-MS (Los Alamos, New Mexico: Los Alamos National Laboratory, March 1987), p.11.

## Detecting High Explosives

*Active Radiation-based Detection of High Explosives*
Several portal-monitoring technologies are under development that use penetrating radiation to detect chemical explosives, including plastic explosives, in luggage or cargoes at commercial airports. These technologies primarily focus on the production of distinctive gamma rays in the explosive through interrogation by thermal or fast neutrons.[25]

*Thermal Neutron Analysis Device.* Thus far, only the thermal neutron analysis (TNA) device, manufactured by Science Applications International Corporation of Santa Clara, California, has been approved by the FAA for use in scanning luggage at commercial airports (see figure 10.5). This machine takes about 6 seconds to scan a piece of luggage.[26]

The TNA device bathes luggage in thermal neutrons produced by slowing down fission neutrons from a californium-252 source. The gamma

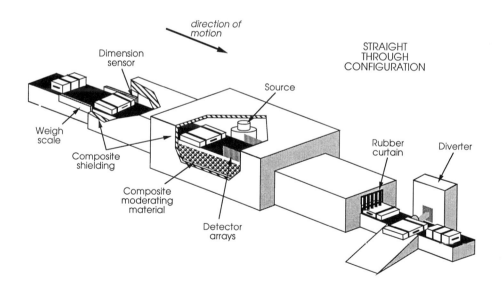

**Figure 10.5:** Thermal neutron analysis: airport baggage/cargo detection system

radiation produced by thermal-neutron capture on elements in the luggage is detected by a large array of sodium iodide detectors, and the data are analyzed by computers for signs of high concentrations of nitrogen, which is found in almost all chemical explosives. The reaction monitored in this device is the thermal-neutron capture on nitrogen-14, producing an excited state of nitrogen-15 that emits a distinctive 10.8-MeV gamma ray as it decays into its ground state.[27] However, the thermal-neutron–capture cross section for this reaction multiplied by the branching ratio for the release of a 10.8-MeV gamma ray is very small—about 10 millibarns. Many other gamma rays, with energies less than 10.8 MeV, are also produced by neutron reactions with the isotopes of other common elements in the luggage.[28] The success of this method therefore depends on distinguishing the 10.8-MeV gamma rays from the very large number of gamma rays produced in other materials.[29] The current device is reportedly designed to detect a bomb containing more than 1 kilogram of chemical explosive.[30]

The TNA machine also monitors for neutron shielding. A significant amount of shielding would set off the alarm.

Since large concentrations of nitrogen are found also in wool, leather, nylon, plastic ski boots, and some foods, the TNA device has a relatively high false-alarm probability of about 4–6 percent per suitcase scanned during normal operation. This rate can be reduced by running the bags through a second time, x-raying the suspect bags, or by visually inspecting these bags.

A more significant problem facing the application of the TNA device at commercial airports is that the bomb which brought down Pan Am Flight 103 over Scotland might have used only one-half kilogram of plastic explosive, which is lower than the detection threshold of this machine.[31] Although the sensitivity of the machine can be increased, setting it to detect a bomb as small as the one believed to have destroyed Flight 103 would significantly increase the TNA's false alarm rate.

Scaling up the TNA device to handle significantly larger containers would be difficult, since the existing device has been optimized to scan objects the size of luggage and benefits from having the target within a cavity that serves to reflect thermal neutrons back into the cavity. However, a multiple source system that irradiates objects from several sides may be applicable to the larger containers.

*Fast-neutron Associated Particle Method.* Argonne National Laboratory is funding DIXCOM Technology Group, Inc. to develop a concept to utilize fast-neutron time-of-flight methods and inelastic neutron scattering to detect high explosives and chemical-weapon agents. This method determines the location of an object within a container by fast-neutron time-of-flight techniques and provides an elemental analysis by inelastic gamma-ray spectroscopy.[32] The gamma rays produced by inelastic scattering off the light elements commonly found in high explosives—carbon, nitrogen, and oxygen—typically have energies of 2–7 MeV and are therefore highly penetrating. This method can also detect and locate hidden fissile or fissionable materials by looking for the distinctive spectrum of the fission-product gamma rays.

DIXCOM's laboratory system can search containers with dimensions of roughly half a meter, determine the location of each object inside the container in three dimensions, and obtain its inelastic gamma-ray spectrum. With this information, the gamma-ray spectrum of an object can be compared to that of a known material, such as a specific explosive. In addition, the ratios of the atomic constituents in an object, such as the

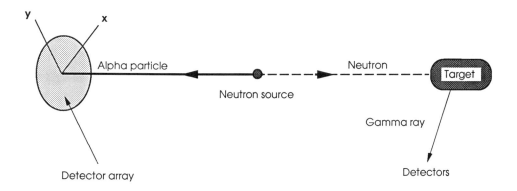

**Figure 10.6:** Associate particle time-of-flight method

Source: DIXCOM Technology Group

carbon-to-oxygen ratio, can be determined.

DIXCOM's system works by producing a beam of timed 14-MeV neutrons in a sealed deuterium-tritium (D-T) neutron generator via the reaction $D + T \rightarrow n + \alpha$. The zero time and direction for each neutron is determined by the detection of the 3.7-MeV recoil alpha particle produced with it (see figure 10.6). Given the direction and speed of the 14-MeV neutrons (about 5 centimeters per nanosecond) and the speed of the gamma rays (30 centimeters per nanosecond), the time between the detection of an alpha particle and a gamma ray in a sodium iodide detector can be related to the location at which the neutron interaction occurred and therefore to the position of the target object relative to the neutron generator. For a source-target-detector geometry with dimensions that are less than half a meter, this time difference is less than about 10 nanoseconds, which ensures a very low background.

The primary limit of the current laboratory system is that the count rate is very low, requiring on order of an hour to obtain the necessary measurements.[33] The count rate on future systems, however, will be increased by the use of more gamma detectors.

Another major stumbling block to all technologies that would use 14-MeV neutron beams in portal monitoring is a lack of small, maintenance-free neutron generators that can operate reliably for extended periods of time. Most of these accelerators can operate only up to a few hundred hours, compared to the several thousand hours required by the FAA.

*Chemical Sniffers*

"Chemical sniffers" collect and analyze distinctive vapors emanating from chemical explosives.

A major limitation of all chemical sniffers is that vapors must first escape from the explosives. Explosives located in large containers could be double sealed, making vapor detection more difficult. And detection is difficult for explosives with low vapor pressures, such as plastic explosives and explosives found in nuclear weapons, such as HMX.[34]

Thermedics, Inc. of Woburn, Massachusetts, has one of the more developed technologies, relying on the phenomenon of chemiluminescence. A molecule of the explosive is reacted with an appropriate oxidizer, such as ozone, resulting in a molecule in an excited state that emits characteristic infrared radiation lines.

To collect a larger sample of the explosive vapor, Thermedic's pedestrian portal system bathes the passenger in infrared radiation to increase the vaporization of any high explosives on that person's clothing, rapidly filters large amounts of air, and concentrates the material into a small volume for analysis.

During 1988, the FAA tested Thermedics' chemical sniffer on passengers at Boston's Logan Airport.[35] The time to scan each passenger, however, was found to be relatively long—about 30 seconds.

Thermedics is also working on a hand-held, battery-powered device for collecting air samples from airline baggage and for screening automobiles. The sample would then be put into a stationary analyzer nearby.

## Portals Using X-ray Radiography

X-ray radiography involves placing an object between an x-ray source and an array of detectors, producing a two-dimensional projection or silhouette of the object. Since high–atomic-number materials, such as fissile materials, metal handguns, and gamma shielding, absorb more radiation per unit thickness than low–atomic-number materials, x-ray radiography can detect these materials inside packages, containers, or large vehicles. X-ray radiography is used in airports to detect weapons in hand baggage and in industry for quality control. As discussed in chapters 6 and 11, it can also be used to detect the dense lumps of fissile material in nuclear warheads.

One way to enhance a two-dimensional radiographic silhouette is to use the "dual energy radiography" technique, which scans the object with both a high- and a low-energy x-ray beam, resulting in 2 two-dimensional projections of the object. The high-energy beam silhouettes the high–atomic-number components of the object and the second shows the structure of the low–atomic-number components, thereby providing more complete information about the object and reducing the clutter associated with just one image.

Although pedestrian portals usually require only small x-ray units, vehicle portals or portals that handle large packages require larger and more sophisticated x-ray units. The Bechtel National Corporation, in conjunction with American Science and Engineering (AS&E), Inc., and Varian Associates, has developed a machine, now available commercially, that can inspect a wide range of cargo. This "Linatron" uses a 15-MeV electron linear accelerator to produce high-energy x-rays that can penetrate several inches of steel with a line of sodium iodide scintillation detectors on the opposite side of the x-ray source (for more information, see appendix 11.D).

According to Varian officials, the system designed for deployment at the US portal-monitoring facility in the Soviet Union utilizes a 9-MeV Linatron to scan the missile containers. This system is capable of distinguishing the SS-20 first-stage booster from the SS-25 first-stage booster on the basis of a difference in their lengths of a couple of tens of centimeters and a difference in their diameters of about 8 centimeters.[36] Argonne National Laboratory has been developing a simpler, less expensive, cobalt-60–source system for the same purpose.

AS&E has developed a High-Energy Computed Tomography (HECT) imaging system for non-destructive testing of large rocket motors. X-ray–computed tomography generates a three-dimensional density image of the object by performing many scans from different angles. A high-speed x-ray–computed-tomography (CT) system has also been developed by Imatron Inc. that can detect small amounts of high explosives in checked luggage.[37] This system operates primarily by identifying the densities of objects within the luggage. Military high explosives have relatively distinctive densities (about 1.6–2.0 g/cm$^3$). However, many commercial explosives have densities that are similar to items commonly found in luggage.[38] Since this machine can provide a three-dimensional reconstruction of an object with resolution as small as 1 millimeter, the company claims that it could reveal blasting caps and detonators. The system is roughly half as fast as the TNA device.

## CONCLUSION

Table 10.4 summarizes the fissile-material, high-explosive, and nuclear-warhead detection capabilities of the major technologies considered above. These systems provide a solid foundation for the development of portals to verify specific arms-control agreements.

Whatever techniques are used at a portal, they must have a high probability of detecting a treaty-limited item along with modest cost, transportability, limited radiation exposure to personnel, minimum intrusiveness, and ease of operation. This entails that the United States and the Soviet Union continue to develop reliable, sophisticated, automatic equipment.

Because of the special difficulty of locating nuclear weapons and fissile material in vehicles or large containers, administrative procedures should restrict the size of packages and vehicles permitted to pass through a portal.

A portal monitor is only as effective as the perimeter controls applied around a facility. Although only briefly addressed in this paper, perimeter controls need to be improved so that they can be applied quickly and economically to sites covered by future arms-control agreements. If carefully done, however, portal monitoring combined with effective perimeter control can make evasion significantly more difficult, costly, and risky.

**Table 10.4:** Comparison of detection capabilities of various portal-monitoring technologies

|  | Fissile material | High explosive | Nuclear warhead |
|---|---|---|---|
| Automatic passive radiation portals | yes | no | yes |
| Hand-held radiation detectors | yes | no | yes |
| Metal detectors | maybe | maybe | yes |
| X-ray radiography | maybe | maybe | yes |
| TNA device | maybe | yes | yes |
| Associated particle | yes | yes | yes |
| Chemical sniffer | no | maybe | maybe |

## NOTES AND REFERENCES

1. Theresa Maggio, "Lab's Device, Staff Heroes in Rebar Event," *Los Alamos Newsbulletin*, 4, 4, 27 January 1984.

2. "FAA's Preboard Passenger Screening Process," statement of Kenneth M. Mead, Associate Director, Resources, Community, and Economic Development Division, General Accounting Office, GAO/T-RECED-87-34, before the US House of Representatives Subcommittee on Government Activities and Transportation, Committee on Government Operations, 18 June 1987.

3. "FAA's Preboard Passenger Screening Process," op. cit.

4. Nell Henderson, "Airlines Must Check for Plastic Bombs at 40 International Airports," *The Washington Post*, 31 August 1989, p.A28.

5. Carl H. Lavin, "New Machines Can Detect Terrorists's Bombs, Usually," *The New York Times*, 12 September 1989.

6. The treaty requires that all trucks or railcars that could contain an SS-20 missile exit from one designated portal. If an exiting vehicle could carry a prohibited missile, the inspected country must declare so before the shipment arrives at the portal, and state whether the vehicle contains such a missile.

7. Interview with Colonel Douglas M. Englund, Director for Portal Monitoring, On-Site Inspection Agency, 5 September 1989. This small amount of traffic is the result of workers traveling to the plant in buses rather than in private vehicles.

8. Ibid.

9. Ibid.

10. Warren Strobel, "Soviets Block U.S. Inspection at Missile Plant," *The Washington Times*, 16 March 1990, p.A8.

11. Englund interview.

12. The weight of the SS-20 without its warhead section but in its canister is 42.7 tonnes (INF treaty, Memorandum of Understanding, December 1987).

13. This section is based primarily on Paul E. Fehlau, *An Applications Guide to Pedestrian SNM Monitors*, LA-10633-MS (Los Alamos, New Mexico: Los Alamos National Laboratory, February 1986); and Paul E. Fehlau, *An Applications Guide to Vehicle SNM Monitors* LA-10912-MS (Los Alamos, New Mexico: Los Alamos National Laboratory, March 1987).

14. The detection sensitivity of a category-2 portal monitor can be illustrated by considering a simple model of such a portal. The sides of the portal are separated by about 60 centimeters, and each side contains two plastic scintillators with dimensions of 90 × 15 × 3.75 centimeters (number and dimensions of scintillators provided by P. Fehlau, Los Alamos National Laboratory, personal communication,

January 1990). The geometric efficiency of this portal is about 20 percent, and the intrinsic efficiency for these detectors is about 2–3 percent, giving a total efficiency of about 4–6 percent. The background for a category-2 monitor at Los Alamos National Laboratory (where the measurements in table 10.1 were made) is typically between 4,000 and 5,000 counts per second (Fehlau, personal communication, January 1990). At an alarm threshold of four times the standard deviation of the average background, the portal will alarm at 180–200 counts over background during a half-second measurement.

The emission rate of gamma rays from a gram of uranium-235 is about $5.5 \times 10^4$ per second, and the rate for plutonium-239 is about $2.3 \times 10^5$ per second. Under unfavorable conditions, a category-2 monitor can detect 10 grams of weapon-grade uranium and 0.3 grams of weapon-grade plutonium. This type of monitor can therefore detect a plutonium-239 source producing about $6.9 \times 10^4$ gamma rays per second and a uranium-235 source producing about 5.5 $6 \times 10^5$ gamma rays per second. These emission rates, however, do not include self-shielding by the source itself, which is significantly higher in the case of uranium-235 which has gamma rays concentrated at 185 keV, than for plutonium-239, which has gamma rays located both near 130 keV and 400 keV. We estimate that only about 5 percent of the gamma rays escape the uranium-235 source, and about one-third escape the plutonium-239. Including self-shielding, the above uranium-235 and plutonium-239 sources have roughly equal emission rates of $2.5 \times 10^4$ gamma rays per second.

If one of these sources is placed in the center of the portal for a half-second measurement, the number of counts above background will be about 500–750 counts since the total detection efficiency is 4–6 percent. These numbers are much higher than the minimum number of counts needed to trigger the alarm in this portal. The major reason is that the unfavorable performance conditions, including attaching the source to the inside of a person's ankle and swinging his foot through the portal, significantly reduce the count rate. As a result, under more normal conditions, the portal can be expected to perform significantly better than stated in table 10.1.

15. Paul E. Fehlau, personal communication, January 1990.

16. See, for example, Paul E. Fehlau, "Rugged, Lightweight, and Long-Operating Hand-Held Instruments for Neutron and Gamma-Ray Verification Measurements," preprint LA-UR-88-2780, submitted to the 22nd midyear topical meeting of the Health Physics Society on Instrumentation, San Antonio, Texas, 4–8 December 1988; and Paul E. Fehlau and Gary Wiig, "Stabilized, Hand-Held, Gamma-Ray Verification Instruments for Special Nuclear Materials," *IEEE Transactions on Nuclear Science* **36**, 1, February 1989.

17. Paul E. Fehlau, "A Low-Cost Safeguards Pedestrian Portal Monitor using Chamber Neutron Detectors," in *Proceedings of the 9th ESARDA Symposium on Safeguards and Nuclear Material Management*, London, 12–14 May 1987.

18. This total detection efficiency for this chamber detector corresponds to a geometric efficiency of about one third and an intrinsic efficiency for fission spectrum neutrons of about 4.5 percent.

19. The source is located at the least sensitive region of the portal, typically near a side. At this location the total detection efficiency for fission spectrum neutrons is about 1.0 percent. Therefore, for a source of 1,000 neutrons per second (about 20 grams of weapon-grade plutonium metal) and an approximate counting time of about

0.8 seconds, the detectors would count about 8 neutrons from the source per second. If the alarm threshold is set at four times the standard deviation of the average background, the background count for this detector at sea level would be about 4 neutrons per second.

20. *Vehicle SNM Monitors*, p.4.

21. Ibid., pp.22–26.

22. Each chamber has exterior dimensions of 122 × 244 centimeters and a depth of 15 centimeters and contains four helium-3 proportional counters.

23. *Vehicle SNM Monitors*, pp.23–24. Five centimeters of lead corresponds to about 6.5 mean free paths for a 600-keV gamma ray, which is more energetic than almost all the gamma rays emitted by weapon-grade plutonium.

24. Ibid., p.24. Plastic scintillation detectors can detect both gamma rays and neutrons. The performance of plastic scintillators, however, is limited by the gamma-ray background, which is significantly larger than the neutron background. For example, assuming a 1-second measurement, over 100 neutron counts per second are required to set off the alarm (4 sigma) in a plastic scintillation detector in which the neutron background is perhaps 50 counts per second and the gamma-ray background is roughly 500 or more counts per second.

25. See for example, Levin, "New Machines Can Detect Terrorists' Bombs, Usually," *The New York Times*, 12 September 1989.

26. For more information on this machine, see I.M. Bar-Nir and R.L. Cole, "Coping with Bomb Threats to Civil Aviation," Science Applications International Corporation, Santa Clara, California, undated.

27. A 10.8-MeV gamma ray has a mean free path of about 1.8 centimeters in lead and about 16 centimeters in aluminum.

28. For comparison, the $(n,\gamma)$ reactions taking place in hydrogen has a thermal-neutron cross section of 0.33 barns.

29. The background in this system at 10.8 MeV is relatively large because of the continuous spectrum resulting from "pile up" of the frequent signals from lower-energy gamma rays.

30. Eliot Marshall, "FAA's Bomb Scanner: An Awkward Goliath?" *Science*, 245, 1 September 1989, pp. 926-7.
   The TNA device is designed to monitor standard luggage that can fit into a cavity with a vertical cross section of 40 × 65 centimeters. Each device contains a few hundred micrograms of californium-252 and produces on order of $10^9$ neutrons per second.
   Although the actual situation and calculations are much more complicated, the following discussion illustrates the operational principles of this system. We assume that the source is located about 20 centimeters above the cavity, is surrounded by several centimeters of moderating material, and is highly collimated. If about 90

percent of the neutrons are lost to absorption or scattering during thermalization, then about $10^5$ thermal neutrons $s^{-1}$ $cm^{-2}$ enter the top of the cavity.

We consider two types of explosives, TNT and HMX. TNT has the chemical formula $C_7H_5N_3O_6$ and a density of about 1.6 $g/cm^3$. HMX is a common explosive found in nuclear weapons, with chemical formula $C_4H_8N_8O_8$ and a density of about 1.9 $g/cm^3$.

We assume the TNT is in the shape of a thin 25 × 25-centimeter sheet, 1 centimeter thick, with a total mass of 1,000 grams. Since the density of clothing commonly found in luggage is only about 0.1 $g/cm^3$, we ignore any absorption in the clothing in the luggage. For this calculation, we assume that the sheet is situated in the luggage so that its surface is parallel to the top of the cavity and the flux of the neutron beam across the width and length of the TNT sheet is assumed to be constant. With these assumptions, the approximate emission rate can be found by integrating over the thickness of the sheet of TNT, and is:

$$(1000 \text{ s}^{-1} \text{ cm}^{-1})(A)[1/R - 1/(R + t)] \text{ 10.8-MeV gamma rays per second,}$$

where $A$ is the area of the sheet, $R$ is the distance from the source to the sheet, and $t$ is the thickness of the sheet. In this example, if the sheet is lying on the bottom of the luggage ( $R$ = 59 centimeters), the emission rate is about 180 10.8-MeV gamma rays per second. In the case of the sheet moving through the center of the cavity ( $R$ = 40 centimeters), the emission rate is about 380 per second.

The sides of the cavity have sodium iodide detectors. If the sheet is situated on the bottom of the luggage, the detectors located underneath the cavity are close to the gamma source, resulting in a total efficiency of about 20 percent for a 10.8-MeV gamma ray, or a total counting rate of about 36 10.8-MeV gamma rays per second, or 220 counts over 6 seconds. In the case of the sheet moving through the middle of the cavity, the total efficiency of the detector array would be lower, about 10 percent, but the gamma-ray emission rate is higher, for a total of 230 counts over 6 seconds.

For HMX, which is denser than TNT, a 1-centimeter thick square sheet with a mass of 1,000 grams is 23 centimeters on each side. Using the same methods as above, HMX results in almost double the number of counts as TNT, or about 440 10.8-MeV gamma-ray counts in 6 seconds when the explosive is situated near the bottom of the cavity.

31. Ibid.

32. Consolidated Controls Corporation, Advanced Systems Division, "The Neutron Diagnostic Probe System for Location and Identification of Concealed Explosives," 24 February 1987; and DIXCOM Technology Group, "Neutron Diagnostic Probe: Multi-Pixel Detection and Analysis of Simultaneous Targets," slide presentation, undated. Most of the gamma rays are produced through pure inelastic scattering, although (n,α) reactions also occur a small percentage of the time. Los Alamos National Laboratory is also developing the associated particle method for producing three-dimensional images. See for example, Larry E. Ussery, Charles L. Hollas, Kenneth B. Butterfield, and Richard E. Morgado, "Three-Dimensional Imaging Using Tagged 14.7 MeV Neutrons," LA-11423-MS, Los Alamos National Laboratory, October 1988; Larry E. Ussery, Charles L. Hollas, Kenneth B. Butterfield, and Richard E. Morgado, "Analysis of Biological Samples Using Prompt Gamma Radiations Induced by 14.7 MeV Neutrons," LA-UR-89-1274, submitted to Proceedings of the International Conference on Nuclear Analytical Methods in the Life Sciences, Gaithersburg, Maryland, April 17-21, 1989; and Larry E. Ussery, Charles

L. Hollas, and Gaetano J. Arnone, "The Associated-Particle Technique for One-Sided Imaging," LA-CP-89-330, submitted to the Arms Control Verification Conference at Los Alamos National Laboratory, 29–31 August 1989.

33. The count rate of the current laboratory system is typically about half a count per second, requiring about an hour to obtain sufficient counts. Unlike the TNA device, the neutron and gamma backgrounds are very low because the time-of-flight method enables measurements to be taken over a very short period of time. Typically, the background rate is about 0.1 counts per second. (Information about the associated particle method is from interviews with Charles Peters and Alfred Aitken at DIXCOM Technology Group Inc, Lorton, Virginia, November and December 1989.)

34. Comparing the vapor pressures of explosives at 373 K (100° C) provides an indication of a chemical sniffer's ability to detect various types of explosives. The vapor pressures for several important explosives are:

TNT:    14.13 pascals
RDX:    0.55 pascals
PETN:   $1.1 \times 10^{-3}$ pascals
HMX:    $4 \times 10^{-7}$ pascals

One pascal $\approx 10^{-5}$ atmospheres. (Brigitta M. Dobrz, "Properties of Chemical Explosives and Explosive Simulants," UCRL–51319, Rev 1, 31 July 1974, table 4-3.)

35. M. Mitchell Waldrop, "FAA Fights Back on Plastic Explosives," *Science*, **243**, 13 January 1989.

36. "Officials Say New Verification Technologies Won't Replace NTM," *Aerospace Daily*, 21 November 1989, p.304. The first-stage booster of the SS-20 is 8.58 meters long and 1.79 meters in diameter (INF treaty, Memorandum of Understanding, Section VI, Technical Data, December 1987).

37. Breck W. Henderson, "High-Speed X-Ray CT Scanner Could Meet FAA's Explosion Detection Requirements," *Aviation Week & Space Technology*, 13 November 1989.

38. Richard Dick et al., *Explosives and Blasting Procedures Manual*, (Washington DC: Government Printing Office, 1983). Commercially available nitroglycerin-based explosives can have densities of about 1.0 g/cm³ or less.

# Chapter 11

*Steve Fetter*
*Valery A. Frolov*
*Marvin Miller*
*Robert Mozley*
*Oleg F. Prilutsky*
*Stanislav N. Rodionov*
*Roald. Z. Sagdeev*

# Detecting Nuclear Warheads

To the best of our knowledge, all nuclear weapons contain at least several kilograms of fissile material—material that can sustain a chain reaction. Such material provides the energy for fission explosives such as those that destroyed Hiroshima and Nagasaki; it is also used in the fission triggers of modern thermonuclear weapons.

The two fissile materials used in US and Soviet warheads are weapon-grade uranium (WgU) and weapon-grade plutonium (WgPu). The compositions of these materials assumed in this study are given in table 11.1.

Fissile materials are radioactive; they are very dense and absorb certain radiations very well; and they can be fissioned. Therefore, there are three basic ways to detect fissile material: "passive" detection of the radiation emitted by its radioactive decay, or "active" detection involving either radiographing ("x-raying") an object to detect dense and absorptive materials or irradiating an object with neutrons or high-energy photons and detecting the particles emitted by the resulting induced fissions.

Passive detection is the preferred technique for verification purposes, because of its simplicity and safety. As we shall see, however, passive detection can probably be evaded. Active detection can overcome some evasion scenarios, but only at added cost, inconvenience, and complexity. In addition,

**Table 11.1:** The compositions of weapon-grade uranium and weapon-grade plutonium assumed in this study, in percentages of total weight

| Weapon-grade uranium | | Weapon-grade plutonium | |
|---|---|---|---|
| Uranium-234 | 1.0 | Plutonium-238 | 0.005 |
| Uranium-235 | 93.3 | Plutonium-239 | 93.3 |
| Uranium-238 | 5.5 | Plutonium-240 | 6.0 |
| Other* | 0.2 | Plutonium-241 | 0.44 |
| | | Plutonium-242 | 0.015 |
| | | Other* | 0.2 |

* Oxygen concentration for both WgU and WgPu set at 0.2 percent to give the observed ($\alpha$,n) production rate from WgPu. WgU may be contaminated with uranium from reprocessed reactor fuel, thus making WgU far more radioactive. See appendix 11.A for details.

the process of irradiating objects may pose a danger to nearby humans and to the objects themselves, and may in some cases reveal sensitive information.

## WARHEAD MODELS

In order to illustrate the possible variations in the detectability of different types of warhead, we have devised some simple warhead models. The detailed design of nuclear weapons is secret, but the general characteristics of fission weapons are by now well known (see appendix 11.A for more details). An implosion-type fission explosive can be represented by a series of concentric spherical shells, with the fissile material on the inside surrounded by a neutron reflector/tamper, a layer of high explosive, and some sort of case. In this paper, we explore using either WgU or WgPu as the fissile material, and either tungsten or depleted uranium* for the tamper, giving four hypothetical models. The models are depicted in figure 11.1. We should emphasize that these models are not intended to be realistic weapon designs, but were

---

* Depleted uranium is the residue of the uranium enrichment process, typically 0.2 percent uranium-235. (Natural uranium contains 0.7 percent uranium-235.)

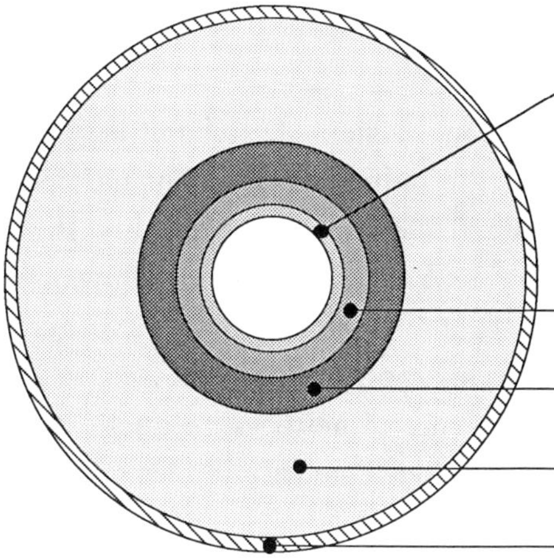

**Figure 11.1:** Hypothetical weapon models used in this study

constructed to define a range of radiation outputs that includes a reasonable *lower bound* on the radiation that would be emitted by actual warheads. Specifically, one of these models (that with a WgU core and tungsten tamper) was deliberately designed to represent a worst case as far as detectability by its radiation emissions is concerned. We doubt that there is today any warhead in the US or Soviet stockpiles that is as difficult to detect as this, but we must acknowledge that such warheads could probably be designed.

## PASSIVE DETECTION

All isotopes of uranium and plutonium undergo radioactive decay. The detectability of this radioactivity varies widely from isotope to isotope, depending on the halflife and the types of radiation emitted. The two types of radiation that might be detectable a few meters or more from a warhead are neutrons and gamma rays (high-energy photons).

## Neutrons

Neutrons are produced primarily by spontaneous fission—fissions of isotopes of uranium or plutonium that occur without the help of an incident particle. Isotopes of plutonium undergo spontaneous fission far more readily than isotopes of uranium, leading to much higher rates of neutron emission. Spontaneous fission occurs at the highest rate in isotopes that have even numbers of both neutrons and protons (for example, plutonium-238, -240, -242, and uranium-238).[*] Table 11.2 gives the rate of neutron production for each isotope and the contribution of each isotope to the neutron-emission rates of WgU and WgPu.

Neutrons are also emitted by light elements such as carbon and oxygen when they absorb alpha particles.[†] These are called "$(\alpha,n)$" reactions. Since isotopes of uranium and plutonium emit alpha particles, and since WgU and WgPu contain small amounts of light-element impurities, $(\alpha,n)$ reactions make a secondary contribution to neutron production in fissile material. Table 11.2 also gives the magnitude of this contribution.

The neutrons produced by spontaneous fission and $(\alpha,n)$ reactions in weapons cause the release of additional neutrons by two mechanisms:

♦ Neutrons will induce additional fissions in the fissile material and so multiply. The shape of the fissile material determines the degree of multiplication. (In our weapon models, each spontaneous fission causes about one additional fission.)

♦ Some materials that may be present in weapons, such as beryllium, can emit two or more neutrons when they absorb a high-energy neutron.

Neutrons are also slowed down in their passage through materials. Here again, two basic mechanisms are involved:

---

[*] Plutonium has 94 protons in its nucleus, uranium has 92.

[†] Alpha particles are helium nuclei that contain two protons and two neutrons.

**Table 11.2:** Neutrons per second per kilogram from spontaneous fission and $(\alpha,n)$ reactions in WgU and WgPu

| | Spontaneous fission $a$ | $(\alpha,n)^*$ $b$ | Fractional composition $c$ | Total $(a+b) \times c$ |
|---|---|---|---|---|
| Uranium-234 | 5.7 | 50. | 0.01 | 0.56 |
| Uranium-235 | 0.30 | 0.012 | 0.933 | 0.29 |
| Uranium-238 | 14. | 0.001 | 0.055 | 0.75 |
| | | | *total WgU* | 1.60 |
| Plutonium-238 | 2,600,000 | 220,000 | 0.00005 | 130 |
| Plutonium-239 | 22 | 630 | 0.933 | 610 |
| Plutonium-240 | 910,000 | 2,300 | 0.060 | 55,000 |
| Plutonium-241 | 500 † | 22 | 0.0044 | 2 |
| Plutonium-242 | 1,700,000 | 33 | 0.00015 | 260 |
| | | | *total WgPu* | 56,000 |

\* Assuming an oxygen concentration of 0.2 percent. See appendix 11.A for details.
† Mostly from americium-241, a decay product of plutonium-241.

**Table 11.3:** The rate of neutron emission at the surface of the four hypothetical weapon designs

| *Weapon model* Fissile material | Tamper material | Multiplication factor* | Emission rate at surface of model *neutrons/second* |
|---|---|---|---|
| 12 kg WgU | tungsten | 1.65 | 30 |
| 12 kg WgU | 79 kg depleted uranium | 1.30 | 1,400 |
| 4 kg WgPu | tungsten | 1.89 | 400,000 |
| 4 kg WgPu | 52 kg depleted uranium | 1.94 | 400,000 |

\* The calculated neutron emission rates from the surfaces of these weapons are greater than the production rate from spontaneous fission. This is due to multiplication from fission and (n,2n) reactions.

♦ Neutrons lose energy like colliding billiard balls by bouncing off the nuclei of light atoms, such as hydrogen or beryllium, without energy being transferred to the internal structure of the struck nucleus ("elastic scattering").

♦ A nucleus of a heavier atom can absorb a fast neutron, keep some of its energy, and emit a slower neutron and a gamma ray through "inelastic scattering."

Finally, slow neutrons are absorbed by the nuclei of most elements, which then release their resulting "excitation" energy in the form of gamma rays or sometimes alpha particles. These are called $(n,\gamma)$ and $(n,\alpha)$ reactions.

We have used the computer program TART to predict the emission of neutrons from our hypothetical weapon models.[1] The results appear in table 11.3.

## Gamma Rays

High-energy gamma rays are released from fissile material primarily as a result of the radioactive decay of isotopes of uranium and plutonium. They are also produced during fission, during the inelastic scattering and absorption of neutrons, and during the decay of radioactive isotopes produced by these reactions. These processes are discussed in more detail in appendix 11.B. Only gamma rays with energies greater than about 0.1 million electron volts (MeV) are penetrating enough to be detectable.*

Unlike neutron emissions, most gamma-ray emissions occur at energies that are determined by the energy-level structure of the parent isotope. And unlike neutron detectors, gamma-ray detectors are available that can determine the energy of a gamma ray with great precision. For these reasons, detection of gamma rays from weapons materials is best accomplished by looking at particular energies where the ratio of the signal from the weapon to the background "noise" from other sources will be highest.

---

* An electron volt is a unit of energy equal to $1.6 \times 10^{-19}$ joules. The gamma rays released during radioactive decay have energies ranging from a few keV (thousands of electron volts) to several MeV.

Our calculations included gamma rays emitted at over 1,000 distinct energies during the radioactive decay of 59 different isotopes of uranium, plutonium, and the nuclei that result from their nonfission decays. We also included a few of the strongest gamma-ray emissions from the radioactive isotopes produced by fission (delayed fission gamma rays). The gamma rays emitted during fission (prompt fission gamma rays) and inelastic neutron scattering are also included, but these have a more-or-less continuous distribution of energies and are therefore less useful with high-resolution detectors.

Using TART, we calculated the fraction of gamma rays produced in the fissile material that escapes from the weapon unscattered—that is, without loss of energy. For the two models with a depleted-uranium tamper, we also calculated the fraction of gamma rays produced in the tamper that emerge from the weapon unscattered. We then multiplied the strength of each gamma-ray emission by the fraction of gamma rays of that energy that escape, and selected the strongest emissions. The results appear in table 11.4.

In all but one case (WgPu with tungsten tamper), the strongest gamma-ray emission is at 1.001 MeV. This gamma ray is emitted by a decay product of uranium-238 in WgU and depleted uranium. This emission is much weaker in the WgU/tungsten model, since there is so little uranium-238 in WgU.

**Table 11.4:** The rate of the strongest gamma-ray emissions at the surface of the four hypothetical weapon designs

| Weapon model | | Emission rate at surface of model | Gamma-ray energy |
| Fissile material | Tamper material | *gamma rays/second* | *MeV* |
|---|---|---|---|
| 12 kg   WgU | tungsten | 30 | 1.001 |
| 12 kg   WgU | depleted uranium | 100,000 | 1.001 |
| 4 kg   WgPu | tungsten | 600 | 0.662 |
| | | 1,000 | ≈1.6* |
| 4 kg   WgPu | depleted uranium | 60,000 | 1.001 |

* There are about 15 neutron-induced gamma rays per keV per second at an energy of about 1.6 MeV. In a low-resolution (sodium iodide) detector, the number of counts per 70-keV channel would be about 1,000 per second, and this may be more detectable than the 0.662-MeV emissions.

In a high-resolution gamma-ray detector, the most prominent emission from the WgPu/tungsten model is probably the 0.662-MeV gamma ray emitted by a decay product of plutonium-241. In a low-resolution detector, however, the neutron-induced gamma rays at energies of about 1.6 MeV may be more detectable. (For more details on neutron and gamma-ray emission and absorption, see appendix 11.B.)

## Detection of Radiation

As one moves away from the weapon, the flux of neutrons and gamma rays (particles per second per unit area) decreases inversely with the square of the distance. For example, the particle flux at 2 meters is four times smaller than at 1 meter; at 3 meters it is nine times smaller, and so on. At some distance, the emissions from the weapon will become undetectable because the flux will be small compared to the flux of natural background radiation.

How far away can our weapon models be detected? To get a rough idea, we first calculate the distance at which the signal is equal to the background, which is given by setting:

$$\frac{A_s \varepsilon_s S}{4 \pi r^2} = A_b \varepsilon_b b \qquad (11.1)$$

where $S$ is the source strength (particles/second), $A_s$ and $A_b$ are the areas of the radiation detector for detecting the signal and the background (square meters), $\varepsilon_s$ and $\varepsilon_b$ are the efficiencies for detecting the signal and background, $r$ is the distance from the source to the detector (meters), and $b$ is the average background rate (particles $m^{-2}$ $s^{-1}$). Solving this equation for $r$ gives (in meters)

$$r = \left[ \frac{\alpha}{4 \pi b} \right]^{1/2} S^{1/2} \qquad (11.2)$$

where $\alpha = (A_s \varepsilon_s / A_b \varepsilon_b)$. This equation is valid for distances of up to about 100 meters.[2]

Here we will consider only hand-held detectors weighing about 10 kilograms and transportable detectors weighing about 100 kilograms. More massive detectors would be more sensitive, but their application to treaty verification will probably be limited to fixed portals through which objects to

**Table 11.5:** The areas, detection efficiencies, and background rates of typical hand-held and transportable neutron and gamma-ray detectors*

| Detector | Energy MeV | $A_s$ $m^2$ | $\varepsilon_s$ | Background[†] $m^{-2}\,s^{-1}$ |
|---|---|---|---|---|
| *Neutron* | | | | |
| Hand-held | - | 0.02 | 0.05 | 50 |
| Transportable | - | 0.3 | 0.14 | 50 |
| | | | | |
| *Gamma-ray* | | | | |
| Hand-held | 0.66 | 0.003 | 0.21 | 100 |
| | 1.0 | | 0.16 | 17 |
| | 1.6 | | 0.10 | 4.4 |
| Transportable | 0.66 | 0.3 | 0.70 | 1,400 |
| | 1.0 | | 0.57 | 860 |
| | 1.6 | | 0.43 | 320 |

* The hand-held gamma-ray detector is assumed to be high-purity germanium, which would be used with a very narrow energy-acceptance window ( ≈ 2 keV). The portable gamma-ray detector is assumed to be sodium iodide with a 10-percent energy resolution.

† Average sea-level terrestrial values

**Table 11.6:** The distance in meters from the centers of the four weapon models at which the neutron and gamma-ray signals equal the background for the detectors in table 11.5

| *Weapon model* | | *Distance (meters)* | |
|---|---|---|---|
| **Fissile material** | **Tamper material** | **Neutrons** | **Gamma rays** |
| 12 kg  WgU | tungsten | 0.2 | < 0.4 |
| 12 kg  WgU | depleted uranium | 1.5 | 3–20 |
| 4 kg  WgPu | tungsten | 25 | < 0.6 |
| 4 kg  WgPu | depleted uranium | 25 | 2–15 |

be inspected would pass.* Table 11.5 summarizes the neutron- and gamma-ray–detector characteristics used here, together with typical background rates; table 11.6 gives the distance from the four weapon models at which the neutron or gamma-ray signal equals the background counting rate in the detector.

As table 11.6 shows, the neutron signal from weapons containing plutonium is greater than the background out to distances of 25 meters. The gamma-ray signal from weapon models that use a depleted-uranium tamper is greater than the background out to 2–20 meters. For the weapon model containing WgU in a tungsten tamper, however, both the neutron and gamma-ray signals are below the background, even at the surface of the weapon.

In our estimates for the performance of detectors, we have assumed that shielding around the detector reduces the background in directions other than toward the source by a factor of 10, making $A_s \approx A_b$ and $\alpha \approx 1$. Gamma rays can be absorbed by relatively thin sheets of heavy elements such as lead and tungsten. Using collimators, it should be possible to reduce the gamma-ray background in a detector by another factor of 10, which would increase the distances given in table 11.6 by a factor of three. In that case, the gamma-ray signal from the models with depleted-uranium tampers would be greater than or equal to the background out to distances of 6–60 meters. In contrast, neutrons are very difficult to collimate, and the low density of neutron absorbers would lead to very large shields.

Signals smaller than the background can be detected if the mean background rate is well known and fairly constant. The size of fluctuations in the background grows with the square root of time, while the signal grows linearly with time. One can simply wait therefore until the signal is larger than normal statistical fluctuations in the background. This criterion can be expressed by the following equation:

$$\frac{A_s \varepsilon_s S t}{4\pi r^2} \geq m(A_b \varepsilon_b b t)^{1/2} \tag{11.3}$$

---

* The helicopter-mounted detector that made the measurements reported in chapter 13 actually had an area of about 2.5 square meters.

where $t$ is the detection time (in seconds) and $m$ is the number of standard deviations in the background that the signal must exceed before we count it as a signal. With $m = 5$, there are fewer than three chances in ten million that a chance variation in the average background could be mistaken for a signal. Using this value of $m$ and solving for the distance, we have

$$r = \left[ \frac{\alpha A_s \varepsilon_s}{400 \pi^2 b} \right]^{1/4} S^{1/2} t^{1/4} \tag{11.4}$$

By substituting the values for $A_s$, $\varepsilon_s$, and $b$ from table 11.5 and the values of $S$ from tables 3 and 4 into the above equation, we can obtain the maximum distance at which each weapon model can be detected by using the assumed hand-held or transportable detector for a given amount of time (see table

**Table 11.7:** The maximum detection range for a given detection time for neutron and gamma-ray emissions from each weapon model if a signal of five standard deviations relative to background fluctuations is required

| Weapon model | | | Distance (meters) | |
|---|---|---|---|---|
| Fissile material | Tamper material | Detection time | Neutrons | Gamma rays |
| 12 kg WgU | tungsten | 1 sec | < 0.1 | < 0.1 |
| | | 1 min | 0.1–0.3 | 0.1–0.2 |
| | | 1 hour | 0.4–0.9 | 0.4–0.6 |
| 12 kg WgU | depleted uranium | 1 sec | 0.3–0.8 | 3–5 |
| | | 1 min | 0.9–2 | 8–15 |
| | | 1 hour | 2–6 | 20–40 |
| 4 kg WgPu | tungsten | 1 sec | 5–15 | 0.2–0.6 |
| | | 1 min | 15–40 | 0.4–2 |
| | | 1 hour | 40–110 | 1–4 |
| 4 kg WgPu | depleted uranium | 1 sec | 5–15 | 2–4 |
| | | 1 min | 15–40 | 6–10 |
| | | 1 hour | 40–110 | 15–30 |

11.7). For a detection time of 1 minute, neutrons from the WgPu models can be detected at distances of up to 40 meters. Gamma rays from depleted-uranium-tamper models are detectable at distances up to 15 meters; if the detectors were collimated, the 1-minute detection distance for these models could be increased to 25 meters. The WgU/tungsten model is almost undetectable.

Since, in many situations, the background may not be well known or constant, the distances reported in table 11.7 should be viewed as theoretical maxima for the weapon models and detectors under consideration. (See appendix 11.C for more details on radiation detectors and backgrounds.)

## Accuracy of Detection

A number of sensitivity tests suggest that our estimates of the detectability of neutron emissions from weapons that use plutonium cannot be far from the mark. First, for warheads of the assumed size and containing plutonium cores, the amount of plutonium cannot be much larger or smaller than 4 kilograms, and the concentration of plutonium-240 in current warheads cannot be much larger or smaller than 6 percent.[3] Making the mass of fissile material more compact by replacing the fissile material shell with a solid core containing the same amount of fissile material could greatly increase neutron emissions because of the greater neutron multiplication in the compact core, but the material would also be more vulnerable to accidental detonation; making it much less compact would decrease neutron emissions by less than a factor of two and would make the warhead unnecessarily large.

Neutrons are not readily absorbed by the materials in our weapon models, but we cannot rule out the use of neutron absorbers designed to protect against effects of nearby nuclear explosions. Such materials would be at most only a few centimeters thick, and could not decrease the neutron flux by much more than a factor of 10. A large amount of lithium deuteride, which is an effective neutron absorber, is used in thermonuclear weapons, but presumably this is not distributed around the fission trigger.[4] Absorption of fission neutrons by air is unimportant at distances of less than 100 meters.

The neutron background is due to cosmic rays and increases with altitude and geomagnetic latitude. At high altitudes or near the poles the neutron flux would be several times greater than the values given in table 11.5, especially during the maximum phase of the solar activity cycle. The neutron flux near a nuclear reactor may be up to 100 times greater than the average background flux.*

Thus, for current warheads with plutonium cores, the distances required for the detection of neutron emissions using portable detectors should not be more than a factor of two less than the values given in tables 6 and 7. If the plutonium is in a more compact configuration, these distances could be significantly greater because of the greater neutron multiplication in the warhead.

The detectability of gamma-ray emissions, on the other hand, is subject to much greater uncertainties. First, the gamma-ray background is less predictable than the neutron background. Second, the rate of gamma-ray emission from a weapon is much more dependent on details of its design than is the rate of neutron emission. For example, some nongovernmental analysts assume that thermonuclear weapons have a casing made out of depleted uranium. If our weapon models each had a uranium case weighing 10 percent of the total mass, the case would be about 1 millimeter thick. Since the mean free path of 1-MeV gamma rays is much greater than that (14 millimeters), about half the gamma rays produced in the case would escape. The resulting gamma-ray flux would be about 10 times greater than that from a depleted-uranium tamper, and would be detectable at a distance three times greater.

As another example, consider a tamper made of beryllium instead of tungsten or depleted uranium. Because beryllium is much less absorptive than these heavy metals (especially at low energies), the gamma rays emitted by the uranium or plutonium in the center of the weapon would be far more detectable. For a WgU core, 186-keV gamma rays would be emitted from uranium-235 at a rate of about 70,000 per second, which would be detectable at a distance of 6 meters (for a counting time of 1 minute). In the case of a

---

* The natural neutron background results in an average yearly radiation dose of roughly 5 millirems (1 rem = $10^{-2}$ sieverts). For comparison, the maximum permissible annual radiation dose for the public is 500 millirems. It is reasonable to assume that reactors will be shielded so that permanently occupied areas meet the 500-millirem criterion, which would result in a neutron flux no more than 100 times greater than the natural background.

plutonium core, 414-keV gamma rays would be emitted from plutonium-239 at a rate of about 500,000 per second, which would be detectable at a distance of about 20 meters.

Moreover, it is possible that the WgU could be mixed or contaminated with uranium from reprocessed reactor fuel (see appendix 11.A for details). If this is the case, the presence of uranium-232 in WgU could make such weapons far more detectable than is indicated by our analysis. Even if this isotope is present at concentrations of less than 1 part per billion, the highly penetrating 2.614-MeV gamma rays emitted during the decay of uranium-232 would be detectable at distances of tens of meters.

In light of these considerations, the estimates presented in tables 11.6 and 11.7 for gamma-ray detection should be considered as approximate lower bounds. In some special cases the detection distances could be somewhat smaller, but in other conceivable circumstances the distances could be many times greater.

## Evading Passive Detection

Passive detection is not foolproof. Some possible weapon types, such as those that contain neither plutonium nor depleted uranium, but use a heavy-metal tamper (like tungsten) and WgU uncontaminated with reprocessed uranium, could be undetectable by portable devices. We presume, however, that plutonium is the preferred core material for situations in which the mass or the size of a weapon is constrained. In the negotiation of future arms-control treaties, persons knowledgeable about weapon design will have to judge whether there is a significant threat of covert deployment of undetectable pure-WgU warheads on certain delivery vehicles could pose a significant military threat. Of course, a ban on nuclear testing would greatly inhibit the development of new warheads designed to evade such detection.

Even weapons with plutonium cores could escape detection by either shielding or purification:

*Shielding.* Assume, for example, that a cheater wanted to be sure that a nuclear weapon could not be detected by a hand-held detector directed at the weapon system for 1 minute from a distance of 1 meter. In our WgPu models, a neutron shielding factor of about 600 would be required to meet this criterion. A blanket of lithium hydride around the warhead 20 centimeters thick and weighing at least 300 kilograms would provide this much shielding. To hide the weapon model from the transportable detector with a detection time of 10 minutes, at least 1 tonne of lithium hydride would be required. Reducing the gamma-ray signal from a depleted-uranium tamper to the same level of detectability would require comparable weights of shielding around the warhead: 600–1,200 kilograms of tungsten. Such large amounts of shielding are impractical for most deployed weapons. However, warheads could be stored in shielded boxes or rooms during an inspection.

*Purification.* Neutron emissions could also be diminished by reducing the concentration of plutonium-240 in WgPu. Reducing the rate of neutron emission to an undetectable level of 100–1,000 per second would require plutonium-240 concentrations 400–4,000 times less than those in WgPu. It appears that this degree of purity is achievable using atomic-vapor laser isotope separation, but only at a cost of one to several million dollars per warhead (see appendix 11.A for further details).

## RADIOGRAPHY

The high density and high atomic number of fissile materials may allow their detection by radiography, which measures the transmission of radiation through various parts of an object with a detection system placed on its far side. An example is the medical x-ray, in which the absorption of x-rays of certain energies is used to indicate the location of bones and other density variations in the body. To be most useful, the probe particles should be sufficiently penetrating so that there is at least a detectable flux through even the most absorptive part of the object under inspection. Although in principle one could simply increase the strength of the source to compensate for a lack of penetrability, in practice this would at some point become destructive to the object and a hazard to nearby humans. But the particles cannot be too penetrating; a measurable fraction of the flux should be absorbed by even the least absorptive parts of the objects of interest to give sufficient contrast in the radiograph.

Since we are interested in detecting the presence of fissile materials, the second criterion for probe particles is that they must discriminate between uranium-235 or plutonium-239 and the materials found in permitted objects. In other words, the particles should be absorbed either much less or much more strongly by uranium and plutonium than by other materials.

Only gamma rays and neutrons both meet these criteria and can be produced in sufficient quantities by portable equipment. Table 11.8 compares the effectiveness of neutrons and gamma rays for radiography. For each particle, table 11.8 gives the ratio of its range in WgU to that in carbon, aluminum, iron, tungsten, and lead. A value of 1.0 would mean that, meter for meter, the particles are absorbed equally by the two materials.

As table 11.8 indicates, "thermal" (very low energy) neutrons offer the greatest discrimination capabilities. What table 11.8 does not show, however, is that some materials that could find legitimate uses would be nearly impossible to discriminate from fissile materials using neutrons. Materials containing lithium or boron would absorb thermal neutrons as efficiently as WgU or WgPu. More important, the high resolution that would be necessary to find relatively small masses of fissile material in large objects would be difficult to achieve with neutrons. High resolution requires either a narrow, well-collimated beam, or a monoenergetic source and the ability to measure the energy of the transmitted particles. Both are difficult to achieve with neutrons.

Both low- and high-energy gamma rays discriminate well between fissile material and light elements such as carbon, aluminum, and iron, but they are very poor at discriminating fissile material from heavy elements such as tungsten and lead. Unlike neutrons, high resolution is easily achieved with gamma-ray beams.

Figure 11.2a shows a diagram of a large x-ray machine, and figure 11.2b shows a radiograph of a car made with a machine being developed for examining truck loads at border crossings. A fan-shaped beam of gamma rays was produced by a linear accelerator and directed at a vertical line of detectors. The car was moved between the source and the detectors and vertical slices irradiated one at a time. By combining many slices with the help of a computer, a picture of the car was obtained.

While high resolution may be necessary to spot small objects, too much detail can easily be revealed. This problem could be dealt with by a jointly

**Table 11.8:** The ratio of the gamma-ray and neutron mean free path (MFP) in carbon, aluminum, iron, tungsten, and lead to that in uranium-235 (WgU)

|  | Energy | Ratio of MFP in element to that in WgU | | | | |
|---|---|---|---|---|---|---|
|  | *MeV* | C | Al | Fe | W | Pb |
| Gamma rays | 0.4 | 22. | 19. | 6.7 | 1.4 | 2.0 |
|  | 10. | 23. | 16. | 4.3 | 1.1 | 1.8 |
|  | 100. | 56. | 27. | 5.5 | 1.1 | 1.7 |
| Neutrons | thermal | 50. | 240. | 24. | 40. | 70. |
|  | 0.001 | 3.0 | 16. | 2.2 | 1.6 | 4.1 |
|  | 10. | 2.2 | 2.4 | 1.5 | 0.94 | 1.5 |

developed computer program that analyzed detailed radiographs to find concentrations of dense material with only this information being passed along to the monitoring party.

An obvious use of radiography would be to try to discriminate between two particular types of object under controlled conditions. For example, an arms-control regime might limit the number of nuclear cruise missiles (or ban them entirely) but not limit non-nuclear versions of the same missile. The inspected party could provide prototype nuclear and conventional missiles, and low-resolution transmission measurements could be made along their entire lengths using neutrons and/or gamma rays (see figure 11.3). Measurements made during on-site inspections could then verify whether a particular missile was armed with a conventional or nuclear warhead. If such measurements were done when the warheads were first mated with the missiles, the missile canisters could be sealed and tagged with a label giving the warhead type. If it was decided to depend primarily on passive means to distinguish between nuclear-armed and non-nuclear weapon systems, radiography could be used to make spot checks for the presence of warheads emitting low levels of radiation or of large amounts of shielding. (See appendix 11.D for more technical information on radiography.)

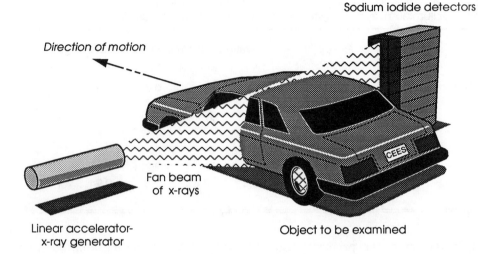

**Figure 11.2a:** Diagram of a large x-ray machine

**Figure 11.2b:** Image of a car generated by a large x-ray machine.
Source: *Linatron: High Energy X-ray Applications for Non-Destructive Testing,*
(San Francisco, California: Bechtel National)

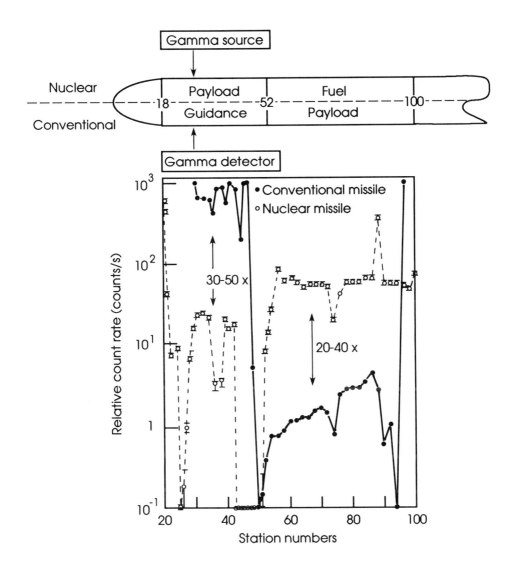

**Figure 11.3:** Comparison of radiographs of sea-launched cruise missiles armed with conventional and simulated nuclear warheads. (John R. Harvey, "SLCM Discrimination," Lawrence Livermore National Laboratory transparency, 6 June 1988.) The purpose of this work was to characterize the radiation-transmission signatures of existing US nuclear and conventional cruise missiles. An important element of verification studies, however, is to examine the degree to which such measures can be "spoofed" in potential future cruise-missile designs. Specifically there may be designs that could reduce the differences in signatures between conventional and nuclear variants illustrated in the figure. This was not addressed in these studies.

Similar results with much weaker gamma sources have been achieved at Argonne National Laboratory (A. De Volpi, private communication).

## INDUCED FISSION

A unique property of fissile isotopes is that they can be caused to fission with low-energy neutrons. If fissions can be induced in an object with slow neutrons, and if the characteristic particles emitted during or after fission can be detected, then the presence of fissile material can be proved conclusively. Higher-energy neutrons and high-energy photons can also be used to induce fission in fissile isotopes, but these are also capable of fissioning other isotopes, such as uranium-238 and thorium-232.

### Neutron-induced Fission

Figure 11.4 shows the number of fissions that would be induced by an isotropic, monoenergetic neutron source located 1 meter from the center of the WgPu/depleted-uranium weapon model. Note that 14-MeV neutrons[*] are 10 times more effective than lower-energy (1 eV–1 MeV) neutrons, and 100 times more effective than thermal neutrons (0.025 eV), in causing fissions in the model. This is because high-energy neutrons can fission uranium-238 in the tamper, and because the larger mean free path of 14-MeV neutrons allows a greater fraction of the neutrons to penetrate to the plutonium in the center of the weapon.

It is important to compare the rates of induced and spontaneous fission, since it would be unprofitable to induce fewer fissions than occur spontaneously. About $2.4 \times 10^5$ spontaneous fissions occur in the WgPu/depleted-uranium weapon model per second.[5] According to figure 11.4, an isotropic source of nearly $2 \times 10^8$ 14-MeV neutrons per second would be required at a distance of one meter to induce the same number of fissions in this weapon model. The strongest portable sources of 14-MeV neutrons (indeed, of neutrons of any energy) generate $10^{11}$ isotropic neutrons per second.[6] At a distance of 25 meters from such a source, the rate of induced fission would equal the rate of spontaneous fission.

---

* 14-MeV neutrons can be produced by a compact source through the fusion reaction $D + T \rightarrow He^4 + neutron$, where D is a deuterium nucleus (containing one proton and one neutron) and T is a tritium nucleus (one proton and two neutrons) and $He^4$ is the nucleus of ordinary helium (two protons and two neutrons).

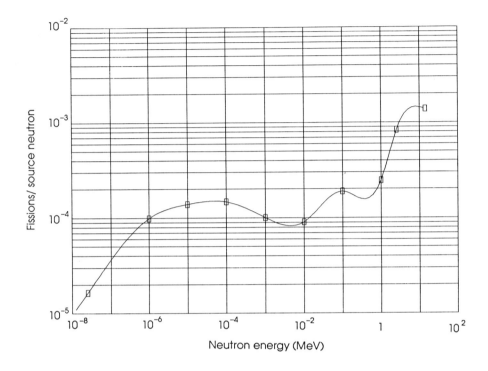

**Figure 11.4:** The number of fissions induced in the WgPu/depleted-uranium model per source neutron as a function of neutron energy, for an isotropic neutron source located one meter from the model

Since a distance of 25 meters is also within the range of passive neutron detection for this weapon model, inducing fission does not seem to be a promising method for increasing the detectability of weapons with plutonium cores. Sources hundreds of times more powerful would be needed to have a significant advantage over passive detection, but these would require large power sources and would pose severe radiation hazards.

Inducing fission is a more useful technique for warheads with cores of uncontaminated WgU or pure plutonium-239 and that do not use a depleted-uranium tamper or case. As has been noted above, such warheads are almost undetectable with passive methods. About the same number of fissions per source neutron can be induced in the WgPu/tungsten and WgU/tungsten weapon models as in the WgPu/depleted-uranium model (except at thermal neutron energies, where the WgU model has an order of magnitude fewer fissions per source neutron). Below we will explore the detection of induced fission in such warheads through the detection of the prompt and delayed particles produced.

*Prompt-radiation Detection*

The number of prompt neutrons per second $S$ from neutron-induced fission that escape from a weapon model is given approximately by

$$S \approx \frac{F(E)}{r^2} S_n v(E) f \qquad (11.5)$$

where $F(E)$ is the number of fissions per source neutron at a distance of 1 meter from an isotropic neutron source of energy $E$ (from figure 11.4) and strength $S_n$ (neutrons per second), $r$ is the distance from the source to the center of the weapon in meters, $v(E)$ is the average number of neutrons per fission induced by neutrons of energy $E$ (see table 11.B.1), and $f$ is the fraction of fission neutrons that escape from the weapon model (about 0.78 for both models). The distance at which the signal equals the background is given in meters by[*]

$$r = \left[\frac{\alpha}{4\pi b}\right]^{1/4} \left[F(E) v(E) f\right]^{1/4} S_n^{1/4} \qquad (11.6)$$

The first quantity in brackets approximately equals 0.2 for the neutron detectors under consideration (see table 11.5). For 14-MeV neutrons, $[F(E)v(E)f]^{1/4} \approx 0.22$. Therefore, as has already been deduced from figure 11.4, a source of $10^{11}$ 14-MeV neutrons per second placed 25 meters from our weapon models would induce enough fissions in the models so that the resulting prompt-neutron flux would equal the background neutron flux at the same distance.

---

[*] The physical origin of the one-fourth power in this equation is the fact that the intensity of the neutron source and the return signal both fall off as $r^{-2}$ with the distance $r$ between the source and the warhead.

The distance for a neutron detector at which the signal is equal to five times the standard deviation in the background is given in meters by

$$r = \left[\frac{\alpha A_s \varepsilon_s}{400\pi^2 b}\right]^{1/8} [F(E)\, v(E)\, f]^{1/4} S_n^{1/4} t^{1/8} \qquad (11.7)$$

The first term is equal to 0.09 for the hand-held detector and 0.15 for the transportable detector (see table 11.5). Therefore, for a source of $10^{11}$ 14-MeV neutrons per second, $r$ ranges from 20 meters for a hand-held detector and a detection time of 1 minute to 50 meters for the larger detector and a detection time of 1 hour.

In the above examples we have, however, ignored a very important effect: 14-MeV neutrons will scatter, both elastically and inelastically, from surrounding materials. These scattered neutrons will create a much larger neutron background than normally exists. Pulsing the source or using an energy-sensitive detector will not help much, since the scattered neutrons will blanket the energy range of the prompt-fission neutrons.

One way to solve this problem would be to use a source of neutrons with energies below those of prompt-fission neutrons. Then the detection of high-energy neutrons would be a unique indicator of fission. Unfortunately, no high-strength source of low-energy neutrons is readily available. For example, alpha particles from polonium-210 can be used to produce $(\alpha, n)$ reactions in lithium, creating about $10^5$ neutrons with an average energy of 0.3 MeV per second per curie of polonium-210. But a source of $10^{11}$ neutrons per second would require $10^6$ curies (220 grams) of polonium; this much polonium would generate over 30 kilowatts of heat. As another example, low-energy neutrons can also be produced by using gamma rays from antimony-124 to produce $(\gamma, n)$ reactions in beryllium. One curie of antimony-124 would produce about $5 \times 10^6$ neutrons per second with energies between 1 eV and 26 keV, but a source of the necessary intensity would require many tonnes of shielding to protect nearby humans. All of these sources use radioisotopes, which, unlike the 14-MeV source mentioned above, cannot be turned off. The hazards posed by these sources make them unsuitable for our purposes.

Using an electron accelerator to produce high-energy gamma rays to produce neutrons with ($\gamma,n$) reactions would eliminate the radiation hazard when inspections were not taking place. Portable linear accelerators are available that, when coupled with a beryllium target, would produce more than $10^9$ neutrons per second with energies of between 0 and 0.33 MeV.[7] These sources are still not strong enough to be useful, however, and the radiation hazard during operation would be very great.

Alternatively, a proton accelerator could be used to produce neutrons of any desired energy using the ($p,n$) reaction with lithium-7. A 150-micro-ampere source of 2.3-MeV protons would produce the equivalent of a $10^{11}$ neutron-per-second isotropic source. Such a machine is commercially available, but it is 1.6 meters long and weighs half a tonne.[8] It also has a duty cycle of only a few percent, which would lead to a corresponding decrease in the average source strength.

A final possible way to generate low-energy neutrons would be to surround a high-energy neutron source with a neutron-moderating material that would slow the neutrons through multiple collisions. A significant fraction of high-energy neutrons would remain, however, so that it would be necessary to produce a short pulse of neutrons and simply wait until all the high-energy neutrons disappeared. Sources are available that can produce from $10^8$ 14-MeV neutrons in a 3.5-microsecond pulse to a total of $5 \times 10^8$ neutrons per second at 10 pulses per second.[9] Ten microseconds after the pulse, all high-energy neutrons would be gone, but the majority of neutrons—those with energies of less than 10 keV—would not have traveled more than 10 meters from the source. It would not be necessary to operate the detector for the entire 100-millisecond period between pulses, since even thermal neutrons travel 15 meters in 1 millisecond. Reducing the operating time of the counters correspondingly, it would be possible to achieve an effective source strength on the order of $10^{10}$ low-energy neutrons per second of counter operating time ($5 \times 10^7$ neutrons per pulse, with roughly half of these in the energy range of interest, spread over a few milliseconds). With such a source, the signal would equal the background out to a source–detector distance of 6–8 meters, and the signal would be more than five times the standard deviation in the background out to 3–16 meters, depending on the detector for detection times ranging from 1 second to 1 hour (see table 11.9).

Detecting the prompt gamma rays released during induced fission would have no advantages over prompt-neutron detection. Although more than twice as many prompt gamma rays are released during fission as prompt neutrons, and although gamma rays can be detected more efficiently than neutrons, a much smaller fraction of gamma rays would escape from the weapon, and the terrestrial gamma-ray background is over 1,000 times greater than the neutron background. The high energy-resolution of gamma-ray detectors does not help, because prompt gamma rays are emitted over a wide range of energies. To make matters worse, even low-energy neutron sources would create a huge background of high-energy gamma rays due to $(n,\gamma)$ reactions with surrounding materials.

### Delayed-radiation Detection

Many of the problems with a 14-MeV neutron source can be solved by detecting delayed neutrons, which have halflives ranging from tenths of a second to one minute. Unfortunately, the number of delayed neutrons is small in comparison to the number of prompt neutrons. The ratio is 0.0064 for uranium-235, 0.0148 for uranium-238, and 0.0020 for plutonium-239 (independent of the incident neutron energy). To use the above equations for delayed neutrons, $v$ should be multiplied by this ratio. Since the average delayed-neutron energy is only about 0.45 MeV, we will assume that $f = 0.5$. Thus, for delayed neutrons and a 14-MeV source, $[F(E)v(E)f]^{1/4} = 0.054$ for WgU and 0.044 for WgPu.

We cannot detect delayed neutrons that are emitted when the source is on; this will thus reduce the effective source strength by a factor of two.* The delayed neutrons are emitted over a long enough time, however, for the $10^{11}$ s$^{-1}$ 14-MeV neutron source to be used in a slowly pulsed mode (for example, on for a second, off for a second). In this way, the delayed-neutron signal would equal the background out to a distance of 4–5 meters, and the signal would be greater than five standard deviations in the background out to 3–11 meters, once again depending on the detector and for detection times ranging from 1 second to 1 hour.

Table 11.9 summarizes the results of our analysis of the detection of

---

* The optimal fraction of time for the source to be on is 0.5.

**Table 11.9:** The maximum detectable distances for the WgU/tungsten and WgPu/tungsten models using prompt neutrons and delayed neutrons from neutron-induced fission

| Detection time | Maximum detectable distances (meters) using | |
|---|---|---|
| | $10^8$ n/pulse, moderated *prompt neutrons* | $10^{11}$ 14-MeV n/s *delayed neutrons* |
| 1 second | 3–6 | 2–4 |
| 1 minute | 6–10 | 3–6 |
| 1 hour | 10–16 | 5–11 |

particles from neutron-induced fission.

Fissions induced by 14-MeV neutrons generate about 250 times more delayed gamma rays than delayed neutrons in both weapon models, but a much smaller fraction of the gamma rays escapes. Overall, about 0.06 delayed gamma rays escape per fission, compared to about 0.005–0.011 delayed neutrons per fission. Although the delayed-gamma signal may be 5–12 times larger than the delayed-neutron signal, the gamma-ray background is 2,000 times larger than the neutron background. Since the detection distance goes as $S^{1/4}/b^{1/8}$, the detection distances for gross delayed-gamma emission will be 30–40 percent less than the corresponding distances for delayed-neutron detection.

Gamma-ray detection might be considerably improved by looking for the line emissions where the signal-to-background ratio is greatest. The strongest delayed-gamma emission line is the 1.597-MeV gamma ray emitted by lanthanum-140. The number of 1.597-MeV gamma rays emitted per fission is about 0.06; the fraction of these gamma rays that escape from the weapon models is about 0.003. For a 14-MeV neutron source at a distance of 1 meter, only about $1.4 \times 10^{-7}$ 1.597-MeV gamma rays are emitted per source neutron; in the notation used above, $[F(E)\nu(E)f]^{1/4} = 0.02$. Combining this with the information on detector characteristics given in table 11.5, we find that the signal would equal the background at a distance of only 1.5–2 meters. Even with the best detector and a counting time of 1 hour, the signal would be equal to five times the standard deviation in the background out to only 3 meters. Looking at several line emissions simultaneously would improve this somewhat, but there are so many delayed-gamma emissions that this is unlikely to be more advantageous than looking at gross delayed-gamma emission. Thus, delayed gamma rays appear to be significantly less detectable than delayed neutrons.

## Photon-induced Fission

High-energy photons can also induce uranium and plutonium to fission. This would appear to be an attractive approach, since the prompt-fission neutrons would not be obscured by a background from the source itself, as is the case when using 14-MeV neutrons to induce fission. Unfortunately, a much smaller number of fissions are generated per source photon than per source neutron for two reasons: a much larger fraction of the photons are absorbed before they can reach the fissile material, and, of those reaching the fissile material, a smaller fraction induce fissions.

Photon-induced fission, or "photofission," has an energy threshold of about 5.3 MeV for uranium-235 and plutonium-239. Only photons with energies greater than the threshold can induce fissions. The photofission reaction becomes much more likely at high energies ( > 14 MeV), but use of such high-energy photons would also lead to a large number of $(\gamma,n)$ reactions in other materials. Among common materials found in weapons, beryllium-9, deuterium, lithium-6, and carbon-13 have $(\gamma,n)$ reactions at or below 5.3 MeV. To eliminate these isotopes as a possible source of confusion, one could measure the various thresholds by varying the gamma-ray energy and noting sudden increases in neutron production.

At a gamma-ray energy of 5.5 MeV, only about 1 percent of the gamma rays headed towards the fissile material penetrate, and only about 0.1 percent of those that penetrate cause fissions. In the notation used above, $[F(E)v(E)f]^{1/4}$ ≈ 0.01. A detection distance of 10 meters would require an isotropic source of nearly $10^{15}$ 5.5-MeV gamma rays per second. A portable electron linear accelerator might be capable of producing such a large intensity of gamma rays, but it is difficult to see what advantages this would have over using a neutron source. (See appendix 11.D for more information on particle sources.)

## CONCLUSION

In the absence of shielding, "ordinary" nuclear weapons—those containing kilogram quantities of ordinary weapon-grade (6 percent plutonium-240) plutonium or uranium-238—can be detected by neutron or gamma counters at a distance of tens of meters. Objects such as missile canisters can be radiographed with high-energy x-rays to reveal the presence of the dense fissile core of any type of nuclear warhead, or the radiation shielding that might conceal a warhead. If subjected to neutron irradiation, the fissile core of any type of unshielded warhead can also be detected by the emission of prompt- or delayed-fission neutrons at a distance on the order of 10 meters.

Devices capable of detecting the presence of nuclear weapons could be useful in verifying compliance with various arms control agreements. Examples discussed in this book include portal monitoring at warhead-dismantlement facilities, verifying warhead stockpile declarations, verifying limits on the number of nuclear warheads on individual ballistic missiles, and verifying limits on the nuclear versions of dual-capable weapons such as sea-launched cruise missiles.

## NOTES AND REFERENCES

1. Ernest F. Plechaty and John R. Kimlinger, "TARTNP: A Coupled Neutron-Photon Monte Carlo Transport Code," UCRL-50400, volume 14, (Livermore, California: Lawrence Livermore National Laboratory, 4 July 1976).

2. The average distance that a fission neutron travels in air before being absorbed is about 250 meters. The average distance that a 1-MeV gamma ray travels before being scattered is 120 meters (100 meters for a 0.66-MeV gamma ray).

3. Some US warheads use "super-grade" plutonium, which is only 3 percent plutonium-240. (Thomas B. Cochran, William M. Arkin, and Milton M. Hoenig, *Nuclear Weapons Databook Volume 1: U.S. Nuclear Forces and Capabilities* [New York: Ballinger, 1984], p.79.) In this case, the distances in tables 6 and 7 would be reduced by 30 percent. The US has successfully tested a warhead using fuel-grade plutonium (which has about three times more plutonium-240 than WgPu), but the superpowers would have no reason to stock such weapons as long as adequate supplies of WgPu are available (Cochran et al., p.24).

4. According to publicly available discussions, the fusion fuel in a standard thermonuclear weapon is contained in a physically separate component. See appendix 11.A.

5. The rate of spontaneous fission is about $10^5$ per second. Each spontaneous fission releases 2.16 neutrons, and each neutron creates a total of 0.62 additional fissions.

6. The A-711 neutron generator produced by Kaman Sciences Corporation in Colorado Springs, Colorado, produces $10^{11}$ neutrons per second in steady-state operation. The cost of the machine in 1989 was about $110,000.

7. Portable 2-MeV electron linear accelerators (x-ray sources) producing 200 rads per minute at 1 meter are available from Varian Associates, Inc., of Palo Alto, California. A bremsstrahlung (braking radiation) spectrum produced when electrons collide with a dense target has a constant power per unit energy; for a beryllium target, only the portion between 1.666 and 2 MeV could be used for neutron production. The machine should be able to produce $10^9$ neutrons per second from a beryllium target.

8. The maximum yield of the $(p,n)$ reaction is about $10^8$ neutrons per steradian per microcoulomb ($6 \times 10^{12}$ protons) in the forward direction. This occurs at a proton energy of 2.3 MeV and results in a neutron energy of 0.5 MeV in the forward direction. Emilio Segre, *Nuclei and Particles* (Menlo Park, California: Benjamin/Cummings, 1977), p.617. Thicker targets could yield about five times as many neutrons at the same proton current. Access Systems of Pleasanton, California, produces a small proton linear accelerator that could produce a 150-microampere beam of 2.3-MeV protons, but at a duty factor of only about 2 percent. The cost of the machine in 1989 was about $600,000.

9. Model A-801 manufactured by Kaman produces $10^8$ neutrons with a pulse length of 3.5 microseconds at half-amplitude. The repetition rate is up to 10 pulses per second with $5 \times 10^7$ neutrons per pulse. The cost in 1989 was about $35,000.

*Steve Fetter*
*Valery A. Frolov*
*Oleg F. Prilutsky*
*Roald Z. Sagdeev*

# Fissile Materials and Weapon Models

This appendix describes the fissile materials that might be detected in a search for nuclear weapons. The composition of the fissile material determines the type and strength of the radiation emitted during radioactive decay, and the weapon design determines the fraction of this radiation that escapes.

## COMPOSITION OF FISSILE MATERIALS

Fissile isotopes are those that can sustain a fission chain-reaction with fast neutrons. These reactions were the basis for fission explosions that destroyed Hiroshima and Nagasaki. Modern nuclear weapons derive a large fraction of their energy from fusion reactions, but a fission explosive is still required to ignite the fusion fuel.

The two fissile isotopes used in US nuclear weapons are uranium-235 and plutonium-239. Several other isotopes are also fissile (for example, uranium-233 and plutonium-241), but they are all more costly to produce and fabricate and also more radioactive than uranium-235 or plutonium-239.

### Weapon-grade Uranium

Naturally occurring uranium contains 0.7 percent uranium-235. Weapon-grade uranium (WgU) is produced by using isotope separation techniques to increase the concentration of uranium-235 to more than 90 percent. The amount of WgU that has been produced for use in US nuclear weapons is estimated to be about 500,000 kilograms.[1] Assuming that the United States stockpile contains about 25,000 nuclear warheads, there is an average of 20 kilograms of WgU per US warhead. For comparison, a moderate-yield fission explosive or thermonuclear primary whose fissile material is WgU might contain 10–15 kilograms of WgU.[2] Since most modern nuclear weapons are believed to have fission primaries with plutonium cores, it is likely that most of the weapon-grade uranium in US nuclear weapons is in the so-called "thermonuclear" secondary stages of the warheads.

We assume here that WgU contains 93.5 percent uranium-235, 5.5 percent uranium-238, and 1 percent uranium-234.[3] If virgin natural uranium is used as the feedstock for the enrichment process, then no other uranium isotopes will be present. If, on the other hand, the uranium feedstock is contaminated with uranium from reprocessed reactor fuel, uranium-232, -233, and -236 will also be present.*

Uranium-233 and uranium-236 do not affect our analysis when they are in small concentrations, but uranium-232 is intensely radioactive: at a concentration of only 0.05 parts per billion (ppb), uranium-232 would emit high-energy gamma rays (energies greater than 1 MeV) at a rate equal to that from all other uranium isotopes in WgU.[4] US Department of Energy specifications for "natural" uranium feedstock to uranium-enrichment plants implicitly permit uranium-232 concentrations 80 times greater than this in WgU.[5]

As the WgU in at least some nuclear warheads is contaminated with uranium-232, the gamma-ray emissions from such warheads are much easier to detect than our analysis suggests. A nation attempting to evade passive detection could, however, use only virgin uranium in the enrichment process.

We assume that the WgU in a stockpiled warhead is at least one year old. Most stockpiled warheads contain WgU much older than this. The average age of WgU in the US warhead stockpile is about 30 years.[6] The WgU in Soviet warheads is presumably somewhat younger. The strength of the most detectable gamma-ray emissions does not, however, vary significantly with the age of WgU.

Even if, after enrichment, WgU contains only isotopes of uranium, radioactive decay of these isotopes will in time produce isotopes of many other elements. Also, in practice, WgU will be contaminated with light elements such as carbon and oxygen. Although these elements are not by themselves radioactive, they can emit neutrons when bombarded by the alpha particles released during the radioactive decay of uranium. If light elements are present in significant concentrations, then the rate of neutron emission from $(\alpha,n)$ reactions can be larger than that from spontaneous fission in WgU. Unfortunately, we have been unable to determine the concentration of light-element impurities in WgU. For the purposes of calculating the $(\alpha,n)$ reaction rate, we have assumed the same oxygen concentration as in WgPu (0.2 percent by weight).

## Weapon-grade Plutonium

Plutonium does not exist in nature except as a contaminant introduced by human activities. It is produced from uranium in nuclear reactors. Weapon-grade plutonium (WgPu) contains about 93 percent plutonium-239. The total amount of WgPu in the US

---

* Uranium-232 is produced in nuclear reactors as follows:

$$U\text{-}235 \xrightarrow[770\text{My}]{\alpha} Th\text{-}231 \xrightarrow[1.1\text{d}]{\beta} Pa\text{-}231 \xrightarrow{(n,\gamma)} Pa\text{-}232 \xrightarrow[1.3\text{d}]{\beta} U\text{-}232$$

About 1 ppb of protactinium-231, which has a halflife of 33,000 years and a thermal-neutron–capture cross section of 260 barns, is produced each year by the decay of uranium-235. Uranium-232 emits about $2.7 \times 10^{11}$ 2.614-MeV photons per second per gram due to the decay of one of its radioactive daughters, thallium-208 (see appendix 11.B).

stockpile is estimated to be about 93,000 kilograms.[7] This gives an average of 3–4 kilograms of WgPu per warhead, an amount that is sufficient for a nuclear explosion.[8]

We have assumed that, after reprocessing (the process by which plutonium is extracted from reactor fuel), WgPu contains only isotopes of plutonium and small concentrations of light elements. Trace amounts of fission products may be present in WgPu, but these should not affect our analysis. Once again, radioactive decay will produce many other isotopes. The average age of WgPu in the US stockpile is about 20 years,[9] and Soviet plutonium should be somewhat younger. The most detectable gamma-ray emission from WgPu is 10 times stronger after 20 years of decay than after 1 year of decay.

The isotopic composition of WgPu assumed here is 93.5 percent plutonium-239, 6.0 percent plutonium-240, 0.44 percent plutonium-241, 0.015 percent plutonium-242, and 0.005 percent plutonium-238.[10] Two other long-lived isotopes, plutonium-236 and plutonium-244, are present in minute concentrations. Although plutonium-244 is of no concern, plutonium-236 is highly radioactive. Plutonium-236 would be the major source of high-energy gamma-ray emissions from WgPu at concentrations of as little as 10 ppb.[11] Since it appears that plutonium-236 is present in WgPu at concentrations of less than 1 ppb, we will ignore its presence in WgPu.[12]

Light elements, such as lithium, beryllium, carbon, and oxygen are also present in small concentrations. Although we have been unable to obtain accurate estimates of the concentrations of these impurities, about 1 neutron per second is emitted from $(\alpha,n)$ reactions per gram of WgPu.[13] An oxygen concentration of 0.2 percent by weight would result in an equivalent $(\alpha,n)$ reaction rate.

*Purifying Weapon-grade Plutonium*
Plutonium-239 is not a strong source of neutrons and gamma rays. Almost all the radiations that are useful for passive detection are emitted by other isotopes of plutonium. For example, 97 percent of the neutron emission from WgPu is due to the spontaneous fissioning of plutonium-240. Therefore, if the concentration of the isotopes other than plutonium-239 could be greatly reduced at reasonable cost, passive detection might be thwarted.

About 400,000 neutrons per second would be emitted from a nuclear warhead containing 4 kilograms of standard WgPu (see table 11.3 of the main article). As the calculations in that article show, a neutron emission rate of 1,000 per second would be barely detectable outside the warhead, and a rate of 100 per second would be undetectable (at least by transportable instruments). This would require a factor of 400–4,000 reduction in plutonium-240 concentrations, and a factor of 20–200 reduction in plutonium-238 and -242 concentrations. The concentration of light-element impurities would also have to be reduced by one to two orders of magnitude to reduce neutron emissions from $(\alpha,n)$ reactions.

Although eliminating the light-element impurities would be relatively easy, removing undesirable plutonium isotopes would be difficult and costly. The only existing method of isotope separation that could achieve the high purities required is atomic-vapor laser isotope separation (AVLIS). In this process, a laser ionizes the unwanted (or, alternatively, the desired) isotope, which is then collected on a charged plate. The proposed but recently canceled US Special Isotope Separation (SIS) plant was expected to turn fuel-grade plutonium into WgPu in a single pass through an AVLIS separation stage by reducing the plutonium-240 concentration by a factor of two to three, and the concentrations of other minor isotopes of plutonium by a factor of two to six.

If a constant threefold decrease in the concentration of plutonium-240 could be achieved for each pass through an AVLIS stage, six to eight consecutive passes through a single stage, or a single pass through a series of six to eight identical stages, would be required to achieve a reduction of 400–4,000. Assuming that the cost per pass or stage is dominated by the throughput, the cost of high-purity plutonium would be approximately six to eight times the cost of the WgPu that would have been produced from the SIS plant. Based on current estimates, the latter would have been roughly $200,000 per kilogram.[14] Equipping a warhead with 4 kilograms of high-purity plutonium-239 would therefore cost perhaps $5 million. Since the cost of a warhead, including the fraction of the cost of the associated "delivery vehicle," is at least this high, a cheater might not view the additional cost of purification as intolerable.[15]

These costs would drop if the fractional decrease in the minor plutonium isotope concentrations per pass could be improved so that fewer passes or stages would be required. It is plausible that this can be achieved, even without advances in current AVLIS technology.[16]

## Depleted Uranium

Depleted uranium, or the "tails" of the enrichment process, is thought to be used in nuclear weapons as a tamper or case material. Since it is radioactive and can be caused to fission by high-energy neutrons or photons, it may also be detectable using active or passive means. We will assume that depleted uranium contains 99.8 percent uranium-238 and 0.2 percent uranium-235.

## WEAPON MODELS

The absorption and scattering of neutrons and gamma rays by the materials that surround the fissile material must be taken into account in any accurate estimate of the detectability of fissile material. Although detailed weapon designs are classified, the general characteristics of nuclear weapons are well known by now. The models that we offer below are not intended to be representative of actual US or Soviet weapon designs; they are merely intended to give a range including a lower bound to the amount of radiation that would be emitted by such weapons.

A fission explosive, or the "primary" of a thermonuclear explosive, can be represented by a series of concentric spherical shells. Innermost is the fissile material. We assume that the outside radius of this shell is 7 centimeters for WgU and 5 centimeters for WgPu, and that the mass of the shell is 12 kilograms for WgU and 4 kilograms for WgPu.[17] Surrounding the fissile material is a neutron reflector, which we assume to

**Table 11.A.1:** Models of fission explosives used in our analysis

| | Outside radius *cm* | Mass *kg* | | Outside radius *cm* | Mass *kg* |
|---|---|---|---|---|---|
| *WgU + depleted uranium* | | | *WgPu + depleted uranium* | | |
| Empty | 5.77 | 0 | Empty | 4.25 | 0 |
| WgU | 7 | 12 | WgPu | 5 | 4 |
| Beryllium | 9 | 3 | Beryllium | 7 | 2 |
| Depleted uranium | 12 | 79 | Depleted uranium | 10 | 52 |
| High explosive | 22 | 71 | High explosive | 20 | 56 |
| Aluminum | 23 | 17 | Aluminum | 21 | 14 |
| | | 182 | | | 128 |
| *WgU + tungsten* | | | *WgPu + tungsten* | | |
| Empty | 5.77 | 0 | Empty | 4.25 | 0 |
| WgU | 7 | 12 | WgPu | 5 | 4 |
| Beryllium | 9 | 3 | Beryllium | 7 | 2 |
| Tungsten | 12 | 81 | Tungsten | 10 | 53 |
| High explosive | 22 | 71 | High explosive | 20 | 56 |
| Aluminum | 23 | 17 | Aluminum | 21 | 14 |
| | | 184 | | | 129 |

be a 2-centimeter-thick shell of beryllium. (A thin shield between these components would prevent alpha particles produced in the fissile material from causing ($\alpha,n$) reactions in the beryllium.) Next is the "tamper" of dense material inside the high explosive, which we assume is 3 centimeters thick. We consider two different tamper materials here: depleted uranium,* which would produce its own characteristic gamma-ray emissions, and tungsten, which would act simply as a radiation shield. Surrounding the tamper is a layer of high explosive 10 centimeters thick. We assume that the high explosive is composed of hydrogen, carbon, nitrogen, and oxygen in the ratio 2:1:2:2, and that it has a density of 1.9 grams per cubic centimeter.[18] Finally, we represent the electronics and packaging materials by a shell of aluminum 1 centimeter thick.

The models of the fission explosives thus derived are summarized in table 11.A.1. The models with WgU cores have outside radii of 23 centimeters and weigh about 180 kilograms; the models with WgPu cores have radii of 21 centimeters and weigh about

---

\* This material might be preferred because it could increase the power of the nuclear explosion through fissions by fast neutrons.

130 kilograms. This is reasonably consistent with what is known about modern US warheads, which are estimated to have outside radii of 14–24 centimeters and to weigh 100–200 kilograms.[19] Warheads exist that are significantly smaller and lighter than our models (for example, the 155-millimeter nuclear artillery shell has a radius of 8 centimeters and a mass of 44 kilograms); the radiations from such warheads should be easier to detect because of the smaller amount of shielding. Heavier nuclear weapons also exist, but the radii of these cigar-shaped weapons are about equal to those in our models; the amount of shielding in directions perpendicular to the long axis should therefore be about equal.[20]

When comparing our models to thermonuclear warheads, one should remember that the fusion "secondary," *which is a physically separate component*, comprises a substantial fraction of the mass of the warhead. The fissile material in the primary might therefore be shielded in some directions (for example, in the direction of the secondary) much more than we assume. Some have speculated that the secondary contains a substantial amount of fissile material in the center of the fusion fuel, or that depleted uranium or WgU surrounds the fusion fuel as a tamper.[21] Moreover, some nongovernment analysts assume that a layer of depleted uranium surrounds thermonuclear warheads. If so, then thermonuclear weapons might be more detectable than suggested by our analysis.

## NOTES AND REFERENCES

1. Frank von Hippel, David H. Albright, and Barbara G. Levi, *Quantities of Fissile Materials in U.S. and Soviet Nuclear Weapons Arsenals*, PU/CEES 168 (Princeton, New Jersey: Princeton University, Center for Energy and Environmental Studies, July 1986); Thomas B. Cochran, William M. Arkin, Robert S. Norris, and Milton M. Hoenig, *Nuclear Weapons Databook Volume 2: U.S. Nuclear Warhead Production* (New York: Ballinger, 1987), p.191.

2. The critical mass of a bare sphere of uranium-235 is about 50 kilograms. If a chemical explosion can double the density of uranium, then the critical mass would be about 12 kilograms. Surrounding the uranium with a tamper to reflect about half the escaping neutrons back into the fissile material would make this compressed mass highly supercritical.

3. Weapon-grade uranium is commonly assumed to be about 93 percent uranium-235, but the presence of uranium-234 is often ignored. The concentration of uranium-234 in natural uranium is about 0.0054 percent. Assuming that the depleted uranium contains 0.2 percent uranium-235 and that nearly all the uranium-234 goes into the enriched product, then WgU should contain about 1 percent uranium-234.

4. Uranium-232 has a halflife of 70 years. After 1 year of decay, 1 gram of uranium-232 and its reactive daughter products emits a total of $2 \times 10^{11}$ photons per second—$9 \times 10^{10}$ photons per second with energies greater than 1 MeV. For comparison, 1 gram of WgU with no uranium-232 emits 2,000 photons per second—only about

4 photons per second with energies greater than 1 MeV. Therefore, at a concentration of only 0.05 ppb, uranium-232 would emit high-energy photons at the same rate as all other isotopes of uranium in WgU.

5. The US Department of Energy recognizes that uranium feedstock may become contaminated with reprocessed uranium. The current specifications for commercial "natural" uranium hexafluoride permit a maximum uranium-232 concentration of 0.02 parts per billion (James C. Hall, "Revision to the Uranium Hexafluoride Feed Specification," AE-97-126, US Department of Energy, 6 April 1987). WgU produced with such feedstock would have a uranium-232 concentration of about 4 ppb.

6. Based on data in Cochran et al., *U.S. Nuclear Warhead Production*, p.184, the mean age of WgU in the US stockpile is 29 years.

7. Von Hippel et al., *Quantities of Fissile Material*, and Cochran et al., *U.S. Nuclear Warhead Production*, p.75.

8. The critical mass of plutonium-239 is about three times less than the critical mass of uranium-235 under similar conditions (see note 2). *Fat Man* (the bomb dropped on Nagasaki) used 6.1 kilograms of plutonium. Modern weapons undoubtedly make more efficient use of plutonium.

9. Based on data in Cochran et al., *U.S. Nuclear Warhead Production*, pp.63–65, the mean age of US WgPu is 21 years.

10. The isotopic concentrations (percent by weight) are estimated from data in M.J. Halsall, "Graphs and Tables of the Isotopic Composition of Plutonium Produced in Canadian $D_2O$-Moderated Reactors," AECL-2631 (January 1967), L.J. Clegg and J.R. Coady, "Radioactive Decay Properties of CANDU Fuel," AECL-4436/1 (January 1977), and in M.S. Milgram and K.N. Sly, "Tables of the Isotopic Composition of Transuranium Elements Produced in Canadian $D_2O$-Moderated Reactors," AECL-5904 (August 1977), for a Pickering-type CANDU for a burnup of about 1.2 gigawatt-days per tonne. The isotopic concentrations from a production reactor should be very similar for a plutonium-239 concentration of 93.5 percent.

11. The halflife of plutonium-236 is 2.8 years. After 1 year of decay, 1 gram of plutonium-236 and its radioactive daughters emits about $2 \times 10^{10}$ photons per second with energies greater than 1 MeV. For comparison, 1 gram of WgPu with no plutonium-236 emits only 150 photons per second with energies greater than 1 MeV. After 5 years of decay, the rate of photon emission from plutonium-236 would be 12 times greater.

12. After a burnup of 7.5 gigawatt-days per tonne, plutonium-236 comprises 0.77 ppb of the total plutonium in a natural-uranium-fueled Candu reactor (Scott Ludwig, Oak Ridge National Laboratory, personal communication, 6 June 1988). Even less plutonium-236 would be expected in a plutonium production reactor, since the burnup would be about six times smaller.

13. Ralph Condit and Mel Coops, Lawrence Livermore National Laboratory, personal communications, December 1988.

14. SIS would have convert six to seven tonnes of fuel-grade plutonium into WgPu in about eight years at a projected start-up cost of about $1 billion and operational costs of $60 million per year. Dan W. Reicher and Jason Salzman, "High-tech Protest Against Plutonium Plant," *Bulletin of the Atomic Scientists*, **44**, 9, November 1988, p.27. Assuming an annual capital cost of 10 percent, the WgPu produced by SIS would have cost roughly $200,000/kilogram.

15. The unit program cost of a sea-launched cruise missile is roughly $5 million, and that of the MX missile (if 200 are deployed) and the B-1B bomber (assuming 20 weapons per aircraft) is roughly $15 million per warhead.

16. For the plutonium isotope concentrations of interest, the relationship between the stripping efficiency $\varepsilon$ (defined as the fraction of minor isotope atoms which are separated from the feed per pass in a single stage) and $\delta$ (the corresponding fractional decrease in the minor plutonium isotope concentration) can be approximated by $\delta^{-1} = (1 - \varepsilon)$. The stripping efficiency, in turn, can be written as the product of three factors: $\varepsilon = f_a f_i f_c$, where $f_a$ is the fraction of atoms in atomic states that can be accessed by the lasers, $f_i$ is the fraction of these atoms that are ionized, and $f_c$ is the fraction of ionized atoms that are collected. As noted by Solarz, $f_i$ and $f_c$ can be increased by increasing the laser power and decreasing the stage throughput, respectively, thus increasing $\varepsilon$ and $\delta$. The resulting higher cost per stage can be more than compensated for by the smaller number of stages required. Richard W. Solarz, "A Physics Overview of AVLIS," UCID-20343 (Livermore, California: Lawrence Livermore National Laboratory, February 1985).

17. The radii are 1.4–1.7 centimeters larger than those of solid spheres of uranium or plutonium with the same masses. The dimensions were chosen so that a total of about one induced fission occurs for each spontaneous fission in the material. (This multiplication includes the effect of the reflector/tamper described in the text.) Smaller radii would result in significantly increased neutron emission due to neutron multiplication and would decrease the nuclear safety of the weapon. Larger radii would simply waste valuable space.

18. Clifford Conn, "Synthesis of Energetic Materials," *Energy and Technology Review*, January–February 1988, p.21. Rough calculations show that this amount of high explosive is more than sufficient to compress the fissile material to a highly supercritical state, although proportionately more explosive should be required for the WgU design to achieve an equivalent compression.

19. Thomas B. Cochran, William M. Arkin, and Milton M. Hoenig, *Nuclear Weapons Databook Volume 1: U.S. Nuclear Forces and Capabilities* (New York: Ballinger, 1984), pp.76, 126, gives dimensions for the Minuteman and the MX warheads that would allow maximum radii of 24 centimeters, and on p.297 gives dimensions for the Pershing II warhead that would allow a maximum radius of 14 centimeters. Robert S. Norris,

"Counterforce at Sea," *Arms Control Today*, **5**, 7, September 1985, p.9, gives masses for the Trident I and Trident II warheads of 100 and 200 kilograms. Cochran et al., p.79, gives a mass of 130 kilograms for the air-launched cruise missile warhead.

20. Cochran et al., *U.S. Nuclear Forces*, p.199, gives the masses of the B-28, B-43, and B-83 bombs as about 1,000 kilograms. Ibid, pp.42, 49, gives radii of 25 and 23 centimeters for the B-28 and the B-43 bombs.

21. See, for example, Howard Morland, "The H-bomb Secret (To Know How Is to Ask Why)," *The Progressive*, November 1979, and his further speculations in Robert Del Tredici, *At Work in the Fields of the Bomb* (New York: Harper and Row, 1987), pp.130–131.

Steve Fetter
Robert Mozley

# Emission and Absorption of Radiation

A full understanding of fissile-material detection requires a solid knowledge of nuclear physics, especially the emission and absorption of radiation. The purpose of this appendix is therefore to give a brief explanation of these processes so that the reader can better understand the assumptions leading to the results in the main text.

## THE NUCLEUS

The nucleus is composed of neutrons and protons, which have about the same mass. The atomic number $Z$ of an atom is equal to the number of protons in the nucleus. All atoms of a given element have the same atomic number. For example, plutonium has $Z = 94$.

The atomic weight is the number of protons plus the number of neutrons. Atoms with the same atomic number but with different atomic weights are called "isotopes." For example, plutonium-239 has 94 protons and 145 neutrons and therefore an atomic weight $A = 94 + 145 = 239$. Most elements have more than one stable isotope and several unstable or "radioactive" isotopes.

The unit of nuclear mass is called an "atomic mass unit" or "amu." One amu is defined as exactly one-twelfth of the mass of a carbon-12 atom, which contains six protons and six neutrons.

The unit of energy most commonly used in atomic and nuclear physics is the electron volt (eV). An electron volt is the energy an electron would obtain in falling through an electric potential of 1 volt. Masses can also be expressed in units of energy using the equivalence between mass and energy expressed in the famous formula $E = mc^2$. A proton has a rest mass corresponding to 938.272 million electron volts (MeV), while that of a neutron is 939.566 MeV (ignoring the $c^2$, or velocity of light squared). An electron's rest mass corresponds to only 0.510976 MeV, and the atomic mass unit is 931.494 MeV. These units can be transformed into metric units of energy and mass as follows: 1 MeV $= 1.602 \times 10^{-13}$ joules, and 1 MeV/$c^2$ $= 1.783 \times 10^{-32}$ kilograms.

The atomic mass unit is less than the average of the masses of the proton and the neutron because it takes energy to break up a nucleus and with that addition of energy comes additional mass.

Using accurately determined masses, one can determine the feasibility of many interactions. For example, one can see that the mass of a neutron is greater than the masses of a proton and an electron combined. Hence, the decay of a neutron into a proton, an electron, and a neutrino* is energetically possible and, in fact, occurs when

---

* A neutrino, which always accompanies beta decay, is a neutral, almost noninteracting, zero-mass particle.

a neutron is outside a nucleus. However, not all nuclear reactions that would conserve energy occur in nature because other quantities, such as electric charge, must also be conserved.

## RADIOACTIVE DECAY

The nuclei of all uranium and plutonium isotopes are unstable in that they each decay into one or more sets of nuclei and other particles, which have a lower total rest mass—the energy difference going into the energy of motion of the resulting nuclei and particles (see figures 11.B.1–4). The isotopes under consideration here undergo three types of radioactive decay: alpha decay, beta decay, and spontaneous fission. Other modes of decay are possible (for example, electron capture and internal transitions), but these will be ignored here.

When an isotope undergoes alpha decay, an alpha particle (helium-4 nucleus) is emitted with an energy of several MeV, the atomic weight of the isotope decreases by four, and the atomic number decreases by two. During beta decay, a beta particle (electron or positron) and a neutrino are emitted with a total energy of up to a few MeV; the integer atomic weight remains unchanged (although the actual atomic mass decreases slightly) and the atomic number is increased or decreased by one. The excess of neutrons in the fission products restricts the beta emissions primarily to electrons, and hence the atomic number generally increases. Gamma rays (high-energy photons) are also emitted during alpha and beta decay if the emitted particle does not carry away all of the excess energy, leaving the residual nucleus in an "excited" state that releases the remaining energy in the form of photons. X-rays (intermediate-energy photons) are often emitted if the atomic electrons are disturbed, and bremsstrahlung ("braking radiation"—photons with a continuous energy probability distribution up the full energy of the electron) is emitted as energetic beta particles slow down in matter.

Spontaneous fission is quite different from alpha or beta decay. In this case, the nucleus is so unstable that it splits into two parts. Several gamma rays and neutrons are emitted instantaneously; the remaining fragments are highly unstable and may undergo beta decay several times. Fission, both spontaneous and neutron- and photon-induced, is described in detail below.

Radioactive decay of all kinds is random in time (the moment of decay for a given atom cannot be predicted), but the rate of decay is proportional to the number of atoms present. This leads to the following relationship: if $N(t)$ is the number of atoms present at time $t$, and $\lambda$ is 1/(mean life), the decay rate $Q(t) = N(t) \times \lambda$, and can be expressed by

$$Q(t) = Q(0)\exp(-\lambda t) \tag{11.B.1}$$

where $Q(0)$ is the initial decay rate. If $t$ and $\lambda^{-1}$ are expressed in seconds or years, the relation gives the number of decays per second or year. $N(t)$ can also be expressed in terms of the mass of the radioactive material present using the fact that $6.02 \times 10^{26}$

(Avogadro's constant) atoms of an isotope with atomic mass $m$ weighs $m$ kilograms:

$$N(t) = K(t) \frac{6.02 \times 10^{26}}{m} \qquad (11.B.2)$$

where $K(t)$ is the mass in kilograms at $t$, given by

$$K(t) = K(0) \exp(-\lambda t) \qquad (11.B.3)$$

$\lambda$ can be expressed in terms of the halflife of a material $t_{1/2}$, the time in which half of it decays, as

$$\lambda = \frac{\log_e(2)}{t_{1/2}} = \frac{0.693}{t_{1/2}} \qquad (11.B.4)$$

This leads to a value of

$$Q(t) = K(t) \frac{6.02 \times 10^{26}}{m} \frac{0.693}{t_{1/2}} \qquad (11.B.5)$$

Dividing the expression above by the number of seconds in a year, $3.16 \times 10^7$, gives the decay rate in number of decays per kilogram per second when the halflife is expressed in years:

$$Q(t) = K(t) \frac{1.32 \times 10^{19}}{t_{1/2}\, m} \qquad (11.B.6)$$

The situation is actually made more complex because (in the terminology of nuclear physics) many "parent" radioactive isotopes decay to "daughter" isotopes that are themselves radioactive, and these decay to other radioactive isotopes, and so on. The decay chains for isotopes of uranium and plutonium contain up to 14 radioactive daughter products. These decay schemes are shown in figures 11.B.1–4. Alpha decays are indicated by a downward arrow; beta (electron) decays by an arrow pointing to the right. Halflives are listed next to each arrow.* When isotopes can decay by either alpha

---

* $\mu s = 10^{-6}$ seconds, ms $= 10^{-3}$ seconds, s $=$ seconds, m $=$ minutes, h $=$ hours, d $=$ days, y $=$ years, ky $= 10^3$ years, My $= 10^6$ years, and Gy $= 10^9$ years.

or beta emission, the probability of each type of decay is noted next to the arrows. Spontaneous fission (sf) is indicated by a forked arrow, and the percentage of decays via spontaneous fission is given. Branches of the decay chain that occur with low probability (less than 0.1 percent) are omitted from these diagrams.

The rate of decay of the $j$th daughter at time $t$ is given by

$$Q_j(t) = Q(0) \prod_{i=1}^{j} \lambda_i \sum_{h=0}^{j} \left[ \frac{e^{-\lambda_h t}}{\prod_{\substack{p=0 \\ p \neq h}}^{j} (\lambda_p - \lambda_h)} \right] \qquad (11.B.7)$$

where $\lambda_i$ refers to the decay constant of the $i$th daughter isotope with $i = 0$ for the isotope at the top of the decay chain.[1] The rate of particle emission per kilogram of the parent isotope is given by

$$S_x(t) = \sum_j \left[ Q_j(t) \sum_i f_{i,j} \right] \qquad (11.B.8)$$

where $S_x$ is the rate of emission of particle $x$ (photons, alpha particles, beta particles, or neutrons), and $f_{i,j}$ is the probability of the $i$th emission* of particle type $x$ per decay of the $j$th radioactive daughter. Values for $f_{i,j}$ can be found in standard references or databases.[2] Our calculations of the photon emissions from WgU and WgPu included over 1,000 emissions from 59 daughter isotopes resulting from decays of various uranium and plutonium isotopes (see figures 11.B.1–4).

The gamma spectrum emitted by WgU is dominated by excited decay products of uranium-238 and uranium-234. The most valuable gamma emission for detection purposes is the 1.001-MeV line of protactinium-234m,† which is a decay product of uranium-238. Higher-energy gamma emissions at 1.399 MeV and 1.832 MeV (from protactinium-234m) or at 1.764 MeV and 2.204 MeV (from bismuth-214, which is a decay product of uranium-234) may also be valuable in situations where a large amount of heavy-metal shielding is present. The decay of uranium-235 results in a negligible emission of high-energy gamma rays; the most prominent emission has an energy of only 0.186 MeV.

The spectrum of depleted uranium at high energies is dominated by the gamma-ray emissions of protactinium-234m at 1.001, 1.399, and 1.832 MeV.

---

* For example, after a particular alpha or beta decay, the residual nucleus may emit gamma rays carrying several different energies.

† Protactinium-234m is a metastable or long-lived excited state of protactinium-234.

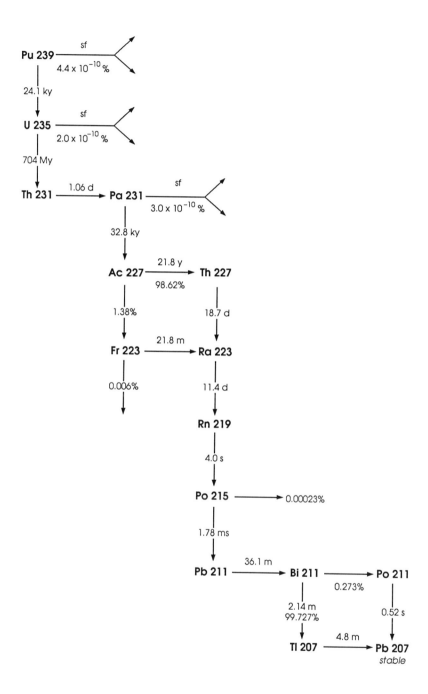

**Figure 11.B.1** The decay chain for plutonium-239 and uranium-235

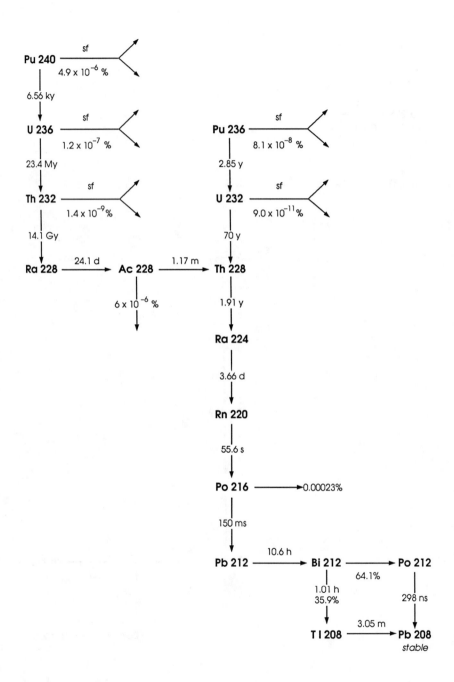

**Figure 11.B.2** The decay scheme for plutonium-240

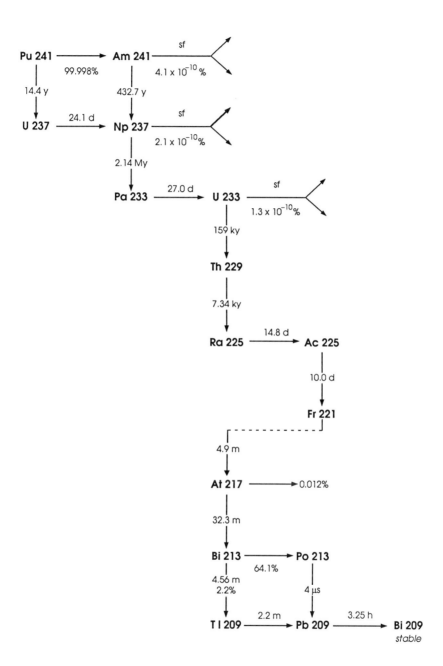

**Figure 11.B.3** The decay scheme for plutonium-241

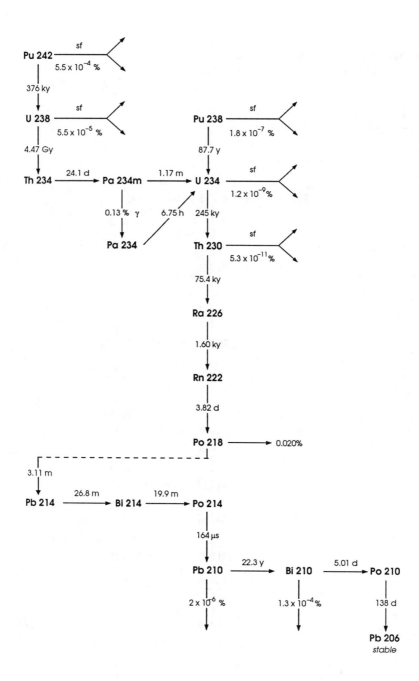

**Figure 11.B.4** The decay scheme for uranium-234 and -238 and plutonium-238 and -242

WgPu lacks intense gamma-ray emissions at energies above 1 MeV. Depending on the degree of shielding, gamma emissions at 0.414, 0.662, 0.722, or 0.769 MeV will be most valuable for detection purposes. The 0.662- and 0.722-MeV gamma rays are emitted by americium-241, which is a decay product of plutonium-241, and the 0.414- and 0.769-MeV gamma rays are emitted during the decay of plutonium-239.

## FISSION

As mentioned above, a heavy nucleus can be so unstable that it fissions, or splits into two nuclei of intermediate weight. There are about 80 different combinations of these intermediate-weight nuclei that can result from the fission of any particular nucleus. Figure 11.B.5 shows the yield of fission fragments as a function of mass for fast-neutron fission of uranium-235 and plutonium-239. Note that the maximum yield for a given mass number is about 7 percent.

Fission occurs spontaneously at a low rate for the heaviest isotopes in the plutonium and uranium decay chains shown in figures 11.B.1–4. Some nuclei, such as the fissile-weapon materials plutonium-239 and uranium-235, can easily be fissioned by incident neutrons of any energy. Others, such as uranium-238, can only be caused to fission with neutrons or photons above a certain energy threshold.

About 200 MeV of energy is released in fission, most of which (about 165 MeV) appears as the kinetic energy of the two fission fragments. The remainder of the energy is associated with prompt and delayed radiations—neutrons, gamma rays, and beta particles. Prompt radiations are those that are emitted during the fission process (actually, within about $10^{-14}$ seconds of fission), while delayed radiations are those emitted by the unstable fission fragments. Since beta particles are stopped by only a few grams/cm$^2$ of matter, we concentrate here on neutron and gamma-ray emissions.

### Prompt Neutrons

Two to five prompt neutrons are released during fission, depending on the isotope and the energy of the incident particle. Table 11.B.1 gives values for $v$, the average number of neutrons emitted per fission, for fissions caused by neutrons of various energies $E_n$. Also given is the halflife* and $v$ for spontaneous fission. The rate of neutron production from spontaneous fission of all of the isotopes in a decay chain can be calculated by multiplying $\Sigma f_{i,j}$ by $v_j$ in equation 11.B.8. Approximate formulas for $v$ are as a function of the energy $E_n$ of the neutron (in MeV) causing fission are:

| | | |
|---|---|---|
| Uranium-235: | $v = 2.432 + 0.066E_n \quad (0 \le E_n \le 1)$ | (11.B.9) |
| Plutonium-239: | $v = 2.874 + 0.138E_n \quad (0 \le E_n)$ | (11.B.10) |

---

* The halflife for spontaneous fission is that which the nucleus would have if *fission* were its *only* decay mode. Since other types of decay occur with a much higher frequency, the actual halflife is much shorter.

The normalized energy spectrum of the neutrons produced by fission is given by

$$\frac{dN(E_n)}{dE_n} = \frac{2}{(\pi\varepsilon^3)^{1/2}} E_n^{1/2} \exp(-E_n/\varepsilon) \qquad (11.B.11)$$

where $N(E_n)$ is the number of neutrons with energies between $E_n$ and $(E_n + dE_n)$, and $\varepsilon$, which is equal to two-thirds of the average neutron energy, can be approximated in MeV by[3]

$$\varepsilon = 0.49 + 0.43\sqrt{v + 1} \qquad (11.B.12)$$

This spectrum peaks at $\sqrt{\varepsilon/2}$ (about 1 MeV), and has a significant tail out to energies of several MeV, which gives a mean neutron energy of about 2 MeV.

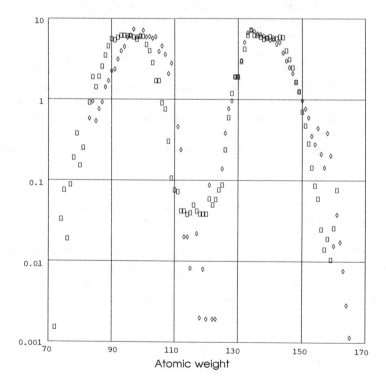

**Figure 11.B.5** The percent fission yield by atomic weight of the fission products of uranium-235 (boxes) and plutonium-239 (diamonds) exposed to a fast-reactor neutron spectrum

## Prompt Gamma Rays

About seven prompt gamma rays are emitted soon after the prompt neutrons; they carry off the excess energy and angular momentum of the excited fragments after neutron emission is no longer energetically possible. Within the accuracy of the available data, the prompt gamma-ray spectrum can be approximated (in $MeV^{-1}$) by[4]

$$\frac{dN(E_\gamma)}{dE_\gamma} = 6.7\exp(-1.05E_\gamma) + 30\exp(-3.8E_\gamma) \qquad (11.B.13)$$

over the energy range $0.3 < E_\gamma < 7$ MeV. Although this equation is for $(n,\text{fission})$ reactions, it should also apply to spontaneous fission since the spectrum is fairly

**Table 11.B.1:** The halflife for spontaneous fission (SF) and the average number $v$ of prompt neutrons released per fission as a function of the energy of the incident neutron

|  | Halflife for SF years | SF | neutrons/fission $v$ thermal | fission | 14 MeV |
|---|---|---|---|---|---|
| Uranium-232 | 8. $\times 10^{13}$ | ≈2. | - | - | - |
| Uranium-233 | 1.22 $\times 10^{17}$ | ≈2. | 2.49 | 2.6 | 4.2 |
| Uranium-234 | 2.04 $\times 10^{16}$ | ≈2. | - | 2.4 | - |
| Uranium-235 | 3.50 $\times 10^{17}$ | 1.86 | 2.42 | 2.5 | 4.4 |
| Uranium-236 | 1.95 $\times 10^{16}$ | 1.91 | - | 2.5 | - |
| Uranium-238 | 8.20 $\times 10^{15}$ | 2.01 | - | 2.4 | - |
| Neptunium-237 | 1. $\times 10^{18}$ | 2.05 | 2.7 | 2.8 | 4.7 |
| Plutonium-236 | 3.52 $\times 10^{9}$ | 2.13 | - | - | - |
| Plutonium-238 | 4.77 $\times 10^{10}$ | 2.22 | 2.92 | - | - |
| Plutonium-239 | 5.48 $\times 10^{15}$ | 2.16 | 2.88 | 3.0 | 4.8 |
| Plutonium-240 | 1.31 $\times 10^{11}$ | 2.16 | 2.26 | 2.9 | 4.9 |
| Plutonium-241 | 2.5 $\times 10^{15}$ | 2.25 | 2.93 | 3.0 | 5.0 |
| Plutonium-242 | 6.84 $\times 10^{10}$ | 2.15 | 2.18 | - | - |
| Plutonium-244 | 6.61 $\times 10^{10}$ | 2.30 | - | - | - |
| Americium-241 | 1.05 $\times 10^{14}$ | 2.27 | - | - | - |

Sources: Edgardo Browne and Richard B. Firestone, *Table of Radioactive Isotopes* (New York: John Wiley & Sons, 1986); Victoria McLane, Charles L. Dunford, and Philip F. Rose, *Neutron Cross Sections, Volume 2: Neutron Cross Section Curves* (Boston, Massachusetts: Academic Press, 1988); R.T. Perry and W.B. Wilson, "Neutron Production from (α,n) Reactions and Spontaneous Fission in $ThO_2$, $UO_2$, and $(U,Pu)O_2$ Fuels," LA-8869-MS (Los Alamos, New Mexico: Los Alamos National Laboratory, 1981).

independent of the incident neutron energy. Integration gives 7.2 photons per fission and 7.2 MeV per fission, which compare well with the measured values of 7.4 photons per fission and 7.2 MeV per fission in this energy range.[5]

## Delayed Neutrons

The fission fragments mostly relieve themselves of excess neutrons by successive beta decays, converting them into protons through the emission of electrons and neutrinos. However, some isotopes decay to an excited state that can decay by the instantaneous emission of a neutron. Since these neutron emissions immediately follow the beta decay from the parent nucleus, the rate of emission appears to follow the exponential law of radioactive decay for the parent nucleus. The energy spectrum of delayed neutrons is complex because they are emitted by many different isotopes; the average energy is about 0.45 MeV.

The number of delayed neutrons is small: for uranium-235 fissioned by thermal neutrons,[*] only about 0.016 delayed neutrons are emitted per fission. The ratio of the number of delayed neutrons to the number of prompt neutrons per fission is independent of the energy of the incident neutron. This ratio is 0.0064 for uranium-235, 0.0148 for uranium-238, and 0.0020 for plutonium-239. To predict the rate of delayed neutron emission, $v$ in the above equations should be multiplied by these ratios.

## Delayed Gamma Rays

The unstable fission fragments emit gamma rays as they beta-decay into stable isotopes. Using the observed distribution of some 80 different fission fragments and data on the gamma ray emissions from them and their many radioactive daughters, computer models have predicted precisely the observed delayed gamma-ray spectrum. Delayed gamma ray emission decreases steadily with time at a rate approximately proportional to $t^{-1.2}$, and the average energy of the gamma rays decreases with time. This spectrum varies with the isotope undergoing fission and the energy of the incident neutron, since these change the distribution of fission fragments.

For uranium-235, the delayed gamma-ray spectrum from neutron-induced fission can be approximated in $MeV^{-1}$ by[6]

$$\frac{dN(E_\gamma)}{dE_\gamma} = 5.1\exp(-1.03E_\gamma) + 3.8\exp(-1.36E_\gamma) \qquad (11.B.14)$$

This equation was fitted to measurements made using sodium iodide detectors, and is

---

[*] Thermal neutrons are neutrons that have been slowed down by successive collisions with the nuclei of some material until on average they are as likely to gain as lose energy in subsequent collisions. Their average kinetic energy is then characteristic of the temperature of the material. At a standard temperature of 293 K, they have an average energy of about 0.025 eV.

accurate to within a factor of two over a wide range of energies. Equation 11.B.14 has been normalized to give the measured value of 6.84 MeV per fission when integrated.

The normalized delayed-gamma spectrum is not very different for other isotopes, but the difference in the total energy released can be significant. For example, while the normalized delayed-gamma spectrum from the fission of uranium-238 does not vary by more than 15 percent from that of uranium-235 for gamma-ray energies less than 2 MeV, the total delayed-gamma energy is 60 percent greater. For simplicity, we assume that the delayed-gamma spectrum is given by the above equation multiplied by the ratio of the total energy released relative to uranium-235. The total gamma-ray energy is 10.8 MeV for thorium-232, 4.24 MeV for uranium-233, 10.9 MeV for uranium-238, and 6.15 MeV for plutonium-239.[7]

The total delayed gamma-ray energy from spontaneous fission should be about the same as that from neutron-induced fission of an isotope with one fewer neutron. The total gamma-ray energy from spontaneous fission of plutonium-240 should, for example, be about the same as that from the neutron-induced fission of plutonium-239.

The large number of delayed gamma-ray emissions makes the gamma-ray energy spectrum from fission appear continuous when viewed with a low-resolution detector. A high-resolution instrument would reveal the line spectra from the radioactive decay of various fission products.

The most detectable delayed-gamma emission is probably the 1.597-MeV emission from lanthanum-140, which is emitted in 5–6 percent of all fissions (fairly independently of the isotope undergoing fission).[8] We have used this line to estimate the detectability of delayed gamma rays with high-resolution detectors.

## INTERACTION OF RADIATION WITH MATTER

### Neutron Interactions

As neutrons pass through matter, they can interact in two distinct ways. Often the interaction is an "elastic" scattering in which the neutron bounces off a nucleus without giving the target nucleus any internal "excitation" energy—as if it were a small billiard ball colliding with a large one. One can use this billiard-ball model to calculate the energy that a neutron would lose in an elastic collision. Using a nonrelativistic approximation, which is valid at the neutron energies with which we are dealing, we obtain the relationship:

$$p_2 = \frac{p_1}{M + m} \left[ m \sin\theta + (M^2 - m^2 \cos^2\theta)^{1/2} \right] \qquad (11.B.15)$$

where $p_1$ is the initial momentum of a neutron with of mass $m$, and $p_2$ is its final momentum after it has scattered elastically at an angle $\theta$ from a nucleus with mass $M$. The result for the case where the scattered neutron bounces back in the direction from which it came ($\theta = 180°$) illustrates the sensitivity of the energy loss to the relative mass of the nucleus with which the neutron collides:

$$p_2 \bigg|_{\theta=180°} = p_1 \left[ \frac{M - m}{M + m} \right]^{1/2} \qquad (11.B.16)$$

If $M = m$ the final momentum is zero; if $M \gg m$, $p_2 \approx p_1$. This is why light elements are better at slowing down neutrons.

The neutron can also bounce off a nucleus inelastically and transfer energy to the nucleus's interior. This nuclear excitation will result in the later emission of one or more gamma rays by the nucleus.

Finally, the neutron can be absorbed by the nucleus, raising its atomic weight by one unit. As a result of the addition of the absorbed neutron, the new nucleus may emit a gamma ray, an alpha particle, a neutron, or it may fission. A neutron's ability to cause these results depends critically on its kinetic energy at the time it is absorbed. These prompt interactions are described by the notation "$(n,\gamma)$," "$(n,\alpha)$," "$(n,n)$," "$(n,\text{fission})$," and so on. Neutron absorption may also create an unstable nucleus that decays with a characteristic halflife by the emission of an electron (beta decay).

A term frequently used to describe the likelihood of various nuclear interactions is "cross section." The probability that an interaction will take place is expressed as proportional to an equivalent cross-sectional area centered on the target nucleus. An effective cross section can represent the probability for elastic scattering, $(n,\alpha)$ interactions, and so on. It is as if the total cross-sectional area for a collision were subdivided into areas for all of the possible interactions. The cross sections change as a function of energy.

The radius of a proton or a neutron is slightly larger than $10^{-13}$ centimeters, which corresponds to geometric cross sections of a few times $10^{-26}$ square centimeters. Because nuclear densities do not vary much (except for the very lightest nuclei), the radii of nuclei generally increase as the cube root of their masses. The unit used to measure nuclear cross sections is the "barn," which is equal to $10^{-24}$ square centimeters. (Atomic radii are much larger than nuclear radii—about $10^{-8}$ centimeters—with corresponding geometric cross sections of a few times $10^{-16}$ square centimeters.)

Of particular interest for our discussion are the total cross section for interaction and the elastic scattering cross section. If we wish to know the mean free path $l$ a neutron can proceed through a material without interacting, we can use the measured total cross section $\sigma$ and the density of nuclei in a material $N$ to calculate $l = 1/N\sigma$. Then the number of neutrons $n(d)$ in a beam originally containing $n(0)$ neutrons decreases with distance $d$ into the material as

$$n(d) = n(0) \exp(-d/l) \qquad (11.B.17)$$

The quantities $d$ and $l$ can be measured either in centimeters or in grams/cm$^2$.

Thus, for example, in a material with a density of 10 grams/cm$^2$, a mean free path of 1 centimeter would equal a mean free path of 10 grams/cm$^2$. The unit of grams/cm$^2$ is used in calculations of shielding masses.

Neutrons are particularly useful for fissile-material detection because their range in matter is quite large. Table 11.B.2 gives the mean free paths of neutrons of different energies in various materials, both in centimeters and in grams/cm$^2$. This table gives the interaction length for undergoing any kind of interaction, including elastic scattering, absorption, or fission. If the neutron undergoes elastic scattering, there will still be a neutron moving through the material in a different direction and at a lower energy. The mean free paths for absorption of the neutron (that is, until it disappears) are therefore generally much longer than those given in the tables.

The fact that neutrons can interact in so many different ways, and that the probability of these many interactions is a complicated function of the neutron energy, makes it almost impossible to estimate the emission of neutrons from the weapon models by hand. We have used the computer program TART to trace the tortuous paths of neutrons emitted by fissile material and to estimate the number of neutrons escaping from the weapon models.[9]

## Gamma Rays

Gamma rays (high-energy photons) are essentially pure energy. A gamma ray can transfer all of its energy to an atomic electron, ejecting it from its bound state as part of the atom. This is called "photoelectric absorption" and is a major cause of energy loss of low-energy gamma rays (below 1 MeV). Since photoelectric effect is most probable when the binding energy of the electrons is comparable to the energy of the absorbed photon, and high-$Z$ atoms contain electrons that are more tightly bound, photoelectric absorption occurs much more readily at high energies in high-$Z$ atoms.

If the gamma-ray energy is not totally absorbed by the ejection of the electron, the interaction is not an absorption but a "Compton" scattering, with the gamma ray proceeding in a new direction and at a lower energy after the interaction. For scattering from an electron that is unbound or whose binding energy may be neglected, the energy $E$ of the scattered gamma ray is given by

$$E = \frac{E_0}{1 + (E_0/mc^2)(1 - \cos\theta)}$$  (11.B.18)

where $E_0$ is the original energy of the gamma ray, $m$ is the electron mass, $c$ the velocity of light, and $\theta$ is the scattering angle. Compton scattering is dominant at energies of about 1 MeV in light elements.

At energies greater than 1 MeV, pair production starts playing a role in absorption and, at 100 MeV, it is the dominant method of absorption. In this interaction the gamma ray ceases to exist—it is transformed into an electron and an anti-electron

**Table 11.B.2:** Mean free paths of neutrons for various elements and isotopes

a: *grams/cm²*

| | Neutron energy eV | | | | | |
|---|---|---|---|---|---|---|
| | $10^{-1}$ | 10 | $10^3$ | $10^5$ | $10^6$ | $10^7$ |
| Hydrogen | 0.042 | 0.042 | 0.042 | 0.083 | 0.42 | 2.4 |
| Helium | 8.9 | 8.9 | 8.9 | 8.3 | 1.0 | 4.2 |
| Lithium | 0.022 | 0.20 | 1.7 | 6.7 | 8.3 | 5.6 |
| Beryllium | 2.5 | 2.5 | 2.5 | 2.8 | 4.5 | 10. |
| Boron-10 | 0.0056 | 0.067 | 0.83 | 3.3 | 7.9 | 14. |
| Boron-11 | 3.7 | 3.7 | 3.7 | 4.2 | 8.3 | 15. |
| Carbon | 4.7 | 4.7 | 4.8 | 4.8 | 7.7 | 17. |
| Nitrogen | 1.9 | 2.1 | 2.6 | 4.7 | 12. | 15. |
| Oxygen | 6.7 | 6.7 | 6.7 | 6.7 | 3.8 | 16. |
| Aluminum | 28. | 30. | 30. | 11. | 15. | 22. |
| Iron | 7.8 | 9.3 | 12. | 23. | 31. | 40. |
| Copper | 13. | 15. | 14. | 21. | 29. | 42. |
| Tungsten | 31. | 61. | 22. | 26. | 44. | 61. |
| Lead | 31. | 31. | 31. | 31. | 72. | 58. |
| Uranium-235 | 0.78 | 9.8 | 13. | 36. | 56. | 65. |
| Uranium-238 | 40. | 26. | 16. | 26. | 44. | 57. |
| Plutonium-239 | 0.44 | 1.3 | 13. | 27. | 57. | 57. |

b: *centimeters*

| | Density g/cm³ | Neutron energy eV | | | | | |
|---|---|---|---|---|---|---|---|
| | | $10^{-1}$ | 10 | $10^3$ | $10^5$ | $10^6$ | $10^7$ |
| Hydrogen *liquid* | 0.071 | 0.59 | 0.59 | 0.59 | 1.2 | 5.9 | 34. |
| Helium *liquid* | 0.125 | 71. | 71. | 71. | 66. | 8.0 | 34. |
| Lithium | 0.534 | 0.42 | 0.38 | 3.2 | 13. | 16. | 11. |
| Beryllium | 1.85 | 1.4 | 1.4 | 1.4 | 1.6 | 2.4 | 5.6 |
| Boron-10 | 2.17 | 0.0024 | 0.028 | 0.35 | 1.4 | 3.4 | 5.9 |
| Boron-11 | 2.38 | 1.6 | 1.6 | 1.6 | 1.8 | 3.6 | 6.5 |
| Carbon | 2.27 | 2.1 | 2.1 | 2.1 | 2.1 | 3.4 | 7.4 |
| Nitrogen *liquid* | 0.81 | 2.4 | 2.6 | 3.2 | 5.8 | 15. | 18. |
| Oxygen *liquid* | 1.14 | 5.9 | 5.9 | 5.9 | 5.9 | 3.3 | 14. |
| Aluminum | 2.7 | 10. | 11. | 11. | 4.2 | 5.6 | 8.3 |
| Iron | 7.9 | 0.99 | 1.2 | 1.5 | 3.0 | 3.9 | 5.1 |
| Copper | 9.0 | 1.5 | 1.7 | 0.47 | 2.4 | 3.2 | 4.7 |
| Tungsten | 19.3 | 1.6 | 3.2 | 1.1 | 1.3 | 2.3 | 3.2 |
| Lead | 11.4 | 2.8 | 2.8 | 2.8 | 2.8 | 6.4 | 5.0 |
| Uranium-235 | 18.71 | 0.041 | 0.52 | 0.69 | 1.9 | 3.0 | 3.4 |
| Uranium-238 | 18.95 | 2.1 | 1.4 | 0.84 | 1.4 | 2.3 | 3.0 |
| Plutonium-239 | 19.84 | 0.023 | 0.07 | 0.70 | 1.4 | 3.0 | 3.0 |

Source: Hildebrand and Leith, *Physical Review* **80**, 842 (1950).

(positron). Below 1.02 MeV, a gamma ray does not possess enough energy to produce a positron/electron pair (the rest mass of an electron or positron is 0.51 MeV). Pair production is greatly enhanced in the strong electric field of a high-$Z$ nucleus. At higher energies, pair production can lead to electromagnetic showers, because the resulting high-energy electrons and positrons generate high-energy gamma rays (which can cause additional pair production) as they interact with the surrounding medium.

The combined cross section of all of these interactions is called the total cross section: table 11.B.3 lists the corresponding mean free paths of gamma rays in various materials.

High-resolution gamma-ray detectors are most effective when they are used to

**Table 11.B.3:** Mean interaction free paths of gamma rays for various elements

a: *grams/cm²*

| | Photon energy MeV | | | | | | | | |
|---|---|---|---|---|---|---|---|---|---|
| | 0.1 | 0.2 | 0.4 | 0.6 | 1.0 | 2.0 | 4.0 | 6.0 | 10.0 |
| Hydrogen | 3.4 | 4.1 | 5.3 | 6.3 | 7.9 | 11 | 17 | 22 | 31 |
| Beryllium | 7.6 | 9.2 | 12 | 14 | 18 | 25 | 38 | 47 | 62 |
| Carbon | 6.7 | 8.2 | 11 | 12 | 16 | 23 | 33 | 41 | 52 |
| Aluminum | 6.2 | 8.3 | 11 | 13 | 16 | 23 | 32 | 38 | 44 |
| Iron | 2.9 | 7.2 | 11 | 13 | 17 | 24 | 30 | 33 | 34 |
| Copper | 2.3 | 6.8 | 12 | 13 | 17 | 24 | 30 | 32 | 33 |
| Tungsten | 0.24 | 1.4 | 5.7 | 9.9 | 16 | 23 | 25 | 24 | 22 |
| Lead | 0.19 | 1.1 | 4.8 | 8.8 | 15 | 22 | 24 | 23 | 20 |
| Uranium | 0.94 | 0.85 | 3.9 | 7.4 | 13 | 21 | 23 | 22 | 20 |

b: *centimeters*

| | Photon energy MeV | | | | | | | | |
|---|---|---|---|---|---|---|---|---|---|
| | 0.1 | 0.2 | 0.4 | 0.6 | 1.0 | 2.0 | 4.0 | 6.0 | 10.0 |
| Hydrogen *liquid* | 48 | 58 | 74 | 88 | 110 | 160 | 240 | 317 | 440 |
| Beryllium | 4.1 | 5.0 | 6.4 | 7.0 | 9.6 | 14 | 20 | 26 | 34 |
| Carbon | 3.0 | 3.6 | 4.7 | 5.5 | 7.0 | 10 | 15 | 18 | 23 |
| Aluminum | 2.3 | 3.1 | 4.0 | 4.8 | 6.0 | 8.6 | 12 | 14 | 16 |
| Iron | 0.37 | 0.92 | 1.4 | 1.7 | 2.1 | 3.0 | 3.9 | 4.2 | 4.3 |
| Copper | 0.26 | 0.76 | 1.2 | 1.5 | 1.9 | 2.7 | 3.4 | 3.6 | 3.7 |
| Tungsten | 0.012 | 0.073 | 0.30 | 0.51 | 0.81 | ·1.2 | 1.3 | 1.2 | 1.1 |
| Lead | 0.017 | 0.098 | 0.42 | 0.77 | 1.3 | 1.9 | 2.1 | 2.0 | 1.8 |
| Uranium | 0.050 | 0.046 | 0.21 | 0.39 | 0.71 | 1.1 | 1.2 | 1.2 | 1.0 |

Source: J.H. Hubbell, *International Journal of Applied Radiation and Isotopes*, **33**, 1269 (1982).

detect emissions at particular energies rather than the emissions at all energies. For this reason, it is relatively simple to use the mean free paths given in table 11.B.3 to estimate the gamma-ray emissions from the weapon models, because only those gamma rays that have not interacted at all with the weapon materials will reach the detector

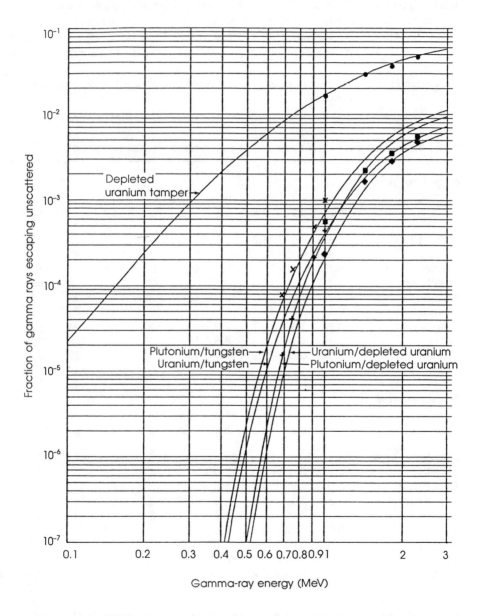

**Figure 11.B.6** The fraction of gamma rays produced in the fissile material or depleted uranium that escapes unscattered from each weapon model as a function of gamma-ray energy

undegraded in energy, and can be counted in the "full-energy peak."

The fraction of gamma rays escaping from the weapon can be given as

$$f = G \prod_k F_k \tag{11.B.19}$$

where $G$ is the self-shielding factor (that is, the fraction of gamma rays that escape from the fissile material unscattered) and $F_k$ is the fraction of gamma rays that pass through the $k$th material between the source and the detector unscattered.

For a spherical shell source of inside radius $r$ and outside radius $R$ inside a concentric spherical shell absorber with inside radius $a$ and outside radius $b$, it can be shown that $F$ is equal to[10]

$$\frac{3}{R^3 - r^3} \int_r^R \int_0^{\pi/2} x^2 \sin\theta \exp(\mu[a^2 - x^2\sin^2\theta]^{1/2} - \mu[b^2 - x^2\sin^2\theta]^{1/2})\, d\theta\, dx$$

$$\tag{11.B.20}$$

where $\mu$ is the linear attenuation coefficient of the absorber (1/mean free path). If the source is thin ($[R - r] << r$), then this can be approximated by

$$\int_0^{\pi/2} \sin\theta \exp(\mu[a^2 - \rho^2\sin^2\theta]^{1/2} - \mu[b^2 - \rho^2\sin^2\theta]^{1/2})\, d\theta \tag{11.B.21}$$

where $\rho$ is the average radius $(R + r)/2$. If the radius of the source is small compared with that of the absorber ($b >> R$) then

$$F \approx \exp(-\mu[b - a]) \tag{11.B.22}$$

Similarly, the self-shielding factor $G$ is given by

$$\frac{3}{2(R^3 - r^3)} \left[ \int_r^R \int_0^{b\cos(r/x)\,+\,\pi/2} x^2 \sin^2\theta \exp(\mu\, x\cos\theta - \mu[R^2 - x^2\sin^2\theta]^{1/2})\, d\theta\, dx\, + \right.$$

$$\left. \int_r^R \int_{b\cos(r/x)\,+\,\pi/2}^{\pi} x^2 \sin\theta \exp(\mu\, x\cos\theta - \mu[R^2 - x^2\sin^2\theta]^{1/2} + 2\mu[r^2 - x^2\sin^2\theta]^{1/2})\, d\theta\, dx \right]$$

$$\tag{11.B.23}$$

Equation 11.B.23 can be approximated by

$$G \approx \frac{1}{\beta\mu\delta r}(1 - \exp[-\beta\mu\delta r])$$  (11.B.24)

where $\delta r = (R - r)$ and $\beta$ ranges between $^4/_3$ and 4, depending on the aspect ratio of the source. For spheres, $\beta = ^4/_3$; for thin shells, $\beta = 4$. An approximate expression for $\beta$ is

$$4 - \frac{8}{3}\exp\left(-0.57\sqrt{r/\delta r}\right)$$  (11.B.25)

Equations 11.B.24 and 11.B.25 are good to within ±10 percent over a wide range of $\mu\delta r$ ($\mu\delta r > 0.01$) and $r/\delta r$ ($r/\delta r < 1,000$).

Figure 11.B.6 gives the fraction of gamma rays escaping from the depleted uranium tamper and from the fissile cores of each weapon model as a function of the gamma-ray energy as predicted by equations 11.B.19, 11.B.20, and 11.B.23. Also shown are the results of TART calculations for several energies. Note that there is good agreement between the analytical expressions and the Monte Carlo results of TART, with TART giving estimates 10–30 percent higher.

## NOTES AND REFERENCES

1. This relationship was first derived by Bateman in 1910; its derivation is well described in Evans, *The Atomic Nucleus* (New York: McGraw-Hill, 1955).

2. The values used in this study are taken from Edgardo Browne and Richard B. Firestone, *Table of Radioactive Isotopes* (New York: John Wiley & Sons, 1986); C. Michael Lederer and Virginia Shirley, *Table of Isotopes*, 7th edition (New York: John Wiley & Sons, 1978); and from an online data library maintained by the National Nuclear Data Center at Brookhaven National Laboratory.

3. G. Robert Keepin, *Physics of Nuclear Kinetics* (Reading, Massachusetts: Addison-Wesley, 1965), p.46.

4. Ibid., p.67.

5. Ibid., p.66.

6. Fit by the authors to data in figure 5-1 of Keepin, *Physics of Nuclear Kinetics*. Data for five decay times ranging from 1.7 to 1,000 seconds were integrated numerically over time for each of 17 energy groups ranging from 0.11 to 6.5 MeV.

7. Keepin, *Physics of Nuclear Kinetics*, p.136.

8. This photon is emitted in 5.74, 5.14, and 5.50 percent of uranium-235, pluton-ium-239, and uranium-238 fast fissions respectively. In some circumstances, other photon emissions may be somewhat more detectable. For example, the 3.317-MeV photon emitted from rubidium-90 in 0.83 percent of uranium-235 and 0.34 percent of plutonium-239 fast-fissions would be about as detectable after passing through 10 centimeters of uranium.

9. Ernest F. Plechaty and John R. Kimlinger, "TARTNP: A Coupled Neutron-Photon Monte Carlo Transport Code," UCRL-50400, volume 14, (Livermore, California: Lawrence Livermore National Laboratory, 4 July 1976).

10. This equation is valid for detectors located at a distance from the source and shielding materials that is large compared to their dimensions. Although equations 11.B.20 and 11.B.23 must have been derived before, we could find no solution to this problem in the standard references on radiation shielding.

Appendix 11.C

Robert Mozley
Oleg F. Prilutsky
# Radiation Detectors and Backgrounds

## RADIATION DETECTORS

Almost all radiation detectors make use of the ionization caused by charged particles as they pass through matter. As a charged particle moves past an atom, its charge can attract or repel the atomic electrons sufficiently to remove them from the atom, creating both positive ions (atoms missing electrons) and negative ions (electrons). The neutron and gamma ray have no charge (one of the reasons they are so penetrating), so their presence can be established only by detecting ionization produced by their transferring energy to charged particles.

A gamma ray, through its electromagnetic field, can give some of its energy to atomic electrons, which in turn can create ions as they move through the detector. As has been explained in appendix 11.B, a gamma ray can interact in three ways. Photoabsorption, in which the photon is totally absorbed and photoelectrons are ejected from atoms, dominates at low incident-gamma-ray energies; Compton scattering, in which the photon loses only part of its energy, producing recoil electrons, predominates at intermediate energies; and pair production, in which the photon energy goes into the creation of an electron–anti-electron (positron) pair, is most prevalent at high energies.

A neutron, on the other hand, loses energy most easily by colliding with a light nucleus, particularly that of ordinary hydrogen (a single proton), giving the recoiling nucleus some or all of its energy. This charged recoiling nucleus can then cause detectable ionization. The neutron can also be absorbed by a nucleus, and the new, unstable nucleus can decay by emitting charged particles that cause ionization.

## PRINCIPLES OF OPERATION

We shall deal with three general types of detector: scintillation counters, proportional chambers, and solid-state detectors.[1]

### Scintillation Counters

When atoms in materials are ionized or excited they often emit light when the electrons settle back in place. In most materials the light emitted is absorbed by the material before it can travel very far. But in some crystals and specially treated plastics the light can travel meters before being attenuated significantly. In scintillation counters, photons in a visible (or slightly higher frequency) spectral band are produced by fast charged particles moving in a transparent "scintillator" (organic or nonorganic crystals, plastics, or organic liquids).

Dense nonorganic crystals are the most convenient scintillator material for gamma-ray detection because the gamma-ray–absorption mean free path is shorter in these crystals than in organic scintillators. Of the nonorganic scintillators, sodium iodide, "activated" by thallium to convert the energy of ionization into light, has been used most often in gamma-ray detectors.

Plastic scintillators are frequently used for charged-particle detection. In plastic scintillators, the length of time required for the ionization energy to be converted into light can be of the order of $10^{-10}$ seconds (in comparison to $2 \times 10^{-7}$ seconds for sodium iodide).

The light emitted by the scintillator travels through the scintillator and "light pipes" to a photomultiplier tube (see figure 11.C.1). Light pipes are rods of highly polished plastic or glass that, like the scintillator itself, contain the light by internal reflection on the polished surfaces. At the end of the pipe, the light hits the photocathode of the photomultiplier. If the photocathode material is properly selected, about one in three photons from the light pipes will cause the ejection of an electron from the cathode into the vacuum of the photomultiplier. This electron is then accelerated by a strong electric field to hit another electrode from which about three electrons are emitted for every incident electron. By going through many stages a strong electric signal can be obtained from a few incident photons. In properly designed counters, the amount of light generated in the scintillator material and the size of the signal from the photomultiplier tube are proportional to the energy deposited by the detected particle in the scintillator. The amount of energy deposited that is needed to generate one photoelectron on the photocathode is about 1 keV.

## Proportional Counters

One of the oldest types of detector is the proportional counter, which consists of a fine

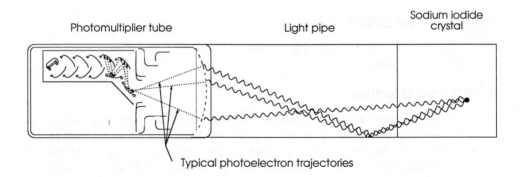

Photomultiplier tube         Light pipe       Sodium iodide crystal

Typical photoelectron trajectories

**Figure 11.C.1** : Basic elements of a scintillator detector and photomultiplier tube
Source: Glenn F. Knoll, *Radiation Detection and Measurement*
(New York: John Wiley & Sons, 1979), p.273.

wire stretched tightly through a gas (see figure 11.C.2). The fine wire is at a high positive voltage with respect to the outside wall of the gas container. If ion pairs are produced in the gas by an energetic charged particle passing through it, the electric field from the positive wire will attract the electrons towards it. As the electrons get closer to the wire they experience an increasingly strong field that causes them to gain appreciable energy between collisions with the gas molecules. At some stage they have sufficient energy to cause more ions to form, the electrons from these ions are in turn accelerated, creating more ions, and so on.

If the voltage on the wire is very high, a huge negative charge develops on the wire and its voltage drops, stopping the whole transaction. The tube would then be operating as a Geiger counter. This was one of the earliest counters used, invented by Hans Geiger in 1928. It has the advantage of giving a very large pulse when a particle is detected, but all information is lost about the number of original ions setting off the interaction.

If the voltage is lower, on the other hand, a smaller signal is produced that is proportional to the original number of ions created and hence the amount of energy deposited. This is a proportional counter. Counters similar to these are used in arrays of thousands in high-energy–physics experiments. The drift time across a proportional counter can be greater than a microsecond, so these counters are not used when a fast signal is required. Although the low density of gases makes proportional counters relatively ineffective as gamma-ray detectors, they provide the heart of one of the most useful neutron detectors.

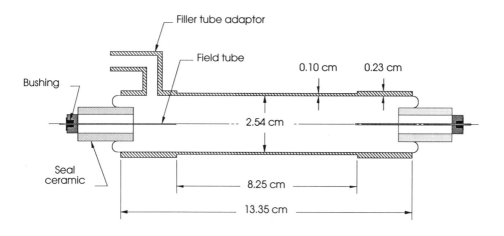

**Figure 11.C.2 :** Cross-sectional view of a specific proportional tube used in fast-neutron detection

Source: Glenn F. Knoll, *Radiation Detection and Measurement*
(New York: John Wiley & Sons, 1979), p.273

## Solid-state Detectors

One of the shortcomings of scintillation detectors is that the energy resolution obtainable is very limited. The problem is a statistical one: the conversion of energy deposited in the scintillator into light and eventually to photoelectrons in the photomultiplier is very inefficient, with only about 1,000 photoelectrons produced per MeV lost in the scintillator.

The energy required for generating a free electron in solid-state detectors is hundreds of times less than the energy required for the generation of one photoelectron from the photocathode of a photomultiplier used in a scintillation detector. This leads to the much better energy resolution of solid-state detectors. Solid-state devices have densities on the order of 2–5 g/cm³, and can be made thick enough to be efficient at stopping gamma rays.

Insulators and semiconductors are similar to gases in that something akin to ionization can be caused in them by charged particles. Atomic electrons are tightly bound, but a passing charged particle can give enough energy to an electron to move it out of its bound position in the atom. (In the case of insulators, the energy required is about 5 eV; with semiconductors such as silicon or germanium it is closer to 1 eV.) This results in a free negative electron and a positive "hole" where the electron was located, similar to the ion pair produced in a gas. Both the free electron and the hole can move (in opposite directions) in an electric field. Since electron-hole pairs can also be produced by the thermal energy present in the semiconductor and the number of such thermally produced pairs increases exponentially with temperature, these detectors must generally be cooled to be effective.

Impurities that have different electronic structures can be added to different layers of semiconductor crystals to provide electrons more easily released by thermal vibrations (*n*-type), but without leaving behind a mobile hole. This type will donate electrons to the conduction band of the crystal. In a similar way, other types of impurities can donate holes to the crystal (*p*-type). If the same crystal has an *n*-type impurity implanted on one side and a *p*-type on the other, it will be very difficult to make electrons flow from the *p*-type side, which has extra holes, to the *n*-type side, which has extra electrons. If, on the other hand, electrons are injected on the *n*-type side, they can easily be made to flow to the *p*-type side, and positive holes can be made to flow in the opposite direction. This produces a circuit element in which an applied voltage will cause current to flow in only one direction. One can apply a large voltage to such a crystal in the other direction without causing current flow (reverse bias). The ionization from a charged particle passing through the crystal would leave behind a series of holes and electrons free to migrate in the applied field and, in the act of moving, can generate a voltage pulse. Since no multiplication takes place (in contrast to the proportional counter) this pulse would be very small, and low-noise amplifiers are needed to produce a detectable signal.

Semiconductor detectors have sensitive regions of only 2 or 3 millimeters depth—just the region near the *p-n* junction. However, if a region of material of a neutral nature is located between these two regions, the entire new region can be sensitive (see figure 11.C.3). The material must, however, be completely neutral (that is, it must contain no extra mobile electrons or holes). One effective method for producing such neutral material is to use a *p*-type material compensated by the implanting of *n*-type material. The *p*-type material can be either silicon or germanium. The implanted material is lithium. The implanting is done by drifting lithium ions into the other material under an electric field. This can be done at an elevated temperature and can produce an almost perfectly compensated material over a region more than one centimeter deep.

The mobility of lithium in germanium is so great, however, that the material must be cooled to liquid nitrogen temperature to prevent the lithium from drifting to form noncompensated regions (regions in which there are different numbers of *p*-type and *n*-type atoms). This cooling must be maintained continuously, or else the detector may have to be recompensated. With lithium-drifted silicon, cooling is necessary to reduce noise when in use but is not invariably needed otherwise.

The neutral region can also be made of very-high-purity germanium, which has a major advantage over lithium-drifted germanium in that it needs cooling only when used.

Solid-state detectors can be made from many different semiconductors, but high-purity germanium detectors are used most often in high-energy (MeV region) gamma-ray spectroscopy because they can be made in large sizes, which results in high efficiency. Silicon detectors, on the other hand, are favored for detecting low-energy gamma rays because silicon has some intrinsic advantages over germanium with regard to absorption by the photoelectric effect and its lower thermally generated leakage current. But, because of the low atomic number of silicon (14) relative to that of germanium (32), its density and stopping power is much lower and it cannot be

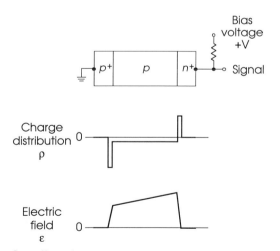

**Figure 11.C.3:** The configuration of an intrinsic germanium detector. The *p*-type central region is made of germanium of the highest available purity, and the *n*⁺-*p* junction is reverse biased.
Source: Glenn F. Knoll, *Radiation Detection and Measurement*
(New York: John Wiley & Sons, 1979), p.495

made effective for higher-energy gamma rays ( > 100 keV).

Since germanium semiconductor detectors must be cooled to liquid nitrogen temperatures when used, "cryostats" (insulated cooling environments) must be a part of the measurement system. The use of liquid nitrogen cooling for these devices complicates their design and it may make it impossible to place the detector as close to the object surveyed as one might wish. In addition, although nitrogen is inert, it must be used in a place with good air circulation, since the vented nitrogen gas might otherwise create a hazard by displacing oxygen in the nearby area. Figure 11.C.4 shows some of the complication of a germanium detector installation.

## Neutron Detection

### Scintillation Counters

Plastic scintillation counters, which contain a high concentration of hydrogen, can measure the light generated by protons (hydrogen nuclei) that are hit by neutrons passing through the scintillator (proton-recoil counters). Because the neutron gives up only part of its energy in such a collision, this does not yield a measurement of the total energy of the neutron.[*]

A proton-recoil neutron counter can be made highly efficient if it is sufficiently large. Heavy shielding around it can make it sensitive to neutrons coming from only one direction. The detector is, however, also sensitive to any other particles that are incident—in particular gamma rays that interact in the detector and produce recoil electrons which produce signals that are difficult to distinguish from those of recoil protons. As the natural gamma-ray background is about two orders of magnitude higher than the natural neutron background, the background-limited sensitivity of such detectors will be an order of magnitude worse than that of detectors not sensitive to gamma rays.[†]

---

[*] It is possible to build a complicated device that can determine the energy and direction of the incident neutron by measuring the energy of the recoil proton and the direction and energy of the scattered neutron. The efficiency of such a device is low and its size is too large for it to be portable. It could be very useful for fixed installations. W. Sailor, R. Byrd, Y. Yariv, A. Gavron, R. Hammock, M. Leitch, P. McGaughey, W. Sondheim, and J. Sunier, *A Neutron Source Imaging Detector for Nuclear Arms Treaty Verification*, LA-UR-90-581 (Los Alamos, New Mexico: Medium-Energy Physics Group, Los Alamos National Laboratory).

[†] Since the background counting rate can, in principle, be subtracted from the combined counting rate of signal and background, the ultimate sensitivity-limit imposed by the background counting rate is its statistical fluctuation, which increases as the square root of the counting rate (see below). Neutron events can, in principle, be distinguished from gamma-ray events by the use of a time-of-flight method (the velocity of neutrons with energies of several hundred keV is 10 times less than the velocity of light), but in this case the detector system would be much more complicated and its efficiency would be lower since it would be necessary to detect two neutron interactions and measure the distance and time between them.

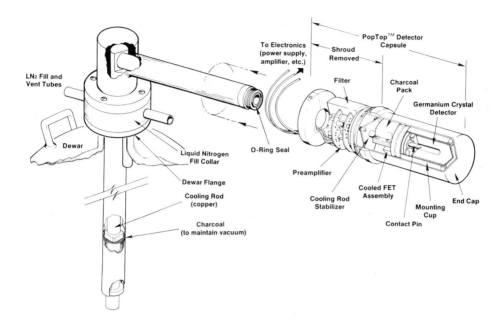

**Figure 11.C.4** : Diagram showing the location of a germanium detector within its vacuum cryostat

Source: courtesy of Sanford Wagner, EG&G Ortec, Oak Ridge, Tennessee

*Proportional Counters*

For the detection of thermal neutrons, it is possible to use a proportional counter filled with a gas that absorbs neutrons. The two most common gases used are helium-3 (cross section for thermal neutrons, 5,700 barns*) and boron trifluoride (cross section of 3,840 barns for boron-10; average cross section 770 barns for the natural mixture of boron isotopes). The reactions involved are:

$$n + \text{helium-3} \rightarrow p + \text{tritium} + 0.764 \text{ MeV}$$
$$n + \text{boron-10} \rightarrow \text{lithium-7} + \text{helium-4} + 2.79 \text{ MeV}$$

The energy released in these reactions goes largely into the motion of the resulting nuclei and results in ionization in the chamber that can be detected as a pulse. The range of all of the charged particles in the gas of the chambers is very small and all of

---

\* A barn is a unit of area equal to $10^{-24}$ square centimeters (see appendix 11.B).

the energy can be readily absorbed.

Since the energy of neutrons from fissile materials is quite high, they have to be slowed to thermal energies before a proportional counter will become sensitive to their presence. For this purpose, neutron detectors are surrounded by a "moderator" to slow the incident neutrons. The usual choice is a material having a high hydrogen content such as paraffin or polyethylene. The necessary thickness of this layer is about 10 grams/cm$^2$. The mean free path of the resulting thermal neutrons in the helium-3 and BF$_3$ gases is 10–100 centimeters, and the efficiency for thermal-neutron detection by these counters can be quite high (10 percent or more).

Proportional counters eliminate the gamma-ray background because the absorption cross section for slow neutrons in the gas is several orders of magnitude higher than the corresponding cross section for interaction of gamma rays.

The energy deposited by the thermal-neutron interactions in the gas chambers is also a great deal larger than would be deposited by any interactions by background gamma rays. In situations with a low counting rate it is therefore possible to clearly identify the incident particle as a neutron.

## Gamma-ray Detection

As noted above, two types of detector are most often used in the gamma-ray energy region around 1 MeV, which is of most interest to us: scintillation counters with nonorganic crystals and solid-state detectors with high-purity germanium crystals.

*Scintillation Counters*
The most frequently used substance is sodium iodide (NaI). Sodium iodide has some disadvantages in that it must be assembled carefully and sealed because it is hygroscopic and can be destroyed by moisture. It also has a relatively slow response, requiring about a fifth of a microsecond to emit most of the light produced. (Some of the light is released with a very long [0.15 second] halflife, which can sometimes be confusing.)

The absorption coefficients of gamma rays in sodium iodide are shown in figure 11.C.5. Photoabsorption of gamma rays is dominant for energies less than 0.2 MeV. Pair production becomes the main absorption process for energies of more than 5 MeV, which is not of much interest here. In our energy band, the main process of gamma-ray interaction with sodium iodide is Compton scattering.

In Compton scattering on electrons, gamma rays lose only part of their energy. Therefore, it is convenient to have two different definitions for the efficiency of detectors: the efficiency of *detection* (the probability of a single gamma-ray interaction with the detector material), and the efficiency for *total absorption* (also called the "photopeak" efficiency). If one is more interested in detecting specific gamma-ray

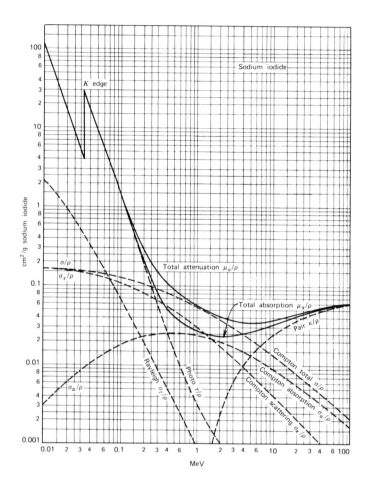

**Figure 11.C.5** : Energy dependence of the various gamma-ray absorption processes in sodium iodide

Source: R.D. Evans, *The Atomic Nucleus* (New York: McGraw-Hill, 1955), p.717

emissions than in detecting gross gamma-ray emission (as we are[*]), then the second efficiency is more important than the first.

For a gamma ray entering a crystal perpendicular to its front face, the probability of an interaction is

---

[*] The detection of particular gamma-ray energy "lines" characteristic of fissile materials is a better indicator of their presence than an increased level of gamma rays of unspecified energy. But in order to detect the lines, the total energy of the associated gamma rays must be absorbed.

$$1 - \exp[-\mu(E)x]$$

where $E$ is the energy of the gamma ray, $\mu(E)$ is called the linear attenuation coefficient, and $x$ is the thickness of the crystal. Typical values of this efficiency are shown in table 11.C.1.

The efficiency for total absorption will be less than the efficiency for interaction and will depend not only on the thickness but also on the transverse dimensions of the detector crystal. Total absorption generally requires a series of Compton scattering events followed by photoabsorption. Values of this efficiency for different crystal sizes, calculated by Monte Carlo (statistical simulation) methods, are shown in table 11.C.2. These values show that the efficiency of total energy absorption can be made quite high for sufficiently large detectors. Maximum sizes of sodium iodide crystals grown for scintillation detectors are about 70 centimeters in diameter and 50 centimeters in thickness.

The energy resolution of scintillation gamma-ray detectors depends on several factors: statistical fluctuations of the number of photoelectrons produced on the photomultiplier photocathode, uniformity of the crystals, and uniformity of light collection. Good scintillation counters using sodium iodide crystals have an energy

**Table 11.C.1**: Gamma-ray detection efficiency in sodium iodide

| Gamma-ray energy MeV | Thickness of crystal cm | | |
| --- | --- | --- | --- |
| | 5 | 10 | 20 |
| 0.45 | 0.85 | 0.98 | 1.00 |
| 1.00 | 0.66 | 0.88 | 0.99 |
| 2.62 | 0.50 | 0.75 | 0.94 |

**Table 11.C.2**: Gamma-ray total absorption efficiency for sodium iodide crystals

| Gamma-ray energy MeV | Size of crystal cm | | |
| --- | --- | --- | --- |
| | $10 \times 10 \times 10$ | $20 \times 20 \times 10$ | $30 \times 30 \times 20$ |
| 0.45 | 0.60 | 0.84 | 0.91 |
| 1.00 | 0.27 | 0.56 | 0.79 |
| 2.62 | 0.13 | 0.30 | 0.59 |

resolution of about 50 keV at 662 keV;* this "line width" increases approximately as the square root of the gamma-ray energy.

It is possible to use other types of inorganic crystal as scintillators. One of the most recently developed materials is bismuth germanate, BiGeO (BGO).

BGO is inferior to sodium iodide in some respects: the scintillation yield (and therefore energy resolution) is lower; its refractive index (a measure of the degree that light is bent, or refracted, on entering the crystal) is higher (a higher value makes light collection more difficult); and the maximum-sized crystals that have been grown are smaller. BGO is superior to sodium iodide in two other respects, however: it is nonhygroscopic, and it absorbs gamma rays better. The shorter absorption length of BGO allows the construction of more compact gamma-ray detectors but, in the absence of size limitations, the use of sodium iodide will generally be preferable.

In conclusion, scintillation detectors for gamma rays can have very high (close to 100 percent) detection efficiency, sufficiently high (close to 100 percent for large detectors) total absorption efficiency, and moderate energy resolution (about 10 percent). These detectors are certainly the best for observing sources with continuous spectra, but their energy resolution may be insufficient for spectroscopy.

### Solid-state Detectors

Good solid-state detectors using high-purity germanium crystals have an energy resolution of about 2 keV at 1.33 MeV (cobalt-60 line) in comparison with the 80-keV resolution of scintillation detectors using sodium iodide crystals at this energy.

The atomic number of germanium is less than that of iodine, and therefore photoabsorption in germanium is less than that in sodium iodide. Photoabsorption in germanium dominates the energy region below 150 keV (compared with 270 keV for sodium iodide) and for that reason the efficiency for total energy absorption in germanium will be less for a given mass of material. But, since the mass density (and therefore the electron density) of germanium is 45 percent greater than that of sodium iodide, the efficiency for a given thickness of material is substantially greater.

Commercial suppliers characterize the efficiency of germanium solid-state detectors by the ratio of the counting rate for total energy absorption of 1.33-MeV gamma rays from a source 25 centimeters distant from the detector to that for a cylindrical sodium iodide detector 3 inches (7.62 centimeters) in diameter and 3 inches thick.[2] The largest commercial detectors have a relative efficiency of about 70 percent. The absolute efficiencies for detection and efficiencies for total energy absorption for germanium crystals of different sizes are shown in tables 11.C.3 and 11.C.4.

The great advantage of solid-state over scintillation detectors is illustrated by figure 11.C.6, in which the detail of spectral lines is revealed only by the germanium detector. In the case of severe backgrounds it may be possible to distinguish fissile materials from background radiation by their characteristic line spectra.

---

* The 30-year halflife isotope cesium-137 emits a 662-keV gamma ray, which is often used as a convenient source for calibrating gamma-ray counters.

## Electronics

All detectors must be supplied with power and must be connected to devices that analyze their output signals. A single-channel analyzer is the simplest of these devices, counting the number of events in which a signal voltage is between two voltage levels. A multichannel analyzer, which counts the number of pulses in each of a large number of contiguous pulse height bins, is more commonly used. While the results of a single-

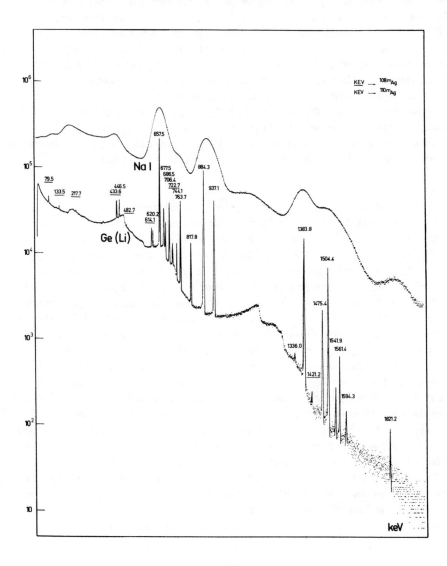

**Figure 11.C.6** : Comparative pulse-height spectra recorded using a sodium iodide scintillator and a germanium detector

Source: J.Cl. Philippot, *IEEE Transactions of Nuclear Science* NS-17, 3 ,1970, p.446

**Table 11.C.3:** Intrinsic gamma-ray detection efficiency in germanium normal incidence

| Gamma-ray energy MeV | Thickness of crystal *cm* | | |
|---|---|---|---|
| | 4 | 5 | 6 |
| 0.45 | 0.85 | 0.91 | 0.94 |
| 1.00 | 0.71 | 0.79 | 0.85 |
| 2.62 | 0.56 | 0.64 | 0.71 |

**Table 11.C.4:** Intrinsic photopeak efficiency for cylindrical germanium crystals normal incidence

| Gamma-ray energy MeV | Size of crystal *cm* | | |
|---|---|---|---|
| | 4 × 4 | 5 × 5 | 6 × 6 |
| 0.45 | 0.19 | 0.24 | 0.30 |
| 1.00 | 0.076 | 0.11 | 0.16 |
| 2.62 | 0.030 | 0.047 | 0.067 |

channel analyzer can be presented as a number, those of a multi-channel analyzer are generally presented on an oscilloscope display or in a computer printout.

The tiny output signal from a proportional counter or a solid-state device must be fed into a nearby amplifier. Often a preamplifier is built into the detector so that the signal, which sometimes is carried a large distance to the electronics package, can be much larger than the noise that may be picked up on the way.

## Cost

All of these devices are commercially available from several suppliers. The total system costs will range from about $10,000 or higher per square meter of proportional counter or plastic scintillator, to $100,000 per square meter for sodium iodide, and to $1,000,000 per square meter for solid state devices.

## COLLIMATORS

When using radiation detectors, it is generally very useful to determine the direction from which the radiation is coming. With simple hand-held instruments, this can be done by moving the detector to find the direction of movement that causes the greatest increase in the signal. This may not be possible with large equipment or there may be multiple signals that have to be distinguished from one another. Here a collimator (that is, an absorbing shield with a hole in it) may be used to allow particles to enter the detector from only one direction.

The most efficient absorbers of neutrons are light elements such as lithium and beryllium; these are more than twice as effective per gram than elements such as lead or uranium. Compared in terms of thickness of material, not absorption per gram, however, the heavy materials may be better, and they may provide a lighter shield in many configurations since the volume of material required to shield a counter increases as the cube of the outside radius of the shield. For gamma rays the heavy materials are much more effective.

In order to shield a neutron counter from 99 percent of a background of 10-MeV neutrons, a 27-centimeter-thick shield of lithium hydride would be required. Such a shield around a 5-centimeter-radius detector would weigh 89 kilograms and it would still not be very effective since the neutrons that interacted in the shield might still penetrate at lower energies. Neutron collimators must therefore generally provide for the absorption of the neutrons by introducing a thermal neutron absorber into the shield. This can be a boron compound or a layer of cadmium. If a hole is cut in such a shield, there will be a much increased sensitivity to neutrons coming from the direction of the hole, but an estimate of the efficiency of a collimator is difficult to make.

Collimation for gamma rays can be much more effective because gamma-ray detectors can measure the energy of the gamma rays and therefore distinguish those associated with particular energy-level transitions that have not scattered. Twenty kilograms of tungsten would be required to shield a 5-centimeter-radius detector to the 1-percent level from 10-MeV gamma rays. Collimators can also be introduced in an opening in a shield to restrict the angles from which gamma rays can hit the counter. Collimators reducing unwanted radiation by much more than a factor of 100 are difficult to build, however, and can become very heavy. A basic problem is scattering in the throat of the collimator, which allows the acceptance of radiation from unwanted angles.

Collimators can diminish the contribution of diffuse background radiation, thus increasing the detection sensitivity. This method of background reduction is limited by several constant factors, however: the natural radioactivity of structural elements (including the collimator and shielding) and the effects of neutrons produced in the collimator by cosmic radiation. Nevertheless, collimation allows the reduction of the background counting rate by more than an order of magnitude relative to an omnidirectional detector.

Collimator masses increase rapidly with the size of the detector, however. For an instrument with a working area of about 0.1 square meters, shielding against gamma rays with energies higher than 1 MeV might exceed 1 tonne (1,000 kilograms).

Image construction with a collimated detector may be done by means of angular or spatial scanning. The angular resolution obtained is approximately equal to the width of the collimator angular resolution response function, and the time needed for image construction equals the time of observation of each angular element multiplied by the total number of elements in the area observed.

An alternative method of image construction in gamma-ray observation can achieve both a large field of view and good angular resolution but has other limitations. This is the "coded aperture" method. In this method, a coordinate-sensitive detector is used: either a scintillation counter of large area and small thickness backed up by a large number of photomultipliers, each sensitive to light from a particular part of the crystal area, or a mosaic of small scintillation counters with separate photomultipliers. Such a detector is exposed to the gamma-ray flux from the source through a mask with a large number of holes distributed in a special pattern. The combination of the detector and coded aperture is sometimes called a gamma-ray "telescope." A bright source throws a "shadow" of the mask on the detector plane. A correlated processing of this distribution allows the determination of the angular coordinates of the source with respect to the axis of the telescope and, given large observational statistics, even permits detecting multiple or extended sources. This technique is most applicable, however, to low-energy gamma rays for which only a thin mask is required.

The angular resolution of the coded aperture system is approximately equal to the ratio of the width of the mask's holes to the distance between the mask and the detector and it may be very good (down to a few milliradians).

However, one pays for increasing the angular resolution of a telescope by the use of a coded aperture with a diminished sensitivity and an increased complexity of instrumentation. The sensitivity decrease is due to several factors: a mask that obscures about half of the working area of the detector; the finite transparency of the mask for high-energy gamma rays, which diminishes image contrast; and an increased contribution of the diffuse component of background radiation in comparison with a collimated telescope. Due to the total of all these factors, the threshold of sensitivity of a telescope with a coded aperture may be three to ten times higher (worse) than that of a collimated telescope with equal geometrical detector area.

For these reasons the use of telescopes with coded apertures as an alternative to collimated telescopes does not seem to be reasonable for remote detection of weak gamma-ray sources unless there are additional tasks for which good angular resolution is needed.

## BACKGROUNDS

To be detected, a signal must at least be larger than statistical fluctuations in the background. For the large numbers considered here, the number of counts in a one-standard-deviation fluctuation of the background counted during an interval of certain length will be equal to the square root of the total number of counts expected from an average background counting rate during that interval. If the background has no systematic trend with time and is random, the chance of an upward fluctuation occurring during any specific measurement interval that is greater than a given number of standard deviations is shown below.

The appropriate number of standard deviations depends on the type of observation. Three standard deviations for a single measurement might seem to be adequate, but

| Number of standard deviations N | Probability of a random upward fluctuation > N standard deviations |
|---|---|
| 1 | 1/6.3 |
| 2 | 1/44 |
| 3 | 1/740 |
| 4 | 1/32,000 |
| 5 | 1/3,400,000 |

consider the following situation: if a detector were making a measurement every 10 seconds, then detection of a three-standard-deviation upward random fluctuation of this size would occur once every two hours. The question then would be "what is the next step?" If the next step were to attempt another type of measurement, the three-standard-deviation criterion is adequate as a trigger of such a response. If the measurement being made is the only one possible, it had better not trigger a serious reaction very often. A less easily fulfilled criterion must be chosen. Hence, in assessing the performance of equipment, it is often wise to use a five-standard-deviation criterion. The signal detected must in that case be greater than five times the square root of the number of background counts expected in the same time interval. To know the background it must have been accurately examined in the absence of a signal. Backgrounds also have systematic fluctuations, so several measurements are generally called for.

If the background is well known and does not have to be measured at the time of the search for a signal, the time required relates only to the signal. If, on the other hand, the background must be measured and is much larger than the expected signal, the same length of time must be spent measuring background and signal. Under these conditions, if the signal is small compared to the background, because the background must be subtracted from the measurement of (signal + background), the error in the signal determination will be increased by the square root of two. To shrink the measurement error down to the level for the case where the background is well known, the background and signal measurements must each take twice as long and the total time for the measurement will therefore be four times as long.

For fissile material detection, the best situation with regard to background is that in which measurements are made in a fixed location, as with portal monitoring. This makes it possible to accurately establish the background, and the background need not be remeasured whenever a search is made for a signal.

There are many sources of unwanted signals in detectors that can simulate the signal we are interested in detecting. The one that affects the beginner most severely is electronic noise. Such noise comes from the electrical environment and can generally be removed by proper electrical shielding and grounding of the detector or its electronics. There is also, however, thermal noise generated within the electronics themselves that can simulate very weak signals; this can only be removed by the use of specially designed components or by cooling. Finally, there are real signals from particles or gamma rays hitting the detector. These signals can come from several sources: radioactive isotopes from materials in the detector, the earth's crust, in the air, or in nearby equipment or reactors, and from cosmic rays passing through the detector.

## Reducing the Background

As pointed out earlier, much of the low-energy background can be removed by collimation. Such collimators can, however, add considerable weight and/or bulk to a detector and hence reduce its portability.

A more effective alternative is often to use active shielding. This generally consists of a layer of scintillator (with associated photomultipliers) covering much of the surface of the detector. Any charged particle passing through the detector that passes through this layer of scintillator will cause a signal that can be used to reject any signal occurring in the detector at the same time with an "anticoincidence" circuit. The design has to be such that the desired interactions to be observed do not also send a charged particle through the anticoincidence detector. If neutrons or gamma rays are being observed, the anticoincidence shield can, if very thin, also cover the sensitive face of the detector.

Using such a shield to reject background neutrons or gamma rays is more complex since additional layers of material must be provided in which the neutral radiation can interact and produce the charged particles required to trigger the anticoincidence shield.

## SENSITIVITY OF A DETECTOR

The sensitivity of a detector may be defined as the lowest level of radiation it can detect in a given time interval. Given the requirement that a signal deviate from the expected background fluctuations by more than five standard deviations, a relationship between signal, background, time, and detector efficiency was derived in equation 11.4 of chapter 11. According to this relationship, the maximum detection distance $r$ is proportional to the square root of the signal $S$, the fourth root of the time $t$, detector area $A_s$, and efficiency $\varepsilon_s$, relative efficiency for detecting signal and background $\alpha$, and inversely proportional to the fourth root of the background $b$:

$$ r \propto \left[ \frac{\alpha A_s \varepsilon_s}{b} \right]^{1/4} S^{1/2} t^{1/4} \tag{11.C.1} $$

Below we consider only hand-held detectors weighing about 10 kilograms and movable detectors weighing on the order of 100 kilograms. More massive detectors would be more sensitive, of course, but their application to treaty verification would probably be limited to fixed portals.

## Neutron Detection Sensitivity

Although some neutrons are produced by the spontaneous fissioning of uranium and thorium in the soil, the only significant natural background source of neutrons is the small component in cosmic rays. There are no neutrons in the cosmic rays when they first enter the atmosphere because neutrons have a short halflife—about 10 minutes. Those present are produced by charged particles interacting in the atmosphere. Neutrons can also be produced by charged particles interacting in the neutron detector itself.

Figure 11.C.7 shows the energy spectrum of cosmic-ray neutrons at sea level and 54° north latitude. Near the air–ground interface there is an enhancement in the near-thermal energy component (not shown in the figure) due to increased production,

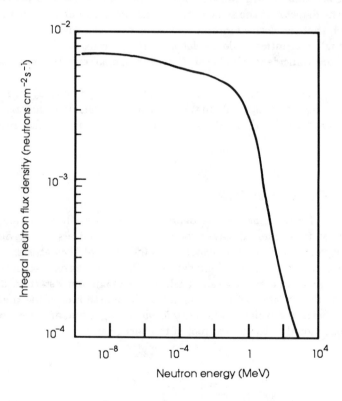

**Figure 11.C.7:** Integral energy spectrum of cosmic-ray neutrons at sea level
Source: *Exposure of the Population in the United States and Canada from Natural Background Radiation*, NCRP Report No. 94 (Bethesda, Maryland: National Council on Radiation Protection and Measurements, 1987), p.14

moderation, and backscatter. At higher altitudes and latitudes the flux increases, although the shape of the spectrum remains nearly constant. At an altitude of 700 meters the neutron flux is twice as great; at 1,600 meters it is five times greater.[3] Near the poles the neutron flux will also be several times greater, especially during the maximum phase of the solar activity cycle. Since few nuclear installations are at high altitudes or near the poles, these considerations are unlikely to be important. Since about half the background flux is in the energy range of fission neutrons, using energy-sensitive neutron detectors will not help much. Eliminating near-thermal neutrons by surrounding a detector with cadmium lowers the background flux to about 50 neutrons per square meter per second.

In this paper, we have based our estimates on two standard neutron detectors: the hand-held SNAP (Shielded Neutron Assay Probe) detector, and a larger but still movable "slab" detector. The SNAP detector shown in figure 11.C.8 consists of one or two 4-atmosphere helium-3 proportional counters 2 centimeters in diameter and 14 centimeters long, embedded in a polyethylene cylinder 7.6 centimeters in diameter and 23 centimeters long. The polyethylene is surrounded by a cadmium sheet to remove thermal neutrons, and may be used with a polyethylene shield (but no additional cadmium) to reduce the signal from one direction, as shown. The total weight of the probe is 9 kilograms, not including the electronics.

Stating an efficiency for such an instrument is dependent on the size attributed to

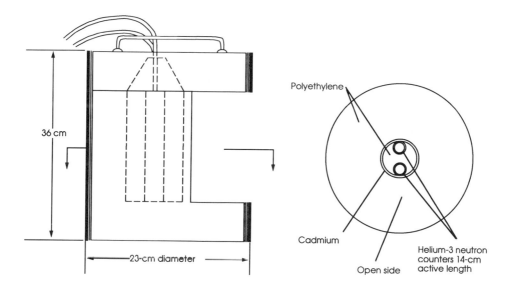

36 cm

23-cm diameter

Polyethylene

Cadmium

Open side

Helium-3 neutron counters 14-cm active length

**Figure 11.C.8**: Drawing of the portable shielded neutron assay probe (SNAP)
Source: R.H. Augustson and T.D. Reilly, *Fundamentals of Passive Nondestructive Assay of Fissionable Material*, LA-5651-M (Los Alamos, New Mexico: Los Alamos Scientific Laboratory, 1974), p.66.

it. If the area is that of the inner polyethylene cylinder (175 square centimeters), the efficiency is about 7 percent at 0.65 MeV, 6 percent at 1.1 MeV, and 4.7 percent at 2.2 MeV. If the area is that of the cross section of the apparatus shown including the shield (830 square centimeters), the efficiency is only about 1.3 percent at 1.1 MeV. The shield makes the detector directional, so that it is sensitive to neutrons from only about one-third of the entire solid angle. Because of this, the area for the signal and for the background are approximately equal. (We ignore the fact that the background is not generally uniform in angle.) Here we shall assume that $\alpha A_s \varepsilon_s = 0.0010$ square meters for the SNAP detector for fission neutrons.

The slab detector shown in figure 11.C.9 has thirteen 6-atmosphere helium-3 tubes embedded in a slab of polyethylene 50 centimeters wide, 61 centimeters long, and 15 centimeters thick. Its weight of 170 kilograms makes it transportable, but not truly portable. It is generally operated with a 10-centimeter-thick shield that leaves one open face (covered only by a thin layer of polyethylene), which increases the area subtended from 0.3 square meters to 0.56 square meters. The efficiency of the slab detector is about 14 percent (assuming an effective area of 0.305 square meters), independent of energy. With the shield, the area and efficiency for signal and background are approximately equal. Here we will assume that $\alpha A_s \varepsilon_s = 0.042$ square meters for the slab detector. Kilogram for kilogram, the slab detector is over twice as effective as the SNAP detector.

Typical values of area, efficiency, background counting rates and sensitivity of neutron detection systems are shown in table 11.C.5. The sensitivity was evaluated for

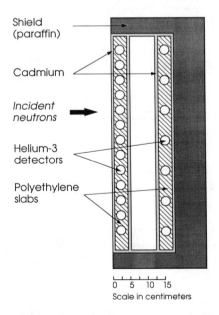

Shield (paraffin)

Cadmium

Incident neutrons

Helium-3 detectors

Polyethylene slabs

0   5   10   15
Scale in centimeters

**Figure 11.C.9** :High-efficiency slab neutron detector
Source: T. Gozani, *Active Nondestructive Assay of Nuclear Materials:*
*Principles and Applications* NUREG/CR-0602

**Table 11.C.5:** Neutron detectors

| Parameter | SNAP | Slab |
|---|---|---|
| $\alpha A_s \varepsilon_s$ $m^2$ | 0.0010 | 0.042 |
| $b\delta E$ neutrons-$m^{-2}$-$s^{-1}$ | 50 | 50 |
| $(\alpha A_s \varepsilon_s / b)^{1/4}$ $m$-$s^{1/4}$-$neutron^{-1/4}$ | 0.067 | 0.17 |
| $r_{max}$ $m$ for $S = 10^5$ n/s, $t = 60$s | 7.4 | 19 |

where $A_s$ = the effective area of the detector in $m^2$, $\varepsilon_s$ = the efficiency for detection of signal, $\alpha$ = $A_s \varepsilon_s / A_b \varepsilon_b$, where $A_b$ and $\varepsilon_b$ are effective area and efficiency for the background, $b$ = background rate per $m^2$, and $\delta E$ = the fractional energy width detected

an exposure time of 1,000 seconds and for a five-standard-deviation level of statistical significance. A maximum distance for neutron detection from sources with an intensity of about $10^5$ neutrons per second would be 7–19 meters.

## Gamma-ray Detection Sensitivity

In the absence of man-made sources of radiation (for example, a nearby nuclear reactor), the principal source of background gamma rays is trace amounts of potassium-40, thorium, uranium, and radium in the soil or in construction materials. The only radiation that causes a serious effect are the gamma rays with energies greater than a few hundred keV. Potassium-40 emits a 1.46-MeV gamma ray. Thorium, uranium, and radium are members of long decay chains (see figures 11.B.1–4 of appendix 11.B) with many decays producing gamma rays with energies of over 1 MeV. In addition to this naturally occurring radiation, there is also a contribution at gamma-ray energies up to 662 keV (the principle gamma-ray emission of cesium-137) from fallout from past atmospheric nuclear weapons tests (and, in some parts of Europe, from the Chernobyl reactor accident). Since naturally occurring radioactive isotopes are located in the greatest concentrations in certain rocks and soils, and since the distribution of fallout is not uniform, the background varies greatly with respect to location and is generally less over the oceans. A typical value for the gamma-ray flux over land at sea level is 1 $\times 10^5$ gamma rays per square meter per second ( > 50 keV).

Figure 11.C.10 shows the background measured by a germanium detector mounted one meter above typical soil. The spikes of certain gamma-ray transitions are easily visible. The flat regions between spikes are background due to less prominent transitions, gamma rays that have been degraded in energy through Compton scattering before entering the detector, and gamma rays that were not completely captured by the detector.

Only about 17 percent of the gamma rays incident on that particular detector

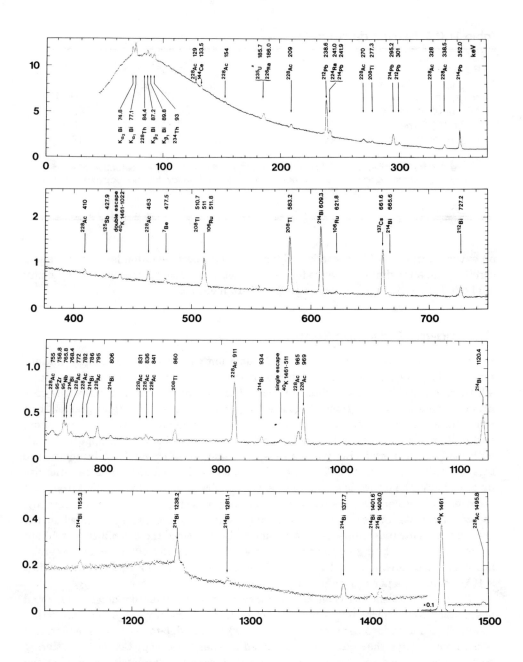

**Figure 11.C.10:** A typical gamma-ray background over land measured with a 60 cm³ germanium detector

Source: R.R. Finck, K. Lidén, and R.B.R Persson,
"In Situ Measurements of Environmental Radiation by the Use of a Ge(Li)
Spectrometer," *Nuclear Instruments & Methods*, **135**, 559 (1976)

would be completely absorbed. About 25 percent would not be detected at all. Given a perfect detector, therefore, the relative areas of the spikes in the actual spectrum would be almost six times as large. Even with a perfect detector, however, the line structure would be reduced relative to that from the source atoms because most of the radiation comes from deep in the soil, and as a result will have scattered at least once before it leaves the soil.

The spectrum associated with natural uranium is different from that which comes from the purified uranium in a weapon. In a natural uranium deposit, very strong radiation comes from elements such as bismuth that are very late in the decay chain and therefore have not had time to build up in the relatively young weapon uranium. Thorium-230, with a halflife of 80,000 years, is an intervening link in the uranium-238 decay chain.

One does not escape radioactive background by going high in the atmosphere. Although the radiation from soil and rock becomes small at high altitudes, there is still background from radioactive gases and aerosols, and the contribution from cosmic rays increases. The cosmic-ray gamma-ray flux is roughly $10^3$ gamma rays per square meter per second at sea level. This increases to a maximum of about 15 times the sea-level value at an altitude of about 15 kilometers.

The cosmic-ray gamma-ray background is only about 1 percent of the background over land at sea level. The 99 percent of cosmic-ray particles that are charged will make such large signals in the detector as to be readily rejected as gamma rays. Only the signals from charged particles hitting the edges of the detector will be indistinguishable from those of gamma rays and even these can be rejected by an anticoincidence shield around the detector. The background from atmospheric radioactivity (mostly due to radon) is about 10 percent of that from the radioactivity of the soil.

In this paper, we have considered two types of gamma-ray detector: a portable high-purity germanium detector with a 30-square-centimeter area and 4 centimeters thick, and a transportable sodium iodide detector with an area of 3,000 square centimeters and 10 centimeters thick.

In the smaller detector, the spikes would be much smaller relative to the flat region since only about 10–15 percent of the gamma rays would be completely absorbed. For the 3,000-square-centimeter sodium iodide detector, on the other hand, about 40–80 percent of the gamma rays would be completely absorbed, but the lines would be spread out because of the low energy resolution in the 1-MeV region.

An ideal detector would combine the size of the sodium iodide with the 2-keV resolution of the germanium, but our choice of a 30-square-centimeter size for the germanium is about as large as is possible for a portable detector. One of 100 square centimeters would be movable but very expensive. Because the size of an individual solid-state detector is limited, such a large-area germanium detector would probably consist of five or six separate germanium detectors with individual read-outs. The multiple detectors could be arranged to gain the aspect of a larger detector by combining the outputs. Depending on the energy of the gamma rays to be detected, the units could be arranged side by side in one layer or in stacked layers.

We assume that background has been reduced by a factor of 10 with shielding, even though such shielding adds considerably to the weight of the detectors. Surrounding the detectors with anticoincidence counters would be more effective. These would detect the charged cosmic rays and reject their signals and, even more important, would detect the showers from the gamma rays that are not completely contained in the detector and reject them as background.

Our background estimates were made using the germanium data in figure 11.C.10 for the germanium detector and the sodium iodide data shown in figure 11.C.11 for the sodium iodide detector. These are not taken at the same place, the former being from a Swedish publication while the latter is described as a typical field spectrum measured near New York City. The data are roughly consistent, however.

In table 11.C.6 we estimate the thresholds for gamma-ray detection by our two

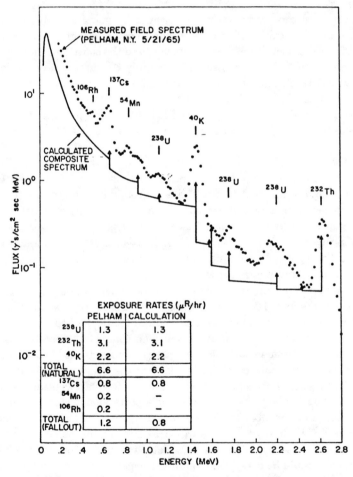

**Figure 11.C.11:** Typical terrestrial gamma-ray background measured with a sodium iodide detector 10 centimeters in diameter and 10 centimeters thick

Source: B.G. Bennett, "Estimation of Gonadal Absorbed Dose Due to Environmental Gamma Radiation," *Health Physics*, **19**, 757 (1970)

**Table 11.C.6:** Gamma-ray detectors

| | Gamma-ray energy *MeV* | | | | | |
|---|---|---|---|---|---|---|
| | Germanium | | | Sodium Iodide | | |
| | 0.66 | 1.0 | 1.6 | 0.66 | 1.0 | 1.6 |
| $\alpha A_s \varepsilon_s$ m$^2$ | $6.3\times10^{-4}$ | $4.8\times10^{-4}$ | $3.0\times10^{-5}$ | 0.21 | 0.17 | 0.13 |
| $b$ m$^{-2}$ s$^{-1}$ keV$^{-1}$ | 70 | 10 | 2 | 30 | 15 | 5 |
| $\delta E$ keV | 1.4 | 1.7 | 2.2 | 46 | 57 | 65 |
| $r_{max}$ m for $S = 10^5$ $\gamma$/s $t = 60$s | 5.6 | 8.1 | 10 | 12 | 13 | 16 |

where $A_s$ = the effective area of the detector in m$^2$, $\varepsilon_s$ = the efficiency for detection of signal, $\alpha$ = $A_s \varepsilon_s / A_b \varepsilon_b$, where $A_b$ and $\varepsilon_b$ are effective area and efficiency for the background, $b$ = background rate per m$^2$, and $\delta E$ = the energy width detected in keV

detectors. We consider energy regions; around 0.66 keV, 1.0 MeV, and 1.6 MeV. The principal line from the uranium-238 series is at 1.001 MeV and provides the best signal for passive detection when a weapon contains a depleted uranium tamper. A gamma-ray energy of 0.66 MeV is about the lower limit for penetration of gamma rays from the fissile material in the model weapons that we have considered and is the site of the strongest line from weapon-grade plutonium (it is also the site of the cesium-137 line); 1.6 MeV is in the midst of a set of neutron-induced gamma-ray emissions. The true sensitivity can be worse than shown because of the variable nature of the background, but the detection range is only weakly dependent on the background. As with the neutron detectors, the background will be incident on all sides of a detector. In the case of the gamma rays it is primarily from the lower hemisphere but with appreciable components in a horizontal direction. We assume the presence of shielding (lead) reduces this background by about a factor of 10 and use the same area for both background and signal. The efficiency for detecting background is greater than for detecting signal because the signal must be completely absorbed in the detector.

Somewhat surprisingly, we find that a 1.6-MeV gamma-ray source can be detected at about the same distance as a fission-spectrum neutron source of equal strength (see table 11.C.5). Compared to neutron detectors, the better energy resolution and efficiency of gamma-ray detectors just compensate for the higher gamma-ray background.

## NOTES AND REFERENCES

1. A more detailed discussion of radiation detection can be found in Glenn F. Knoll, *Radiation Detection and Measurement* (New York: John Wiley & Sons, 1979).

2. Ibid.

3. *Exposure of the Population in the United States and Canada from Natural Background Radiation*, NCRP Report 94 (Bethesda, Maryland: National Council on Radiation Protection and Measurements, 1987), p.14.

*Robert Mozley*
# Particle Sources and Radiography

The transmission of any kind of particle through an object can be used to give information about the interior of the object, just as a medical x-ray uses the absorption of photons to indicate the location of bones and density variations in the body. To yield the most information, the particles should be sufficiently penetrating that there is a detectable flux through the most absorptive part of the object under inspection. The characteristics of various beams will be examined here to understand which may be best for this purpose. The availability of sources of this radiation will also be considered.

The projected material densities* of the nuclear-weapon models described in this report are about 150 g/cm² at their thickest points, with over 90 percent of this due to heavy metals such as tungsten, uranium, and plutonium. For comparison, a conventional warhead of about the same size would be less than 100 g/cm² thick, with nearly all of this due to light elements such as nitrogen and oxygen. Because the fraction of particles passing through a material decreases exponentially with increasing thickness, probe particles should have average ranges of not much less than 10 g/cm², and not much more than 100 g/cm², in the materials of interest.

As we are interested in detecting the presence of fissile materials, the second criterion for probe particles is that they must discriminate between uranium or plutonium and the elements found in permitted objects. In other words, the particles should be absorbed either less or more strongly by uranium or plutonium than by common materials.

It is important to understand how much absorption information is needed to detect the fissionable material in a complicated object. A similar problem is faced in computer-aided tomography when applied to medical examination: how does one use a series of transmission measurements through an object to determine the size and location of internal parts, and how can it be done with a minimum of radiation exposure to the object?

## ALTERNATIVE BEAMS FOR RADIOGRAPHY

In principle, there are several possible choices of probe particles for use in radiography: protons and heavier nuclei, electrons, muons,† neutrons, and photons. For radiographic purposes, protons and heavier nuclei can be considered as similar but inferior to neutrons. They would lose energy due to the ionization caused by the

---

* The mass of material in a column of cross section 1 square centimeter passing through the object.

† A muon is a short-lived (2 microsecond) massive relative of the electron that is produced in the atmosphere by high-energy cosmic rays. It can also be produced with a high-energy particle accelerator.

charges that they carry, and higher-energy beams (which would cause more radiation damage) would generally be required. For the same reasons, electrons, which, like photons, do not interact strongly with nuclei (only electrically), would cause more radiation damage than photons because they are charged. Muons do not interact with nuclei and, because they are 200 times heavier than electrons, lose much less energy than high-energy electrons in collisions with atomic electrons and nuclei. Hence high-energy muons could in principle be used effectively for examining thick objects. However, the cost of producing a useful muon beam and of analyzing it is too great for the technique to be considered practical. Therefore we shall concentrate here on the use of neutrons and photons.

## Neutron Beams

Neutrons possess no charge and, as a result, their only interaction is by nuclear scattering. At low energies (a few electron volts) the cross section (a measure of the likelihood of nuclear interaction—see appendix 11.B) may vary rapidly with energy; at high energies (a few hundred MeV) the cross section is relatively constant. Figure 11.D.1 shows total neutron cross sections for aluminum, tungsten, and uranium-235. Table 11.B.2 of appendix 11.B lists the mean free paths of neutrons in various materials at a series of energies, indicating that strong absorption would take place at low energies and that neutrons with energies of 10 MeV or higher are of greatest interest for transmission measurements.

To show the effect of the mean free path, consider the task of examining a space-launch cargo to see if it has a nuclear warhead concealed in it. Here we assume a payload weighing about 20 tonnes and ask whether a weapon can be concealed in the interior. We consider a simplified case of a 20-tonne sphere that is nominally solid aluminum with a density of 2.7 g/cm$^3$ (certainly an extreme example), a radius of 120 centimeters and a maximum thickness of 650 g/cm$^2$. A 10-MeV neutron would have a mean free path in aluminum of 22 g/cm$^2$; hence the maximum thickness would be $650/22 \approx 30$ mean free paths, and the fraction penetrating the load without a collision would be about exp(–30) or $10^{-13}$. For 100-MeV neutrons, this fraction would be $2 \times 10^{-6}$. (The neutron mean free path becomes longer at higher energies.) The energy deposited by the large number of incident neutrons needed to allow a single one to penetrate the diameter of the sphere is $8 \times 10^{13}$ MeV for 10-MeV neutrons and $5 \times 10^7$ MeV for 100-MeV neutrons. About two-thirds of this energy would be deposited in the first mean free path.

The large beam attenuation seen in this example also reveals a difficulty with transmission techniques. Even if the radiation intensity required does not cause damage to the material examined, any very large attenuation implies that a very unlikely scattering path could become the source of most of the particles detected. For example, if there were any infinitesimal tubular holes or partial holes parallel to the beam direction, of such small size as not to be resolved, they could dominate the actual transmission.

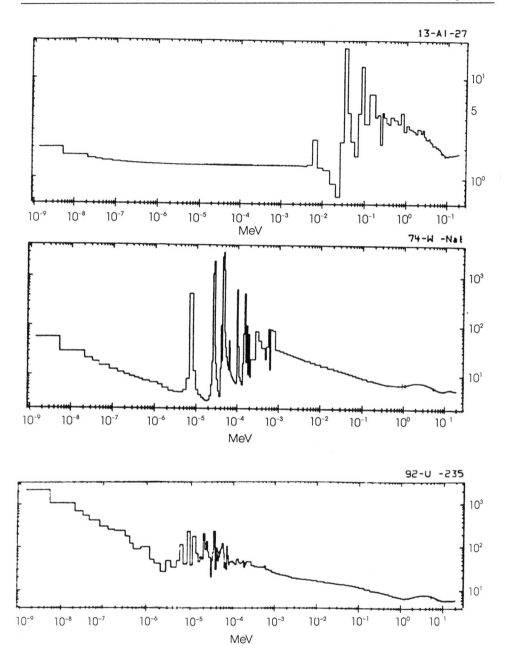

**Figure 11.D.1:** Total cross-sections (barns ($10^{-24}$ cm$^2$)) for aluminum, tungsten, and uranium-235 as a function of neutron energy

Source: E.F. Plecnaty, D.E. Cullen, R.J. Howerton, and J.R. Kimlinger, *Tabular and Graphic Representation of 175 Neutron Group Constants*, UCRL-50400, (University of California, Livermore National Laboratory, and Department of Energy, 1986).

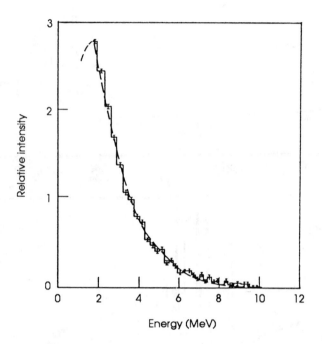

**Figure 11.D.2:** Neutron energy spectrum from the spontaneous fission of californium-252
Source: E.A. Lorch, *International Journal of Applied Radiation and Isotopes* **24**, 585 (1973)

In making estimates of radiation damage to sensitive electronics, the fact that neutrons produce possibly 10 times as much damage per unit energy deposited as photons must be taken into account. This is due to the more concentrated ionization taking place along the track of slower heavy particles (protons or heavier nuclei) recoiling from collision with a neutron relative to that deposited by the faster electrons created by the Compton scattering of high-energy photons.[1]

*Neutron Sources*
For low-energy neutrons, radioisotope sources are available. The only ones that emit neutrons directly as a result of radioactive decay are those that undergo spontaneous fission. Although uranium, thorium, and plutonium isotopes do this, their rates of decay are low; the most useful source is californium-252, which has a halflife of 1.8 years and decays 3 percent of the time by spontaneous fission. Californium sources with intensities of up to $10^{10}$ neutrons per second are available.

Figure 11.D.2 shows the energy spectrum of the neutrons emitted by such a source.

A problem with any source is the background that it produces of unwanted radiation, generally gamma rays. Although californium emits several gamma rays during each fission, many more are emitted during the 30-times more frequent alpha decays of californium. Furthermore, because of the penetrating nature of neutron radiation, a californium-252 source cannot be pulsed except by inserting a massive absorber in a channel in the shielding through which the neutrons pass; this precludes fast pulsing.

The only other radioisotope neutron sources are those in which alpha or gamma emitters are combined with other elements that release neutrons when bombarded with alpha particles or gamma rays through $(\alpha,n)$ and $(\gamma,n)$ reactions. Examples of such reactions are the following:

$$\alpha + \text{beryllium-9} \rightarrow \text{carbon-12} + n$$
$$\gamma + \text{beryllium-9} \rightarrow \text{beryllium-8} + n$$

The alpha and gamma sources are quite different in character because of the very short range of alpha particles. A 5-MeV alpha particle has a range of only about 20 microns (the thickness of a human hair) in silicon. The alpha-source element and the neutron-producing element of the neutron source must therefore be intimately mixed. However, the process of alpha emission and neutron production is almost invariably accompanied by the emission of gamma rays. This will create a neutron background for the neutron detector, one that can be partly eliminated by interposing gamma-ray shielding in the channel through which neutrons pass. The hazard from the sources will be primarily from the neutrons that are emitted and are very difficult to shield. The alpha emitters pose no direct hazard unless they are ingested or inhaled. Gozani[2] compares various $(\alpha,n)$ sources. Spectra from four sources are shown in figure 11.D.3.

Photoneutron $(\gamma,n)$ sources can be constructed with the neutron emitter separate from the gamma source, which allows the neutron source to be pulsed or stopped by interposing a shield between the gamma emitter and the target. There remains, however, the problem of shielding the very penetrating gamma-rays.

Gozani also lists the characteristics of many photoneutron sources, and figure 11.D.4 shows the neutron spectra expected from several of these sources. Photoneutron sources show a more peaked structure than $(\alpha,n)$ sources; this is in large part due to the alpha particles having a very short range, for even though the alpha particles may be emitted at a single energy there will be a broad energy width of alphas hitting the neutron-emitting target.

A powerful photoneutron source such as antimony-124/beryllium, which produces 24-keV neutrons, can be very hard to shield. A source producing $10^8$ neutrons per second would require 20 curies of antimony-124 surrounded by 10 centimeters of beryllium. Over a tonne of shielding would be required to bring the radiation level

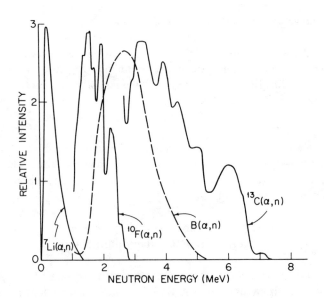

**Figure 11.D.3:** Neutron energy spectra from various (α,n) sources
Source: Lithium-7, K.W Geiger and L.K. Van Der Zwan, *Health Physics*, **21**, 120 (1971), remainder from Lorch, 1973.

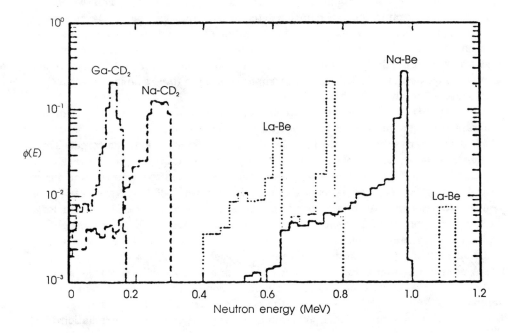

**Figure 11.D.4:** Photoneutron energy spectra from various sources
Source: T. Gozani, *Active Nondestructive Assay of Nuclear Materials: Principles and Applications*, NUREG/CR-0602 (Washington DC: US Nuclear Regulatory Commission, 1981)

down to 30 millirads per hour at the surface.* The size of the beryllium target would also make difficult the production of a narrow neutron beam.

Accelerator sources, although generally more expensive than radioisotope sources, have the great advantage that they produce almost no radiation when they are turned off. Compact and reasonably inexpensive Cockcroft-Walton D–T sources are available producing up to $10^{11}$ 14.3-MeV neutrons per second.[3] For lower energies, a moderator can be used to slow down the 14-MeV neutrons, although this will not completely eliminate the higher-energy neutrons. Higher-energy beams or beams of a lower but more specific energy can be produced using an electron, proton, or deuteron (d) beams from an accelerator to produce $(\gamma,n)$, $(p,n)$, or $(d,n)$ reactions.[4]

A neutron source can be very dangerous when operating since the neutrons are very penetrating and can bounce around corners. The sources described here produce neutrons in all directions; unwanted neutrons must be removed by shielding. A point source of $10^{11}$ neutrons per second would require about a tonne of water shielding (approximating the inside radius of the spherical shield as zero) to bring the radiation at the surface of the shielding down to a level of 20 neutrons $cm^{-2}$ $s^{-1}$ (a level assumed safe for 1-MeV neutrons).[5]

## Gamma-ray and X-ray Beams

The absorption of photons can be understood as a combination of three effects. The photoelectric effect dominates at energies below 0.1 MeV. The Compton effect is most important in the 1-MeV region, while electron–positron pair production, with a threshold at about 1 MeV, is the main absorption mechanism in the high energy region.

The photoelectric effect for high-Z (high atomic number) materials causes such strong absorption at energies lower than 100 keV that photons below this energy are not useful for examining thick materials. The Compton effect is absorption by scattering from the electrons of the material; the electron recoils and the photon loses energy and goes off in a different direction. The fact that the electrons are bound to the atom is only important at low energies. As a result, absorption is proportional to the mass of material as long as the proton–neutron ratio of the nuclei of the atoms making up the material remains constant (deviation from this constant ratio occurs only for hydrogen and the heavy elements). Pair production, on the other hand, varies as $Z^2$. This implies that, at high photon energies, it may be possible to distinguish between elements of equal density, separating uranium and plutonium from elements such as tungsten and lead.

Table 11.B.3 of appendix 11.B gives photon mean free paths for selected elements and energies. One can see the effect of the pair production starting to show itself at 10 MeV, where the uranium has a slightly smaller absorption length

---

* 500 millirads is the recommended maximum *annual* accumulated radiation dose for nonradiation workers.

than tungsten, which is more dense. These tables do not give an accurate picture of the penetration of photons through thick targets, however, since they give the mean path length before an interaction, not complete absorption. Compton scattering results in a photon of lower energy. As a result, a beam of 10-MeV photons after passing through 10 absorption lengths of lead has about a four times greater energy flux than would be calculated if the absorption mean free path was assumed equal to the interaction mean free path alone.

### Photon Sources

There is a wide choice of gamma-ray sources: Gozani[6] lists a number. Some of the less intense ones are used for calibrating gamma-ray detectors. Some have sufficient intensity for transmission measurements, but none produces gammas with energies above 2 MeV.

The best source of higher-energy photons therefore is a beam of high-energy electrons colliding with a target (generally of a heavy metal). The x-ray beam produced contains a broad continuum of energies—a bremsstrahlung spectrum. The energy distribution is approximately constant per energy interval of the x-rays, extending from zero to the energy of the electron beam. Expressed in terms of numbers of photons per energy interval, the number of x-rays produced per unit x-ray–energy varies inversely with the energy of the x-rays produced (see figure 11.D.5).

It would be desirable to use monochromatic x-ray beams together with energy-sensitive detectors in order to distinguish those x-rays that have not been scattered.

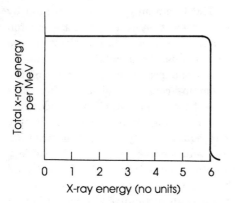
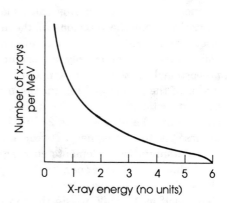

**Figure 11.D.5:** Bremsstrahlung photon energy spectrum (approximate)

There are methods of producing almost monochromatic beams using collimated beams from crystal targets[7] or by scattering a laser beam from a beam of high-energy electrons.[8] Both of these techniques require very accurate collimation of the incoming electron beam and the outgoing x-ray beam, and the radiation from the collimators must be kept very small. The alignment is so difficult that these beams are unsuitable for operational use except in physics experiments. Moreover, inside a very short distance in the material examined, the monochromatic beam would be degraded by Compton scattering into a broad spectrum.

Since monochromatic beams are not practical, the best way of ensuring that the beam has not been scattered is to measure the energy of the transmitted photons and select only the high-energy portion of the transmitted spectrum. The lower energy part of the transmitted beam will be dominated by lower-energy photons resulting from the scattering of the showers made by higher-energy photons.

High-energy x-ray beams are generally produced by linear accelerators, commercial units being available with a peak x-ray energy of 15 MeV.[9] One version is presently the tool of choice for medical radiation therapy. These machines generally use thick targets in which multiple interactions take place, thus enhancing the low-energy end of the spectrum. The opposite effect—removing the low-energy end of the spectrum without depleting the high-energy portion—can be accomplished by passing the beam through some low-Z material that will selectively absorb low-energy photons and "harden" the beam. The use of a "hardener" is complicated by reradiation from electrons produced in the absorption process. As a result such hardeners usually operate in a strong magnetic field with well-designed collimation.

To estimate the energy deposited per x-ray traversing the object examined, we again assume that the object to be radiographed is a 20-tonne aluminum sphere with density 2.7, radius of 120 centimeters and a maximum thickness of 650 g/cm². If 10-MeV photons are used as typical of those from the upper part of a 15-MeV peak energy bremsstrahlung spectrum, one x-ray in 3 million would penetrate the center. In a continuous spectrum with a 15-MeV peak energy there will be about 15 MeV in the incident x-ray beam for every photon between 5 and 15 MeV. The total energy put into the system in order to get one photon through would therefore be about $4.7 \times 10^7$ MeV. For a 6-MeV peak-energy bremsstrahlung beam, the corresponding number is $3.6 \times 10^9$; and at 3 MeV it is $5.7 \times 10^{12}$ MeV.* As with neutron beams, about two-thirds of this energy would be deposited in the first interaction length.

## RADIOGRAPHY

The type of examination being considered here gives information on the total projected mass in the path of a beam of particles or x-rays. A great deal of

---

* Also, at low energy, multiple Compton scattering in the electron target would spread the x-ray beam.

information can be obtained from a single projection. An effective system for such radiography is available commercially.[10] It comprises a linear accelerator producing a fan beam of x-rays that is projected at a vertical line of sodium iodide scintillation counters. The object to be inspected is pulled between the x-ray source and the detectors and the attenuation produced in a series of slices is observed. By combining the slices (with the aid of a computer) an image can be produced (see figure 11.2a).

Some sort of stereoscopic imaging is required to obtain information on the location of materials in the object examined. Two projected views at different angles give depth information but there are ambiguities that cannot be resolved, as illustrated in figure 11.D.6. The addition of a third view can resolve that particular ambiguity, but additional more complex ones remain. To avoid these problems, multi-angle viewing is used.

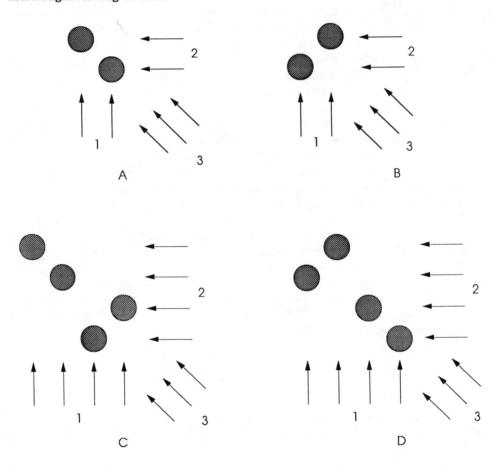

**Figure 11.D.6:** Projections (1) and (2) give the same profile of A and B; projection (3) resolves the ambiguity. Projections (1), (2), and (3) give the same profile of C and D; a fourth projection is needed to resolve the ambiguity

## Multi-angle Viewing

A tomographic scan is an example of multi-angle viewing. Such a scan views a slice of an object (*tomos* means "section" in Greek). In medical tomography, a fan beam of x-rays penetrates a slice of the body being examined. Detectors on the far side measure the intensity of the beam that emerges. The beam source and detectors are then rotated slightly around an axis perpendicular to the slice, and the procedure is repeated. This may take place a few hundred times to give a few hundred views of the slice. The data are then analyzed with a computer to give a portrait of the interior of the slice. As viewed by doctors using this technique, the information is presented as an axial view through the thin slice. To examine a large solid, many slices are examined. Tomography is therefore well suited to the examination of a long cylindrical object such as a typical space payload for a shuttle orbiter. Other projections could also be made, however; the source could for example be moved to many points on a spherical surface rather than the circumference of a slice.

The resolution obtained in tomography is determined by the sizes of the beam source and detectors on the far side. If the beam source is a point on the circumference of the object being examined, the resolution will be about half the width of the individual detectors. The best depth information would come from having views from all parts of the circumference. In the cases of neutron or photon beams, where there would be a great deal of absorption near the points of entry with possibly as many as one million particles incident for every one penetrating the thickest part of the object examined, moving the source around the circumference would also minimize radiation damage near the surface.

Since it is desirable to expose the object to as little radiation as possible, it is necessary to understand the relationship between the accuracy of density measurements and the number of particle trajectories through a resolvable part of the object. What follows is not a proof of the relationship but a plausibility argument that indicates that the accuracy of the density measurement in a region is related primarily to the number of trajectories passing through the region.

Consider a particular small volume in the interior of the slice. There is a group of particle trajectories that have in common the fact that all pass through this volume. If the density of the material in the selected volume is increased, all trajectories penetrating it will be affected and the number passing through reduced. Other interior regions will have only a few of their trajectories affected, those that they share with the subregion that has the increased absorption.

More explicitly, for a small volume with thickness $t$ along the beam direction irradiated with particles of mean free path $L$, the ratio of the number penetrating the volume without interacting, $N_p$, to the number incident is $\exp(-t/L)$. Differentiating, we obtain $dt/t = -(L/t) \times dN_p/N_p$.

The precision of the density measurement depends both on the fractional absorption taking place in the volume cell and on the number of particle trajectories passing through the entire object. Since the precision of a count is effectively $(N_p)^{-1/2}$, one would expect that, for a volume with dimensions on the order of a mean free path for the beam particle (thickness $t$ equal to $L$), 1,000 trajectories that penetrate both the specified volume and the entire object would give about a 3-percent result.

This description considers a hypothetical increase or decrease of the absorption of a region as isolated. A similar change in the trajectories passing through it could be caused by absorbers or voids in other parts of the object so that all the special trajectories passing through that particular region would be affected. This would indeed cause the same rise or decrease but, to do this, the amount of material introduced or removed at each of the other volumes would have to be the same as that required for the special location. If 10 percent extra absorption were to be produced for the trajectories passing through a 10-square-centimeter region in a particular slice, one could increase the absorption of that one special region or increase it for possibly 10–50 other locations intercepting the same special trajectories. However, the additional absorption associated with the second approach would affect many other trajectories and therefore the whole optimization procedure, and a radical change in the calculated density would be introduced for the other regions. It is here that the need for many different beam-incidence angles can be seen. The more angles measured, the more locations must have material added to them to simulate a change in the special zone.

## Radiation Doses

In order to understand whether detailed examination by a tomographic scan might cause radiation damage to parts of the material examined, we consider once again the example of an examination with the 15-MeV peak-energy x-ray beam of the hypothetical 120-centimeter-radius solid aluminum payload. Our calculation illustrates the factors that must be taken into account in any estimate of radiation doses.

For the examination of a circular disk of diameter $D$, $\exp(D/L)$ photons of mean free path $L$ must be incident normal to the circumference of the disk for every one that penetrates through the center of the disk to the other side without interacting. We assume that the beam illuminates all points on the circumference uniformly and that the number of particles incident at different angles relative to the perpendicular would be programmed to allow the same number to penetrate at each angle. (A much smaller number of incident particles would be required for trajectories that pass through only the edges of the object.) For 10-MeV photons and the maximum-thickness slice, this optimization results in the total energy deposited in the disk being about 4.8 times less than if all the photons had to pass through the center of the disk. There will be a similar increase in transmission as the scan goes to cross sections of the sphere with smaller diameters.

The average number incident for each one passing through is then $\exp(D/L)/F$, where $F$ is the reduction factor of 4.8. It is assumed that the interior region to be observed is divided into cubes $d^3$ cm$^3$ in volume and the number of tracks that should pass through a specific cube in one plane and leave the sphere is $N$. If $N$

tracks pass through the four sides of the cube that are perpendicular to the disk, the average number passing out through each of the sides is $N/4$.

If the disk being examined is $D$ centimeters in diameter and $d$ centimeters thick, the circumference can be considered as occupied by the outsides of $\pi D/d$ boxes $d^2$ square centimeters in area through each of which $N/4$ particles must exit if the disk is to be filled with a uniform number of penetrating tracks. Therefore $(N/4) \times (\pi D/d) \times \exp(D/L)/F$ photons must be incident on the circumference.

An estimate of the radiation dose received by the surface region of the sphere can be made by assuming that two-thirds of the energy of these photons will be deposited in the first absorption length. Since a bremsstrahlung spectrum is assumed, and since the lower energy portion of the spectrum has a shorter interaction length, we assume that the maximum dose would be about twice that expected from the total number of high-energy photons incident.

The energy deposited in an annulus of the disk $L$ centimeters deep is therefore equal to

$$(\pi/3) \times E \times N \times (D/d) \times \exp(D/L)/F$$

where $E$ is the energy of the photon in MeV. This will be deposited in $\pi DdL$ cubic centimeters of material. The amount deposited per cubic centimeter is then $(E \times N \times \exp(D/L) \div (3 \times d^2 \times L \times F)$ MeV/cm$^3$. In our example, $D = 240$, $d = 10$, $E = 15$, $L = 16.2$, $N = 1,000$, and $F = 5$. This results in an energy deposition density of $2 \times 10^6$ MeV/cm$^3$ or $7 \times 10^5$ MeV/gram. One MeV/gram is equal to $1.6 \times 10^{-8}$ rad, so the maximum exposure will be about 0.01 rads. This is insufficient to have any effect on sensitive electronics.[11] In the case of a space-launch payload examination, it is worthwhile to note that the daily radiation dose to be expected in low earth orbit is about 0.4 rads.[12] For lower-energy photons, the absorption length is much smaller so the radiation exposure increases exponentially. The same calculation for a 3-MeV bremsstrahlung beam would give a radiation exposure of 2,000 rads.

For neutrons, the situation would be much worse. For 10-MeV neutrons, the energy deposited would lead to a radiation exposure of $3 \times 10^4$ rads even without multiplying by the probably appropriate factor of 10 for the greater damage per rad of neutron dose. Even using 100-MeV neutrons with their greater absorption length, the radiation exposure would be about 20 rads or, with the factor of 10, equivalent to 200 rads from an x-ray beam. Neutrons might, on the other hand, be useful for examining a less absorbing object.

If the multi-angle viewing of a tomographic scan were replaced with the two views of normal stereoscopic viewing, the total amount of radiation exposure could undoubtedly be much reduced but the amount of information given would also be much reduced.

## Effects of Poor Resolution

In the case of large attenuation, other ways for the probe particles to reach the detector without passing through the absorber may become significant. Particles can scatter around the object in numbers that might be comparable to those that pass through, or might pass through small holes or cracks in the object itself. In the above examples we have considered attenuations of over one million. If our area of resolution is 10 × 10 centimeters, a 0.1-millimeter hole would double the transmission, and a 1-millimeter hole would increase the transmission by a factor of 100. There may be some way of analyzing the pattern of transmitted particles to determine that all of the penetration is from one direction, but the analysis would be much more complicated. It will be difficult therefore to use interacting beams such as neutrons or photons beyond attenuations of $10^4$ unless very high resolution is achieved. The type of absorption calculations done above apply best to continuous shields (for example, hiding fissionable material in a tank of hydrocarbons).

## CONCLUSION

The only two probe particles worth considering for ordinary applications are high-energy photons and neutrons. Except for muons, which are very expensive to produce and cannot be made with portable equipment, other particles are not as effective. Table 11.8 in chapter 11 gives the ratio of the range of neutrons and of photons in carbon, aluminum, iron, tungsten, and lead to that in weapon-grade uranium (WgU). The energies of the particles were chosen so that their ranges or mean free paths in uranium and common materials are 4–50 g/cm². A value of 1.0 means that, centimeter for centimeter, the particles interact equally in the two materials.

The table shows that discrimination between WgU and most other elements is possible using either photons or neutrons but that for tungsten the difference is very small, except for low-energy beams that do not have much power of penetration.

Hydrogenous materials, such as plastics, or materials containing lithium or boron, would absorb thermal neutrons as efficiently as weapon-grade uranium or weapon-grade plutonium, but hydrogen, lithium, and boron absorb far fewer gamma rays per centimeter than WgU.

An examination of mean free paths above 10 MeV (see tables 11.B.2 and 11.B.3 in appendix 11.B) shows that the penetrabilities of neutrons and photons are to some extent complementary, neutrons having longer mean free paths in the high-Z elements and photons in the low-Z elements. Neutron beams of 14 MeV would be very much cheaper than electron-beam x-ray sources of the same energy due to the exothermic nature of the D–T interaction. (Most of the energy of the neutron comes from the nuclear reaction, not the accelerator.) In the region above 14 MeV, the electron accelerator sources become cheaper.

The major advantage of high-energy x-rays over lower energy ones would appear to be their ability to distinguish between different heavy elements and the fact that they would be affected less by Compton scattering. Commercial sources between 2–15 MeV are available.

Muon beams, although very expensive and in some ways hard to direct at the object being studied, would be superior to all others if great penetrability were required in a laboratory setting.

## NOTES AND REFERENCES

1. See, for example, G.D. Messenger and M.S. Ash, *The Effects of Radiation Damage on Electronic Systems*, (New York: Van Nostrand Reinhold, 1986).

2. T. Gozani, *Active Nondestructive Assay of Nuclear Materials: Principles and Applications*, NUREG/CR-0602 (Washington DC: US Nuclear Regulatory Commission, 1981), p.82.

3. Kaman Sciences Corporation, Colorado Springs, Colorado produces several of these devices in which deuterons and tritons are accelerated before striking a target containing deuterium and tritium. Model A-711, which cost about \$110,000 in 1989, produces $10^{11}$ 14.3-MeV neutrons per second continuously. Model-801, which cost about \$35,000 in 1989, produces $10^8$ 14.3-MeV neutrons in a 3.5-microsecond pulse with pulses 50 percent as intense at repetition rates of 10 pulses per second. These models can also be supplied with deuterium targets producing 2.5-MeV neutrons via the reaction D + D = helium-3 + $n$, although the neutron yield is about 100 times less.

4. The maximum yield of the $(p,n)$ reaction is about $10^8$ neutrons/s-sr-$\mu$amp in the forward direction. This occurs at a proton energy of 2.3 MeV and results in a neutron energy of 0.5 MeV in the forward direction (Emilio Segre, *Nuclei and Particles*, 2nd edition [Menlo Park, California: Benjamin/Cummings, 1977], p.617). Thicker targets could be used to yield about five times as many neutrons at the same proton current but at the expense of an increased neutron energy spread. Access Systems, Pleasanton, California, produces a small proton linear accelerator that can produce a 150-microampere beam of 2.3-MeV protons, but at a duty factor of only 2 percent. This machine cost about \$600,000 in 1989.

5. An equivalent in rads can be estimated as follows: if we assume that each neutron deposits 1 MeV of energy in about 10 centimeters of tissue, 1 neutron $cm^{-2}$ $s^{-1}$ will deposit $1.6 \times 10^{-7}$ ergs $g^{-1}$ $s^{-1}$. Since 1 rad = 100 ergs/gram, this is equal to $6 \times 10^{-6}$ rads per hour; thus, 20 neutrons $cm^{-2}$ $s^{-1}$ $\approx$ 0.1 millirads per hour. Since neutrons produce 10 times as much damage as photons or electrons per unit of energy deposited, this translates to a dose of about 1 millirem per hour. The maximum permissible dose to members of the public is 500 millirems per year.

6. Gozani, p.117.

7. Mozley and DeWire, *Nuovo Cimento*, **27**, 1281 (1963).

8.  Milburn, *Physical Review Letters* **10**, 75 (1963); Arutyunian et al., *Journal of Experimental and Theoretical Physics* **18**, 218 (1964); Murray and Klein, SLAC-TN 67-19 (Stanford, California: Stanford Linear Accelerator Center, 1967).

9.  Varian Associates, Palo Alto, California, produces a portable 2-MeV electron linear accelerator producing about $10^9$ photons $cm^{-2}$ $s^{-1}$ at 1 meter in the upper 20 percent of its spectrum. Varian also produces a 15-MeV accelerator with greater capacity, but this is not portable.

10.  This machine is produced by Varian Associates and the Bechtel Corporation. A 15-MeV electron linear accelerator is used as a source of x-rays and a line of sodium iodide scintillation counters is used as the detector.

11.  See, for example, Messenger and Ash.

12.  G.R. Woodcock, *Space Stations and Platforms* (Malabar, Florida: Orbit Book Co., 1986), p.16.

# Chapter 12

*Roald Z. Sagdeev*
*Oleg F. Prilutsky*
*Valery A. Frolov*

# Passive Detection of Nuclear-armed SLCMs

Effective procedures have been developed, using national technical means (photoreconnaissance satellites, radiointercept stations, etc.), for verification of reductions in land-based intercontinental ballistic missiles, submarine-based ballistic missiles, and strategic bombers. However, there is no agreement on procedures for verifying limitations of numbers of long-range nuclear-armed cruise missiles. The difficulties in developing such procedures are sometimes regarded (by opponents of nuclear disarmament) as a reason why cruise missiles based on ships and submarines ought not to be limited by future arms-reduction treaties. This chapter considers the detectability of nuclear-armed cruise missiles through the penetrating radiation emitted spontaneously from their warheads.[1]

## BACKGROUND[2]

Cruise missiles are small unmanned jet aircraft with automatic high-precision control. The V-1, a World War II German projectile, was the first cruise missile used in battle on a large scale. Long-range nuclear-armed cruise missiles were first manufactured in the US during the 1950s, but were inaccurate, unreliable, bulky, and vulnerable. After ballistic missiles were developed, long-range cruise missiles were abandoned.

In the 1970s, however, technological advances made it possible to create a new generation of small highly accurate long-range cruise missiles carrying high-yield nuclear warheads that could penetrate a sophisticated air-defense system. Modern cruise missiles are compact enough to be launched from submarine torpedo tubes or from launchers carried on or embedded in the decks of surface ships. They have ranges of thousands of kilometers. The fuselage of a long-range nuclear-armed sea-based cruise missile carries a navigation system, a nuclear warhead, fuel tanks, and a turbofan jet engine. During launch, the missile is accelerated initially by a solid-fuel booster engine. Incoming cruise missiles are difficult to detect because the effective nose-on "cross section" for generating radar echoes has been reduced to a very small value (0.01 square meters or less), and flight altitudes as low as several tens of meters can be achieved.

In 1972–73 the US started to develop a sea-launched cruise missile, the Tomahawk, and air-launched cruise missiles (ALCMs). In 1977, it was decided to develop a Tomahawk derivative as a ground-launched cruise missile (GLCM) to be deployed in Western Europe. Subsequently, the USSR also developed similar cruise missiles. The first modern US cruise missiles went into service with the US Air Force, US Army, and US Navy in 1981, 1983, and 1984 respectively.

According to the 1987 INF treaty, both US and Soviet GLCMs are being dismantled and destroyed. Nevertheless, both sides retain the right to deploy thousands of nuclear-armed long-range cruise missiles aboard strategic bombers, surface ships, and submarines. In the case of ALCMs, the strategic bombers carrying the cruise missiles are to be limited and controlled under the Strategic Arms Reduction Treaty (START), so the problem of controlling their missiles is secondary. For SLCMs, however, the objects of control cannot be the carriers on which missiles are based but must be the missiles themselves. Surface ships and attack submarines that carry nuclear-armed cruise missiles have conventional as well as nuclear roles that, as of today, are not limited. Ships and submarines also carry a great number of short-range "tactical" nuclear weapons from which cruise missiles must be distinguished. All these factors make sea-based cruise missiles difficult to control.

## VERIFICATION

Detailed analyses of the general aspects of nuclear-warhead detection have been given elsewhere.[3] This chapter therefore focuses on issues specific to the verification of limits on sea-based cruise missiles. We consider below detection of nuclear-armed cruise missiles by their neutron and gamma emissions, taking into account the observational conditions (backgrounds, operational constraints, and so forth) and the possibilities for blocking or masking the emissions.

Radiation detection is considered for two types of inspection: remote and on-site. In a remote inspection, the warhead-detection apparatus is not on the ship, but might, for example, be carried by a helicopter flying above or near the inspected ship at a distance of tens to one hundred meters. In an on-site inspection, the apparatus is brought onto the ship. It might be operated either by inspectors or perhaps under their supervision by the crew of the ship being inspected.

On-site inspection requires more cooperation between the inspecting and inspected sides. Such inspection might be opposed out of concern that the inspectors might acquire sensitive information not relevant to the control of nuclear cruise missiles. Nevertheless, this type of inspection may be necessary because of the limitations of remote inspection.

## Detection of Neutron Emissions

The only significantly detectable neutron emission from nuclear warhead material is that from the spontaneous fission of plutonium-240, a minor fractional constituent of weapon-grade plutonium. The rate of spontaneous fission of this isotope is several orders of magnitude higher than those for the naturally occurring uranium nuclei or for plutonium-239 (the main plutonium isotope in weapons). The neutron-emission rate of warheads containing a few kilograms of ordinary weapon-grade plutonium will be several hundreds of thousands of neutrons per second, independent of other aspects of the warhead design, and the typical energies of the neutrons will be about 1 MeV.[4] Reducing the plutonium-240 content of warheads is, in principle, possible, but would be costly.

Neutrons—like gamma rays—can only be detected through their

transfer of energy to charged particles. Neutrons generate fast protons and other nuclear particles through scattering and by neutron-absorption interactions.

The collimation of high-energy neutrons is very difficult, and systems capable of measuring the direction of incoming neutrons are extremely complicated and are not sensitive enough to use in inspection systems.

The mean free path in air of neutrons with an energy of about 1 MeV is several tens of meters. The energy lost in one scattering in air is typically quite small (less than 10 percent). Therefore, for omnidirectional detectors, the flux will drop with distance $r$ as $r^{-2}$ to distances of more than one hundred meters.

Constraints on neutron-detection systems will arise mainly from the inspection regime. In a remote inspection regime, when detectors are installed on a helicopter, the weight and overall dimensions of the system can be quite large (the system weight can be larger than 1 tonne and total detector area can be several square meters). Such a detection system would be constructed using many commercially available proportional counters filled with beryllium fluoride or helium-3, each with a cross-sectional area of about 100 square centimeters. These counters, which are sensitive to low-energy neutrons, are embedded in polyethylene, which will "moderate" the energy of the fast neutrons as they successively collide with the hydrogen in the plastic. For on-site inspection with hand-held neutron detectors, the total weight would be limited to less than 15–20 kilograms, and the system might consist of one or two counters embedded in a relatively thin moderator.

Typical values for the areas, efficiencies, background counting rates, and sensitivities of neutron detection systems are shown in table 12.1.[5] The detection threshold was estimated for an exposure time of 100 seconds and for a five-standard-deviation level of statistical significance. Under optimal conditions, for exposure periods of about 1 hour, the maximum distance for remote detection of a neutron source with an intensity of about 100,000 neutrons per second will be about 100 meters.[6]

The sensitivity estimates in table 12.1 were obtained assuming statistically random background fluctuations. Possible nonrandom variations of neutron background can, in principle, change the results—especially for the

**Table 12.1**: Neutron-detection system parameters

| Inspection regime | Area $m^2$ | Detection efficiency % | Background counts $s^{-1}$ | Detection threshold $n\ m^{-2}\ s^{-1}$ |
|---|---|---|---|---|
| On-site hand-held | 0.02 | 5 | 0.05 | 110 220%* |
| Remote helicopter-mounted | 0.3 | 20 | 3.0 | 14 30%* |

* signal-to-background

inspection of inner compartments of ships.

The largest neutron backgrounds will be observed on ships with nuclear-power systems. Even if the neutron background outside the shielding of the nuclear reactor corresponds to radiation safety standards, it can be two orders of magnitude higher than the natural neutron background. Moreover, the search for nuclear warheads will be made more difficult because the spatial distribution of the neutron background will not be homogenous.

Neutron radiation can be blocked by introducing thick shielding. Neutrons lose energy by scattering in a shield material and, when their energy is sufficiently small, are absorbed. Scattering cross sections for high-energy neutrons are similar for all materials but the highest energy loss per scattering is in materials with a large hydrogen content. Slow neutron absorption by the nuclei of some isotopes can be several orders of magnitude higher than for others. Lithium-6 in lithium hydride can provide very effective shielding from slow neutrons because neutron capture by lithium-6 produces no gamma rays. Materials routinely carried by a ship—for example, fuel or water—could also be used to shield warheads from neutron detectors.

## Detection of Gamma Emissions

Two types of detector are most often used for detection of gamma rays in the energy region of interest (from several hundreds of keV to several MeV): scintillation counters made of inorganic crystals and solid-state detectors made of high-purity germanium crystals.[7]

Scintillation counters are based on the detection of visible and ultra-violet photons that are produced by fast charged particles moving in certain transparent crystals, plastics, or organic liquids. Inorganic scintillator crystals are the most convenient material for gamma-ray detection because their gamma-ray absorption is higher than that of organic scintillators.

To measure small fluxes of gamma rays, the best material is thallium-doped sodium iodide. Sodium iodide crystals grown for scintillation detectors can be as large as 70 centimeters in diameter and 50 centimeters in thickness. Good scintillation counters, using thallium-activated sodium iodide crystals, have an energy resolution of 8–9 percent at the 662-keV cesium-137 line. At higher energies, this percentage resolution width narrows approximately as the inverse square root of the energy.

Large sodium iodide detectors combine close to 100-percent detection efficiency, close to 100-percent total energy absorption, and moderate energy resolution. These detectors are certainly the best for the observation of sources with continuous spectra but, for the detection of the monoenergetic gamma lines characteristic of different radioactive isotopes, their energy resolution may be insufficient.

Better energy resolution can be obtained with solid-state detectors. In these detectors, a signal is generated by electron-hole pairs produced in a semiconductor by fast charged particles. The amplitude of this signal is proportional to the energy that is absorbed in the detector material.

Solid-state detectors, in principle, can be made from many different semiconductors, but high-purity germanium detectors are used most often in gamma-ray spectroscopy because of the possibility of making relatively large-volume and therefore relatively high-efficiency detectors. Germanium semiconductor detectors must be cooled to liquid nitrogen temperatures (77 K) when used, however, which means that cryostats must be a part of the measurement system.

The energy required for generating one electron-hole pair in a solid state detector is hundreds of times less than the energy required for the generation of one photoelectron from the photocathode of a photomultiplier used in a scintillation detector. This circumstance, together with favorable fluctuation properties (small value of the Fano factor[8]), leads to the much better energy resolution of solid-state detectors. However, for systems of similar area, the cost of a semiconductor detector will be tens of times greater than that of a scintillation detector.

Both scintillation and semiconductor detectors will register photons coming from any direction. Collimating an instrument's field of view with shielding will therefore lessen the diffuse background radiation and provide some capability for image construction. This method of background reduction is limited by the natural and induced radioactivity of a detector's structural elements (including the collimator and shielding). Nevertheless, collimation and shielding can in theory reduce the background counting rate by factors of tens compared with an omnidirectional detector. Unfortunately, however, the large penetration capability of gamma rays with energy higher than 1 MeV requires a large mass of shielding and collimator. For an instrument with a working area of about 1,000 square centimeters the mass of shielding required may exceed 1 tonne.

Image construction with a collimated detector may be done by means of angular or spatial scanning with an angular resolution approximately equal to the width of the collimator angular response function. The time needed for image construction equals the time of observation of each element multiplied by the total number of elements in the area observed.

The natural gamma-ray background is caused by cosmic rays (this source dominates at sea) and the natural radioactive decay of elements in the upper layers of soil or rock (their content in sea water is much less). The background radiation is relatively uniform over the surface of the sea, but, above the deck of a ship, its intensity may change because of the heterogenous distribution of material in the ship's structure and the ship's chemical composition. Once again, the presence of a nuclear reactor can sharply change background conditions.

Even with on-site inspection, where the distance to the source might be only about 10 meters, the background from a nuclear power installation could cause serious problems. Therefore, it would be important to discriminate against this background through:

♦ Spectral discrimination—using equipment with good energy resolution and looking for strong emissions from fissile materials in energy regions where the background spectrum intensity is low

♦ Brightness discrimination—using equipment with good angular resolution. The angular size of a warhead with a diameter of about 40 centimeters[9] is about 2 degrees at a distance of 10 meters.

The flux of radiation from the source decreases as the distance from the source becomes large, both because of the distance itself and because of the interaction of gamma rays with the environment. When the distance from the source is small, most photons arrive at the detector without interacting with the air. Their direction of movement will be the direction from the source to the detector; their energy will be their initial energy; and their flux will decline with distance $r$ as $r^{-2}$. When the distance becomes so large that it is comparable to the mean free path of photons in air—about 100 meters—the flux of unscattered gamma rays will begin to decrease much faster. If the telescope has a narrow field of view or high energy resolution, scattered gamma rays will not be counted, and the counting rate will decrease accordingly.

The gamma rays from a nuclear warhead—like the neutron radiation—can be blocked by introducing additional material between the detector and source. The interaction of gamma rays with shielding is a complex combination of scattering, energy decrease, and absorption. The monoenergetic "lines" associated with unscattered gamma rays are therefore shielded out more rapidly than the continuum.

For low-energy gammas, the absorption in dense substances, such as lead and tungsten, is much larger per weight unit than in less dense ones. For photons with energy of more than 1 MeV, however, the shielding properties of different substances are not very different per weight unit (see table 11.B.3). At such energies, quite large amounts of material (a few hundred $kg/m^2$) are needed for an effective shield. However, in the interior of a ship, such massive shielding is quite possible. A search for such shields could be conducted with the help of active methods (neutron and gamma radiography and backscattering) but such a search would require a very high degree of cooperation during inspection.[10]

## CONCLUSION

The above discussion makes clear that we must consider the use of radiation-detector techniques as only one of a whole complex of verification techniques; they are not capable of solving all problems.

## NOTES AND REFERENCES

1.  Chapters 13 and 14 describe measurements of the gamma and neutron radiation from a Soviet cruise missile warhead. These measurements were made after the present chapter and chapter 11 were completed.

2.  Based on Thomas B. Cochran, William M. Arkin, and Milton M. Hoenig, *Nuclear Weapons Databook Volume 1: U.S. Nuclear Forces and Capabilities* (New York: Ballinger, 1984). See also chapter 8 and its appendix for further information and references on both US and Soviet SLCMs.

3.  Roald Z. Sagdeev, Oleg F. Prilutsky, and Valery A. Frolov, "Problems Relating to the Control of Sea-Based Cruise Missiles with Nuclear Warheads" (preprint, Space Research Institute of the Academy of Sciences of the USSR, 1988). Much of the material in this preprint has been included in this book in chapter 11.

4.  Ibid.

5.  See appendix 11.C for more details and references on neutron detectors.

6.  See Sagdeev et al. and chapter 11.

7.  For a discussion of gamma detectors and references, see appendix 11.C.

8.  U. Fano, *Physical Review*, **72** (1947), p.26.

9.  For a description of simplified "model" nuclear warheads, based on public information, see chapter 11.

10.  For a discussion of such active detection methods, see Sagdeev et al. and chapter 11.

# Chapter 13

*Steve Fetter*
*Thomas B. Cochran*
*Lee Grodzins*
*Harvey L. Lynch*
*Martin S. Zucker*

# Measurements of Gamma Rays from a Soviet Cruise Missile

## MEASUREMENTS

To explore the usefulness of various radiation detectors for verification purposes, a series of simple experiments was carried out on 5 July 1989 on the Black Sea near Yalta under the auspices of the Natural Resources Defense Council (NRDC) and the Academy of Sciences of the USSR. The Soviet Navy provided the cruiser *Slava*, which is shown in figure 13.1.[1] We were informed that the *Slava* was armed with a single nuclear-armed SS-N-12 SLCM in the outside forward launcher on the starboard side, and that no other nuclear weapons were on board during the experiment. Teams of scientists from the NRDC and the USSR used seven different types of detectors. The characteristics of these instruments are summarized in table 13.1. Instruments 1–4 were portable devices; all except number 7 detected gamma rays.

This article discusses measurements we made with detector number 1, a coaxial high-purity germanium detector. The 151-cm$^3$ sensitive volume was cylindrical, 5.9 centimeters in diameter and 5.9 centimeters long.[2] It had an energy resolution of about 2 keV (full-width at half-maximum) at an energy of 1,000 keV. The detector pulses were analyzed with a portable multichannel

This chapter appeared previously in *Science* **248**, 18 May 1990, pp.828–834
© 1990 by the AAAS

**Figure 13.1:** The Soviet cruiser *Slava*

Source: Congressman Robert Carr

analyzer (MCA) with 4,096 channels.[3] Only those gamma rays with energies between 30 and 2,670 keV were recorded.

We made the following measurements on the *Slava*: three measurements, totaling about 24 minutes, on the launch tube directly above the warhead;[4] one 10-minute measurement on the adjacent empty launch tube; and two background measurements lasting 60 and 10 minutes on the deck of the ship about 27 and 32 meters in front of the launch tube. The total count rates in these four locations were 393.2 ± 0.5 counts per second (cps), 36.3 ± 0.3 cps, 11.23 ± 0.06 cps, and 11.14 ± 0.14 cps (errors are from counting statistics only).

The three measurements on the launch tube were combined to form a single 24-minute measurement;[5] the combined spectrum is shown in figure 13.2. The spectra were analyzed with three different peak-finding and peak-fitting programs to determine the location and intensity of the peaks. The

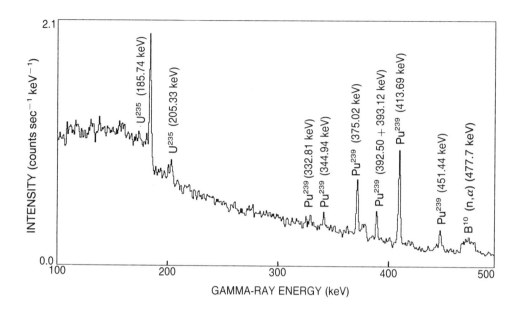

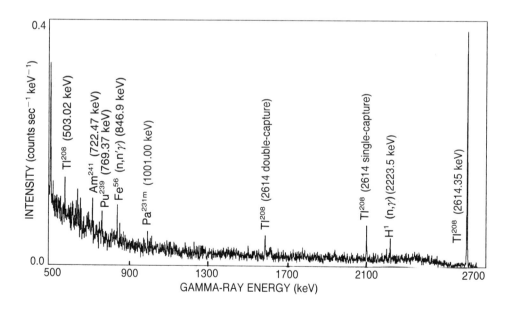

**Figure 13.2:** The combined gamma-ray spectrum recorded by the germanium detector on the launch tube directly above the warhead

results of the three programs were in excellent agreement; for consistency we will only use those given by the HYPERMET program, which gave the most complete results.[6]

The energy calibration of the detector proceeded in two steps. First, a cobalt-60 source, which produces strong gamma-rays at 1,173 and 1,332 keV, was used to give a linear relationship between channel number and gamma-ray energy. This linear calibration was used to identify 16 prominent gamma-ray emissions in the spectrum shown in figure 13.2 that were then used in a quadratic least-squares fit.[7]

Table 13.2 lists 32 peaks whose statistical significance exceeded three standard deviations ($3\sigma$) above the background. The identity of the parent radionuclide was identified in every case but one.[8] Many of the lines were due to uranium-235 or plutonium-239; the presence of either of these materials suggests the presence of a nuclear warhead. In addition, we identified gamma rays emitted by protactinium-234m (a decay product of uranium-238), bismuth-212 and thallium-208 (decay products of uranium-232),[9] and americium-241 (a decay product of plutonium-241). A list of the gamma rays that can be attributed to the presence of uranium and plutonium is given in table 13.3.

The presence of uranium-232 is noteworthy since it is not a naturally occurring isotope of uranium; it is, however, produced in nuclear reactors.[10] The USSR must therefore have used uranium from reprocessed reactor fuel as the feedstock for the uranium enrichment process; uranium-232 would then be enriched along with uranium-235. It should be noted, however, that uranium-232 would not be present in highly enriched uranium that is produced entirely

**Table 13.1:** Detectors used during the experiment

|  | size $cm^2$ |  |
|---|---|---|
| 1 High-purity germanium detector | 27 | US |
| 2 Lithium-drifted germanium detector | 14 | USSR |
| 3 Sodium iodide detector | 100 | US |
| 4 Sodium iodide detector | ≈10 | USSR |
| 5 Ship-based sodium iodide detector | 2,500 | USSR |
| 6 Truck-based sodium iodide gamma-ray telescope | 440 | USSR |
| 7 Helicopter-based helium-3 neutron detector | 2,500 | USSR |

from virgin natural uranium.

The remaining peaks are due to background radiation, neutron reactions, and pair production. For example, the broad peak centered at 478 keV is probably due either to a $(n,\alpha)$ reaction with boron-10 or to inelastic scattering with lithium-7.[11] Other neutron-induced gamma rays were emitted at 846.76 keV, from $(n,n'\gamma)$ reactions with iron-56 in steel, and at 2,223.25 keV, from $(n,\gamma)$ reactions with hydrogen in the fuel of the missile or the high explosive of the warhead. By comparing the spectrum with that taken on the adjacent empty launch tube, we can attribute several peaks to natural background radiation—at 609.31 keV (bismuth-214) and at 1,460.83 keV (potassium-40)—and radioactive fallout—at 604.71 keV (cesium-134) and at 661.66 keV (cesium-137). The peaks at 511, 1,592, and 2,103 keV are due to pair production.[12]

## MAXIMUM DETECTION DISTANCE

For a radioactive source to be identified, its signal at the detector must exceed statistical fluctuations in the background. To minimize the probability of false alarms, a signal is not recorded until a deviation of three to five standard deviations above the mean background occurs.[13] Here we examine two cases: (i) where the signal represents the total count rate integrated over the entire recorded energy spectrum, and (ii) where the signal represents the count rate of discrete gamma-ray emissions.

The simplest way to search for radioactive material is to record the total count rate. One first establishes a value for background count rate for the survey area, and then looks for a count rate that is significantly greater. In our case, however, an analysis of this type suffers from a lack of background measurements taken at many different locations.

Let $C_i$ and $B_i$ equal the total count rate and the background count rate at point $i$, $r_i$ equal the distance between the warhead and the detector at this point, and $F_i$ equal the relative shielding factor along this path. Assuming that the signal decreases as the inverse square of the distance,[14]

$$\frac{(C_i - B_i)\, r_i^2}{F_i} = \text{constant} \tag{13.1}$$

**Table 13.2:** The observed intensity, branching ratio, and decay rate of gamma-ray emissions observed in figure 13.2 that are due to isotopes of uranium and plutonium or their daughters

| Parent radionuclide | Expected energy keV | Observed intensity counts/s | Branching ratio percent/decay | Decay rate $g^{-1} s^{-1}$ |
|---|---|---|---|---|
| Uranium-232[†] | 583.02 | 0.190 ± 0.018 | 86. | $2.31 \times 10^{11}$ |
| | 727.72 | 0.058 ± 0.018 | 6.65 | $4.97 \times 10^{10}$ |
| | 860.30 | 0.071 ± 0.013 | 12.0 | $3.22 \times 10^{10}$ |
| | 1620.66 | 0.030 ± 0.012 | 1.51 | $1.13 \times 10^{10}$ |
| | 2614.35 | 1.369 ± 0.031 | 99.79 | $2.68 \times 10^{11}$ |
| Uranium-235 | 143.79 | 0.118 ± 0.051 | 10.5 | 8,400 |
| | 163.38 | 0.103 ± 0.038 | 4.7 | 3,800 |
| | 185.74 | 1.870 ± 0.074 | 53. | 42,000 |
| | 205.33 | 0.361 ± 0.062 | 4.7 | 3,800 |
| Uranium-238[†] | 1001.00 | 0.082 ± 0.014 | 0.65 | 81 |
| Plutonium-239 | 332.81 | 0.137 ± 0.046 | 0.000505 | 11,600 |
| | 344.94 | 0.191 ± 0.061 | 0.00057 | 13,100 |
| | 375.02 | 0.862 ± 0.160 | 0.00158 | 36,300 |
| | 380.17 | 0.131 ± 0.071 | 0.000307 | 7,040 |
| | 382.68 | 0.160 ± 0.073 | 0.00026 | 6,000 |
| | 392.99 | 0.373 ± 0.090 | 0.00056 | 12,800 |
| | 413.69 | 1.582 ± 0.064 | 0.00151 | 34,600 |
| | 422.57 | 0.139 ± 0.027 | 0.000119 | 2,730 |
| | 451.44 | 0.318 ± 0.036 | 0.000192 | 4,410 |
| | 640.15 | 0.083 ± 0.025 | 0.0000079 | 181 |
| | 645.98 | 0.113 ± 0.025 | 0.0000145 | 333 |
| | 652.18 | 0.075 ± 0.024 | 0.0000064 | 147 |
| | 756.42 | 0.051 ± 0.018 | 0.0000034 | 77 |
| | 769.37 | 0.158 ± 0.020 | 0.0000110 | 252 |
| Plutonium-241[†] | 662.43 | 0.116 ± 0.043[†] | 0.00036 | 174,000 |
| | 722.47 | 0.131 ± 0.026 | 0.00013 | 92,000 |

\*    The error in the observed intensity is the error in the curve fit to one standard deviation. The branching ratios are taken from E. Browne and R.B. Firestone, *Table of Radioactive Isotopes* (New York: John Wiley and Sons, 1986). For gamma rays emitted by a radioactive daughter, the decay rate depends on the age of the material; the decay rates given here are for a decay time of 10 years (i.e., 10 years after starting with 1 gram of the pure parent isotope). For a decay time of 5 years, the production rate from thallium-208 is 10 percent smaller; that from americium-241 is 1.8 times smaller. For a decay time of 20 years, the production rate from thallium-208 is 10 percent smaller; that from americium-241 is 1.6 times greater.

†    After subtracting the background from the decay of cesium-137 at 661.660 keV measured on the adjacent empty launcher.

‡    The presence of this isotope is inferred from the observed daughter activities.

**Table 13.3:** The observed energy, expected energy, and suspected origin of the peaks in figure 13.2 whose significance exceeded three standard deviations*

| Observed energy keV | Expected energy keV | Suspected origin |
|---|---|---|
| 185.7 ± 0.2[†] | 185.74 | uranium-235 |
| 205.3 ± 0.3[†] | 205.33 | uranium-235 |
| 332.4 ± 0.3[†] | 332.81 | plutonium-239 |
| 344.8 ± 0.3[†] | 344.94 | plutonium-239 |
| 375.1 ± 0.3[†] | 375.02 | plutonium-239 |
| 380.3 ± 0.9 | 380.17 | plutonium-239 |
| 382.6 ± 0.9 | 382.68 | plutonium-239 |
| 392.9 ± 0.9 | 392.99[†] | plutonium-239 |
| 413.7 ± 0.2[†] | 413.69 | plutonium-239 |
| 422.6 ± 0.3 | 422.57 | plutonium-239 |
| 451.5 ± 0.2[†] | 451.44 | plutonium-239 |
| 478.0 ± 0.6 | 478.4 | boron-10 $(n,\alpha'\gamma)$ or lithium-7 $(n,n'\gamma)$ |
| 511.0 ± 0.2[†] | 511.00[§] | annihilation radiation |
| 583.4 ± 0.2[†] | 583.02 | thallium-208 (from uranium-232) |
| 604.5 ± 0.3 | 604.71 | cesium-134 *background* |
| 609.4 ± 0.3 | 609.31 | bismuth-214 *background* |
| 639.9 ± 0.5 | 640.15 | plutonium-239 |
| 646.1 ± 0.3[†] | 645.98 | plutonium-239 |
| 652.5 ± 0.5 | 652.18 | plutonium-239 |
| 661.7 ± 0.9 | 661.84[□] | cesium-137 and americium-241 |
| 721.8 ± 0.3 | 722.47[#] | americium-241 (from plutonium-241) |
| 727.7 ± 0.3 | 727.25 | bismuth-212 (from uranium-232) |
| 769.0 ± 0.3[†] | 769.37 | plutonium-239 |
| 846.4 ± 0.3 | 846.75 | iron-56 $(n,n'\gamma)$ |
| 860.2 ± 0.3[†] | 860.30 | thallium-208 (from uranium-232) |
| 1,000.8 ± 0.3[†] | 1,001.00 | protactinium-234m (from uranium-238) |
| 1,460.6 ± 0.2[†] | 1,460.83 | potassium-40 *background* |
| 1,591.7 ± 0.3 | 1,592.35 | thallium-208 *double escape* |
| 1,942.7 ± 0.5 | ? | |
| 2,103.1 ± 0.2[†] | 2,103.35 | thallium-208 *single escape* |
| 2,223.2 ± 0.3 | 2,223.25 | hydrogen-1 $(n,\gamma)$ |
| 2,614.4 ± 0.2[†] | 2,614.35 | thallium-208 (from uranium-232) |

* The error in the observed energy is the combined error of the peak fit and the energy calibration to one standard deviation. Expected energies are from E. Browne and R.B. Firestone, *Table of Radioactive Isotopes* (New York: John Wiley and Sons, 1986) and F.E. Senftle, H.D. Moore, and D.B. Leep, *Nuclear Instruments and Methods*, **93**, 425.

† Lines used in the least-squares energy calibration.

‡ Weighted average of two lines from plutonium-239: 392.50 keV (0.000116 percent) and 393.12 keV (0.000444 percent).

§ Combined with a weaker line from thallium-208 at 510.606 keV.

□ Weighted average of cesium-137 line at 661.660 keV and americium-241 line from decay of plutonium-241 at 662.426 keV.

# Weighted average of two lines from americium-241: 721.962 keV (0.000060 percent) and 722.70 keV (0.00013 percent).

where $i = 1$ and 2 refer to the measurements made on the loaded and empty launch tubes, and $i = 3$ and 4 refer to those made on the deck. From photographs (for example, figure 13.3) and a measurement of the launch tube diameter, we estimate that $r_1 = 0.73 \pm 0.03$ meters, $r_2 = 2.94 \pm 0.14$ meters, $r_3 = 27 \pm 2$ meters, and $r_4 = 32 \pm 2$ meters.

As can be seen from equation 13.1, each measurement made along a different path brings with it two new variables: $B$ and $F$. Only measurements 3 and 4 were made along the same path (and therefore have about the same value of $F$). If we assume that $B$ is equal at these two points, then the background is given by

$$B_3 = B_4 = \frac{C_4 r_4^2 - C_3 r_3^2}{r_4^2 - r_3^2} = 10.9 \pm 0.5 \text{ cps} \qquad (13.2)$$

In other words, the count rate on the deck is due almost entirely to background radiation.

Most of the terrestrial gamma-ray background flux is due to radionuclides in soil and rock: potassium-40 and decay products of thorium-232 and uranium-238.[15] It is reasonable to assume that these radionuclides would also account for most of the gamma-ray flux above a 10,000-tonne ship, since steel is contaminated with potassium, thorium, and uranium impurities present in iron ore. Since emissions from these radionuclides were $2.0 \pm 0.2$ times more intense in the launch-tube spectra than in the deck spectra, we will assume that the total background count rate on the launch tubes is twice as great as that on the deck: $B_1 = B_2 = 22 \pm 2$ cps.[16]

The maximum distance in direction $i$ at which the warhead could have been detected is given by

$$r_{max} = r_i \left[ \frac{C_i - B_i}{C_{min} - B_i} \right]^{1/2} \qquad (13.3)$$

where $C_{min}$ is the minimum count rate that would indicate the presence of a warhead. If the background was perfectly uniform, $C_{min}$ would be determined

**Figure 13.3:** A view of the launch tubes. The launcher on the left has been opened, revealing the missile inside

by counting statistics; for a counting time of 10 minutes and a significance level of $3\sigma$, $C_{min}$ could be as little as 4 to 5 percent above $B_i$, leading to $r_{max}$ of 13 to 20 meters.[17]

The background was not uniform, however. Our analysis suggests that the background above the ship varied by about a factor of two over a distance of 30 meters. If, as would seem prudent in view of this variability, $C_{min}$ must be 30 to 100 percent above $B_i$, then $r_{max}$ would only be 2 to 5 meters, and $r_{max}$ could not be improved by increasing the detector area or counting time.

The preceding discussion made no use of the high energy resolution of germanium detectors. Detecting the characteristic emissions of uranium-235 and/or plutonium-239 represents far more convincing evidence of the presence of a nuclear warhead than an increase in the total count rate, and avoids confusing warheads with other radioactive sources (for example, depleted-uranium bullets) that may be on a ship.

Since only minute concentrations of uranium-235 and plutonium-239 are found in common materials, emissions from these radionuclides can be attributed entirely to the warhead (except for gamma rays from higher-energy peaks that are Compton-scattered into the energy range of the peak under examination).[18] The same assumption does not hold for emissions from thallium-208, however, since thallium-208 is a decay product of both uranium-232 in the warhead and thorium-232 impurities in the steel (see figure 11.B.2). Using the measured intensity of the 911-keV line from actinium-228, which is a decay product of thorium-232 but not of uranium-232, we estimated that 79 ± 17 percent of the thallium-208 decays detected on the deck are due to background.[19]

Applying equation 13.3 to each line, we find that the most detectable emissions are the 186-keV line from uranium-235, the 414-keV or the 769-keV line from plutonium-239, and the 2,614-keV line from the decay of the thallium-208 (from the uranium-232). In direction 1 (above the warhead), $r_{max}$ = 4 meters for uranium-235 and plutonium-239 and 5 meters for uranium-232—all assuming a counting time of 10 minutes and a significance level of $3\sigma$. In direction 2 (to the side of the launch tube), $r_{max}$ = 1.5 meters for uranium-235, 3 meters for plutonium-239, and 4 meters for uranium-232. In direction 3 (in front of the launch tube), $r_{max}$ = 6 meters for uranium-235, 12 meters for plutonium-239, and 6 meters for uranium-232; the lid of the launch tube apparently provides less shielding than its sides.[20] It is apparent that one must be fairly close to launch tube to be certain of detecting fissile materials.

## WARHEAD CONCEALMENT

Even if nuclear weapons can be detected as they are normally deployed, they could be concealed by placing shielding around the weapon or by moving the weapon to a part of the ship that is not open to inspection. It may even be possible to produce special nuclear weapons that emit very little radiation (see appendix 11.A).

For the weapon in our experiment, the most stringent requirement for gamma-ray shielding is set by the highly penetrating 2,614-keV line. A 100-

fold reduction in the intensity of this line would have been required to make it undetectable outside the launch tube; a layer of tungsten at least 6 centimeters thick placed between the missile and the launch tube would have been required. Adding this much shielding is feasible in principle for the launch tubes we examined (which had a 12-centimeter space between the top of the missile and the inside of the launch tube), but the existence of such shielding could be detected by visual inspection or with a few simple gamma-ray transmission measurements.

Concealing a cruise missile in another part of the ship appears to be rather difficult—at least for the US.[21] Little is known about Soviet equipment; however, in the case of the *Slava* it did not appear possible to remove the missiles from the launch tubes while at sea—at least not without the help of a crane from a neighboring ship. In theory, it would be possible to remove the warhead from the missile, conceal it in a shielded box during an inspection, and reinstall it afterward. It is not considered credible, however, to install a US SLCM warhead at sea without seriously compromising the reliability of the missile (see chapter 8).

## WHAT CAN BE LEARNED ABOUT WARHEAD DESIGN?

Concerns are sometimes expressed that measurements of the gamma radiation from a nuclear warhead with a high-resolution gamma-ray detector might reveal sensitive details about the design of nuclear warheads. To investigate this possibility we constructed various models of the warhead on the *Slava*, with the thicknesses of the various components adjusted to give the best possible agreement with our measurements.

The observed intensity of a particular gamma-ray emission $C$ is equal to the product of the decay rate per gram of the parent isotope $S$, the mass of the parent isotope $M$, the self-shielding factor $G$ (that is, the fraction of gamma rays that exit the source unscattered), the external shielding factor $F$ (the fraction of gamma rays exiting the source in the direction of the detector that arrive unscattered), the efficiency of the detector $\varepsilon$ (the fraction of unscattered gamma rays that are fully absorbed in the detector), and the solid angle subtended by the detector $\Omega$:

$$C = SMFG\varepsilon\left(\frac{\Omega}{4\pi}\right) \qquad (13.4)$$

## Self-shielding Correction

If the radioactive material is in the shape of a sphere or an empty spherical shell, then the self-shielding factor far from the sphere can be approximated by

$$G \approx \frac{1}{\beta\mu r}(1 - \exp[-\beta\mu r]) \qquad (13.5)$$

where $\mu$ is the linear attenuation coefficient and $r$ is the thickness of the shell; $\beta$, which depends on the radius ratio of the shell, ranges from 4/3 for solid spheres to 4 for thin shells (see appendix 11.B).

## External Shielding Correction

The external shielding factor $F$ provided by a series of flat parallel absorbers is given by

$$F = \prod_i \exp(-\mu_i x_i) \qquad (13.6)$$

where $\mu_i$ is the linear attenuation coefficient and $x_i$ is the thickness of the $i$th absorber along the path between the source and the detector. The equation for a series of spherical shells in which the thickness is not much smaller than the radius is much more complicated.[22]

## Efficiency Correction

The detector was calibrated in the laboratory using 17 gamma-ray emissions from six radioactive sources of known strength (cobalt-57, yttrium-88, barium-133, cesium-137, thorium-228, and americium-241). The sources were placed in approximately the same position relative to the detector as was the warhead in the measurement on the loaded launch tube. The efficiency of the detector for the $i$th line is given by

$$\varepsilon_i = \left[ \frac{C_i - B_i}{Q f_i \left( \dfrac{\Omega}{4\pi} \right)} \right] \tag{13.7}$$

where $C_i$ is the total count rate and $B_i$ is the background count rate (cps), $Q$ is the activity of the source (decays per second), $f$ is the branching ratio (gammas per decay), and $\Omega$ is the solid angle. Figure 13.4@ shows the results.

## Solid Angle

Accurate evaluation of the solid angle $\Omega$ requires knowledge of the shape and size of the source—information not provided by the Soviets. As mentioned above, we estimated that the distance from the center of the missile to the center of the detector was $73 \pm 3$ centimeters. Since the detector was mounted horizontally on the launcher, $(\Omega/4\pi) = (5.9/73)^2/4\pi = 0.00052 \pm 0.00004$, where 5.9 is the length and diameter of the detector crystal in centimeters.

## Analysis of the Data

We begin our analysis with plutonium-239, for which we observed gamma-ray emissions at 14 different energies from 333 to 769 keV. The plutonium was assumed to be in the form of an empty spherical shell in the center of the weapon, surrounded by low-, medium-, and high-$Z$ ($Z$ is atomic number) materials (see appendix 11.A). High-explosive was chosen to represent low-$Z$ materials (the relative attenuation caused by other common low-$Z$ materials— beryllium, boron, and aluminum—is very similar). Medium-$Z$ materials, such as the steel launch tube, were represented by iron. Uranium represented high-$Z$ materials. The variables in the least-squares fit were the thicknesses of these three materials and the mass and the outside radius of the plutonium shell. The concentration of plutonium-239 was taken to be 96 percent (see below).

The fact that the low-energy gamma rays from uranium-235 were seen

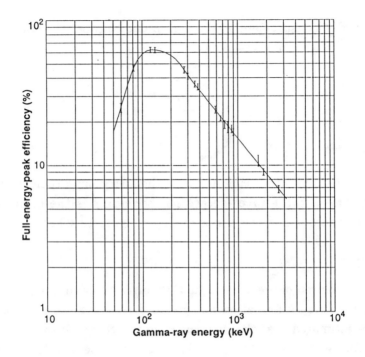

**Figure 13.4:** The full-energy-peak efficiency of the detector as a function of gamma-ray energy

implies that there is very little high-$Z$ material outside the uranium (the mean free path of 186-keV gamma rays is 0.36 millimeters in uranium). Our initial assumption was a shell of uranium-235 surrounding a shell of plutonium-239 (as in a composite-core fission weapon), but we found it impossible to obtain a good fit to either the plutonium or the uranium data with this model.

If, on the other hand, we assumed that the plutonium-239 was immediately surrounded by low-$Z$ material, it was surprisingly easy to obtain acceptable fits ($\chi^2 \lesssim 1$ per degree of freedom) with many combinations of these variables. The combinations that resulted in acceptable fits included plutonium radii of 5.4 to 8.0 centimeters, plutonium masses of 3 to 6 kilograms, high-explosive thicknesses of 3 to 10 centimeters, iron thicknesses of 6 to 8 centimeters, and uranium thicknesses of 0.1 to 0.4 centimeters. In general, changes in the value of one variable could be offset by a combination of changes in other variables. While these values may seem reasonable, it is apparent that this type of analysis cannot uncover sensitive design details.

Using the values given by the above analysis and the count rates of the two americium-241 gamma rays, we estimate the percentage of plutonium-241 to be 0.20 ± 0.10 percent, which corresponds to a plutonium-239 concentration of 96 ± 1 percent and a plutonium-240 concentration of 4 ± 1 percent.[23] For comparison, US weapon-grade plutonium typically contains 6 percent plutonium-240; super-grade plutonium (used in some US warheads) contains 3 percent plutonium-240.[24]

The analysis of the uranium data is necessarily much less precise because we have only 10 lines from all three isotopes and two additional variables: the concentrations of uranium-232 and uranium-238. Moreover, the four uranium-235 lines, which cover a narrow range of low energies (144 to 205 keV), are statistically decoupled from the five uranium-232 lines at much higher energies. The intensity of the single uranium-238 line can only be used to estimate the concentration of that isotope. The data are roughly consistent with a 7–15-kilogram uranium shell with a radius of about 10 to 15 centimeters not surrounded by a thick layer of low-$Z$ materials. The data are also consistent with a uranium-232 concentration of 0.1 to 0.2 ppb and a uranium-238 concentration of 4 to 6 percent (US weapon-grade uranium is 5.5 percent uranium-238.[25]

Thus, we do not believe that such measurements are capable of revealing sensitive information about the design of the warhead. But even if sensitive details could be revealed, there are ways protecting such information. In general, there seem to be three types of worry: (i) that the inspecting nation could learn new weapon design techniques; (ii) that something could be learned about the general technical sophistication of the other nation; and (iii) that the information revealed might aid possible proliferators. The latter problem could be solved simply by keeping the data confidential. The other two worries could be ameliorated by designing special detection equipment that would only collect data in narrow energy bands of interest (for example, around 186, 414, and 2,614 keV).

## CONCLUSIONS

The measurements we made with the germanium detector on the *Slava* provided valuable information for building a verification regime for SLCMs. At close range it is clearly possible to detect the fissionable materials in at least one type of warhead, even when it is shielded by a thick launch tube. The ability to clearly identify either uranium-235 or plutonium-239 would provide prima facie evidence that a nuclear warhead was contained in a launcher.

Detecting line emissions from uranium-235 and plutonium-239 is a more certain and, in the absence of extensive background measurements, a more efficient method of searching for nuclear warheads than looking for an increase in the total count rate. The most intense lines from uranium-235 and plutonium-239 could have been detected through the launch tube at a distance of 4 to 5 meters.

The warhead could have been concealed by placing a thick layer of tungsten inside the launch tube, but such shielding could be revealed by simple gamma-ray transmission measurements. Alternatively, the warhead could be removed to a shielded box, although this is not possible for current US SLCMs. Moving the entire SLCM below deck did not appear possible on the *Slava*.

Our analysis indicates that passive radiation detectors—even those with high energy resolution—cannot be used to reveal sensitive weapon design information, at least if such measurements are constrained to a few locations and counting times of less than 1 hour. There simply is too little information in the spectra to constrain the many possible variables in a realistic warhead design.

Finally, it should be emphasized that passive radiation detection is only one tool of many that may be useful in future arms control agreements. Some types of agreements—such as a ban on nuclear weapons on certain naval vessels, or one in which a particular missile must be identified as conventional or nuclear—might be facilitated by such techniques, while other types of agreement might not benefit at all.

## ACKNOWLEDGEMENTS

We would like to thank the Academy of Sciences of the USSR, the Soviet Navy, and the Natural Resources Defense Council for making this experiment possible, and the Plowshares Fund, the Rockefeller Foundation, and the Union of Concerned Scientists for their support. The authors participated in this experiment as private citizens and not as representatives of their institutions or of the United States government.

## NOTES AND REFERENCES

1. The *Slava*, which became operational in 1982, is about 200 meters long and has a crew of about 600. The turbine-powered ship can launch up to 16 SS-N-12 SLCMs, 64 SA-N-6 surface-to-air missiles, and has ten 21-inch torpedo tubes.

2. A portable Dewar filled with liquid nitrogen keeps the detector (Princeton Gamma-Tech, Inc., model IGC3520) at operating temperatures for up to two days.

3. The portable Davidson model 2056-B 4,096-channel analyzer also provided power for the detector and amplifier.

4. The Soviets suggested that we place the detector about 3.4 meters from the front end of the launch tube to maximize the count rate. We verified this by making measurements along the tube with a small hand-held detector.

5. Because of a small increase in amplifier gain between the first and third measurements, combining the raw data increased the peak width. This had a negligible effect on the energy calibration or the estimates of the areas under peaks, however.

6. G.W. Phillips and K.W. Marlow, "Program HYPERMET for Automatic Analysis of Gamma-Ray Spectra from Germanium Detectors," NRL 3198 (Washington DC: Naval Research Laboratory, 1976). The other two programs were MINIGAM II (sold by EG&G ORTEC) and PEAKFIT (written by Steve Fetter).

7. The least-squares fit was given by $E_\gamma = 1.02 + 0.65103n + (4.25 \times 10^{-7})n^2$, where $n$ is the channel number and $E_\gamma$ is the gamma-ray energy in keV.

8. We could not identify the gamma-ray emission at 1,942.7 keV, which was just significant at the $3\sigma$ level in the combined spectrum.

9. Bismuth-212 and thallium-208 are also decay products of naturally occurring thorium-232, but if thorium-232 was the parent we also would have observed intense gamma rays at 911.16, 968.97, and 1,588.23 keV from the decay of actinium-228.

10. Uranium-232 is produced in nuclear reactors in the following way:

$$U\text{-}235 \quad \xrightarrow[770 \text{ My}]{\alpha} \quad Th\text{-}231 \quad \xrightarrow[1.1 \text{ d}]{\beta} \quad Pa\text{-}231 \quad \xrightarrow{(n,\gamma)} \quad Pa\text{-}232 \quad \xrightarrow[1.3 \text{ d}]{\beta} \quad U\text{-}232$$

About 1 part per billion (ppb) of protactinium-231, which has a halflife of 33,000 years and a thermal neutron-capture cross section of 260 barns, is produced every year by the decay of uranium-235.

11. The excited lithium-7 nucleus created by either reaction would be moving at high speed when it emits the gamma ray, leading to a Doppler-broadened line with a maximum width of about 16 keV, just as we observed. Boron-10 could be in the missile fuel or possibly in the warhead; lithium-7 could be in the thermonuclear fuel.

12. If a gamma ray has an energy greater than 1,022 keV it may create electron–positron pairs in the detector. After coming to rest, the positron interacts with an electron, and both are annihilated, leading to the emission of two 511-keV gamma rays. If both of these gamma rays escape from the detector, a "double escape peak" is registered at 1,022 keV below the energy of the original gamma ray; if only one escapes, a "single escape peak" is recorded. Thus, the strong 2,614-keV gamma ray leads to single and double escape peaks at 2,103 and 1,592 keV. The peak at 511 keV arises from pair production in materials outside the detector.

13. The probability of a 3σ deviation occurring at random is 0.13 percent; for 4σ, it is 0.0032 percent (one chance in 30,000); for a 5σ deviation, it is 0.000029 percent (one chance in three million).

14. The mean free path of even low-energy gamma rays in air is large compared to the distances under consideration (for example, 65 meters at 186 keV). Moreover, the finite size of source does not cause significant errors even at short distances: $r^{-2}$ scaling underestimates the solid angle by less than 4 percent for a source-to-detector distance of 73 centimeters and a source radius of 10 centimeters.

15. These three sources, in roughly equal amounts, typically account for over 90 percent of the background gamma-ray flux; cosmic rays, cosmogenic radionuclides (for example, carbon-14 in the atmosphere), fallout, airborne radon, and other primordial radionuclides account for less than 10 percent. National Council on Radiological Protection, *Exposure of the Population in the United States and Canada from Natural Background Radiation*, NCRP Report 94 (Bethesda, Maryland: NCRP, 1987).

16. We observed five lines from bismuth-214 and lead-214 (daughters of uranium-238), one line from actinium-228 (a daughter of thorium-232), and one line from potassium-40. The ratio of the intensity of these lines on the launch tubes to their intensity on the deck was $1.8 \pm 0.2$, $2.8 \pm 1.2$, and $2.5 \pm 0.3$, respectively. If the three sources contribute roughly equal amounts to the background flux, the average ratio is $2.0 \pm 0.2$. We should note, however, that the presence of nearby humans, each of whom contains over 100 grams of potassium, makes the potassium-40 ratio of doubtful utility.

17. Let $c$ and $b$ be the total number of counts and the total number of background counts recorded in a given time $t$. Then $s = c - b$, $\sigma_s^2 = \sigma_c^2 + \sigma_b^2$, and $s = m\sigma_s$ for a signal significant at the $m\sigma$ level. Solving for $s$, we have

$$s = \frac{m^2}{2}\left[1 + \left(1 + \frac{8b}{m^2}\right)^{1/2}\right]$$

If $b = Bt = (10.9 \text{ cps})(600 \text{ seconds}) = 6{,}540$ counts, and $m = 3$, then $s = 348$ counts, which is 0.58 cps or 5.3 percent of the background rate. Similarly, if $B = 22$ cps, $s$ would be 0.82 cps or 3.7 percent of $B$.

18. In this case, equation 13.3 must include in $B_i$ the background generated by the warhead itself due to Compton-scattered gamma rays. HYPERMET produces estimates of $s$ and $\sigma_s$ for the measurements taken directly over the warhead, where the background, under the lines is dominantly due to these Compton-scattered gammas from the warhead. Since $\sigma_s^2 = \sigma_c^2 + \sigma_b^2 = c + b = s + 2b$, $b = (\sigma_s^2 - s)/2$. Therefore, the Compton-scattered component of the background at $r_{max}$ is given approximately by $(\sigma_s^2 - s)(r_i/r_{max})^2/2$.

19. By comparing the intensity of the 911-keV actinium-228 line to the intensity of the 583- and 2,614-keV thallium-208 lines as measured on the deck, we estimate, after correcting for differences in detector efficiency and self-absorption by the steel, that 79 $\pm$ 17 percent of the thallium-208 counts on the deck arise from the decay of thorium-232. The 2,614-keV background count rate was therefore $0.0058 \pm 0.0017$ cps on the deck; since the ratio of 911-keV count rate on the launch tube to that on the deck was $2.8 \pm 1.2$, the background count rate on the launch tube was $0.016 \pm 0.008$ cps.

20. If we adjust the thickness of steel in the launch tube lid so that the 186-keV and 2,614-keV lines have the correct intensity at point 3, then we would predict the 769-keV line to be much lower in intensity (and the 414-keV line to be greater in intensity) than we observed at this point. The results are, however, roughly consistent with a model in which the plutonium is shielded by heavy metal in direction 3 but that the uranium is not (as might be the case in a thermonuclear weapon with the secondary facing forward).

21. G. Lewis, S. Ride, and J. Townsend, "Dispelling Myths About Verification of Sea-Launched Cruise Missiles," *Science* **246**, 765 (10 November 1989).

22. Ibid.

23. For a Pickering-type CANDU reactor. The concentrations of plutonium-239 and plutonium-240 for a given plutonium-241 concentration are very similar in a graphite production reactor.

24. Thomas B. Cochran, William M. Arkin, Robert S. Norris, and Milton M. Hoenig, *Nuclear Weapons Databook Volume 2: U.S. Nuclear Warhead Production* (New York: Ballinger, 1987).

25. Ibid.

# Chapter 14

S.T. Belyaev
V.I. Lebedev
B.A. Obinyakov
M.V. Zemlyakov
V.A. Ryazantsev
V.M. Armashov
S.A. Voshchinin

# The Use of Helicopter-borne Neutron Detectors to Detect Nuclear Warheads in the USSR–US Black Sea Experiment

The Soviet Navy used a helicopter/ship system called "Sovietnik" to detect a cruise-missile warhead in the joint USSR–US experiment on the Black Sea, which took place on 5 July 1989. The system consists of a ship-based helicopter, equipped with a neutron detector and processing equipment, and associated ship-based equipment. The system was developed at the I.V. Kurchatov Institute of Atomic Energy. Its operation is based on detecting the neutrons emitted from the spontaneous fission of the plutonium-240 contained as an impurity in weapon plutonium. The Sovietnik neutron detector is based on helium-3 counters in a moderator. It is designed to detect the neutron flux from a single nuclear warhead at distances up to 100–150 meters. At these distances, the flux does not exceed 10 percent of the natural background and is comparable with its variations. Therefore when developing the Sovietnik system, great attention was given to factors affecting the level and behavior of the background radiation.

The neutron detector, together with instrumentation for recording, preliminary processing, and analysis of initial information, is located on the helicopter, which carries out the measurement while flying slowly past the ship in question. The presence of fissile material on the ship is considered to be established if the measured neutron-radiation signal in the ship zone is higher than a "threshold" defined as a level that exceeds expected background fluctuations by three standard deviations, allowing for possible background variations and other uncertainties of statistical and nonstatistical character.

The instrumentation system includes a helicopter altimeter and a laser range finder. The data from these instruments are recorded synchronously with those from the neutron detector. The laser range-finder measures the distance to the ship, which makes it possible to estimate the absolute neutron flux and thus, in some cases, if information about the isotopic composition of the plutonium is available, to estimate the amount of plutonium. The helicopter is also equipped with a videorecorder, making possible the visual correlation of information from the neutron detector with a particular region of the ship.

Additional equipment is located on an accompanying ship, where detailed processing and analysis of the data recorded on the helicopter takes place. This system includes a computer, database, and video-reproduction system. Operators on the ship see the image recorded by the helicopter video recorder on a TV monitor in real time. The ship instrumentation also includes a neutron detector to measure the background during the helicopter system's operation. A block diagram of the Sovietnik complex is presented in figure 14.1.

The Sovietnik system operates in several modes: background measurements, measurements taken from the helicopter when it is hovering over the ship, and measurements taken from the helicopter when it is not close to a ship. The measurement results, after preliminary analysis on the helicopter, are displayed on the helicopter control panel for the pilot-operator to decide whether a repeat-flight is required. Detailed processing is done later on the accompanying ship, using either an integrating algorithm that calculates the average neutron radiation from the entire ship, or a zone-processing algorithm that calculates the neutron radiation from individual zones along the length of the ship. The zone algorithm can distinguish between multiple radiation sources if the distance between them is at least comparable with distance between the ship and helicopter. If this is the case, the Sovietnik system can detect nuclear weapons even on ships with a nuclear reactor.

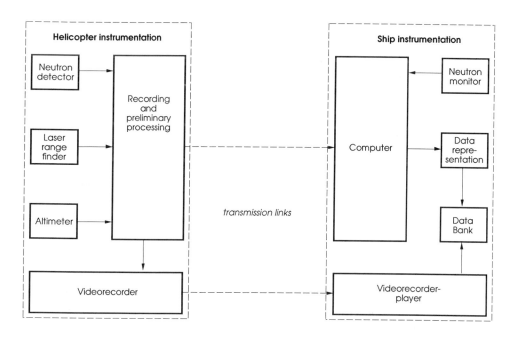

**Figure 14.1:** The Sovietnik complex

The ability of the Sovietnik system to detect sea-based nuclear weapons has been verified by tests carried out under natural conditions in the Black Sea and the Mediterranean during 1978–79. During the tests several NATO ships, including some from the US Navy, were inspected. Nuclear weapons were detected on some US ships in the Mediterranean. For example: the frigate *Truett*, was inspected on 21 January 1978 in the Mediterranean at the point 32° 35′ N, 25° 50′ E; the aircraft carrier *America* was inspected on 7 February 1978 at the point 38° 31′ N, 18° 35′ E; the destroyer *Barry* was inspected twice, on 30 January 1978 at the point 33° 58′ N, 21° 00′ E, and on 11 May 1979 at the point 42° 27′ N, 06° 16′ E; and the weapons and special equipment transport ship *Mount Baker* was inspected on 16 April 1979 in the Mediterranean at the point 40° 17′ N, 13° 39′ E.

In the experiment on the Black Sea, the missile cruiser *Slava* was inspected by two helicopters equipped with the Sovietnik system and flying from the base ship *Apserhon*. One helicopter flew by the *Slava* at a distance of about 30 meters, the other, at a distance of about 80 meters. The results are shown in tables 14.1 and 14.2, and in figure 14.2. They give the distribution of the neutron radiation in 10 spatial zones along the length of the ship. It follows from the integral processing of the data obtained from the 30-meter flight, that with a probability not lower than 0.95, the missile cruiser *Slava* had a neutron source on board. The results of zone processing of the data obtained at both distances show that this source is in spatial zone number 7, corresponding to the location of the launcher containing a cruise missile known to have a nuclear warhead. The detection time in this zone did not exceed 10 seconds.

The Sovietnik system should be considered only as one part of a monitoring system to provide remote detection of nuclear weapons on board ships. There are a number of issues which have not been considered here, such as the problems relating to concealment of nuclear weapons. It also appears that the identification of nuclear-weapon types during the measurements can only be accomplished by closer inspection. Nevertheless, the Black Sea experiment using the Sovietnik system has proven the possibility of confidence building through cooperative remote monitoring.

We hope to be able to publish a fuller description of the Sovietnik neutron detector in the near future.

Table 14.1: Helicopter about 30 meters from ship

| Bin | Distance to stern *meters* | Measurement length *second⁻¹* | Counts *second⁻¹* | Count minus background *second⁻¹* | $3\sigma$ detection threshold* *second⁻¹* | Above threshold? |
|---|---|---|---|---|---|---|
| 0 | 0–17.59 | 14 | 14.07 | 1.28 | 2.25 | no |
| 1 | 37.45 | 10 | 14.20 | 1.40 | 2.60 | no |
| 2 | 56.17 | 11 | 13.36 | 0.41 | 2.49 | no |
| 3 | 74.85 | 10 | 14.90 | 1.85 | 2.60 | no |
| 4 | 92.88 | 8 | 13.63 | 0.66 | 2.89 | no |
| 5 | 111.83 | 10 | 13.90 | 1.00 | 2.60 | no |
| 6 | 130.11 | 13 | 13.54 | 0.71 | 2.32 | no |
| 7 | 150.29 | 11 | 20.18 | 6.87 | 2.49 | yes |
| 8 | 166.91 | 7 | 16.71 | 2.58 | 3.07 | no |
| 9 | 187.00 | 13 | 15.85 | 2.33 | 2.32 | maybe |

Table 14.2: Helicopter about 76 meters from ship

| Bin | Distance to stern *meters* | Measurement length *second⁻¹* | Counts *second⁻¹* | Count minus background *second⁻¹* | $3\sigma$ detection threshold* *second⁻¹* | Above threshold? |
|---|---|---|---|---|---|---|
| 0 | 0–17.77 | 8 | 13.50 | 1.95 | 2.71 | no |
| 1 | 37.76 | 9 | 12.22 | 0.41 | 2.57 | no |
| 2 | 55.99 | 8 | 11.63 | 0.86 | 2.72 | no |
| 3 | 74.39 | 8 | 10.63 | −0.03 | 2.72 | no |
| 4 | 92.89 | 8 | 10.88 | 0.30 | 2.72 | no |
| 5 | 111.39 | 8 | 11.75 | 1.21 | 2.72 | no |
| 6 | 131.47 | 9 | 11.44 | 0.88 | 2.58 | no |
| 7 | 149.24 | 8 | 13.25 | 2.92 | 2.73 | yes |
| 8 | 169.23 | 9 | 11.56 | 1.21 | 2.58 | no |
| 9 | 187.00 | 8 | 11.50 | 1.26 | 2.72 | no |

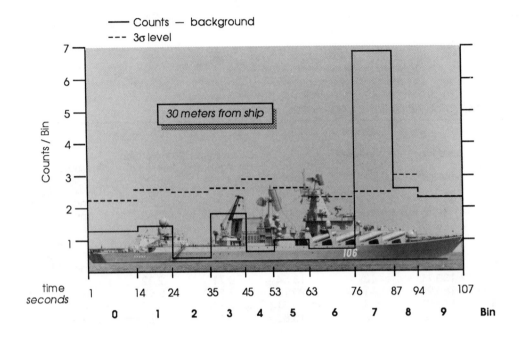

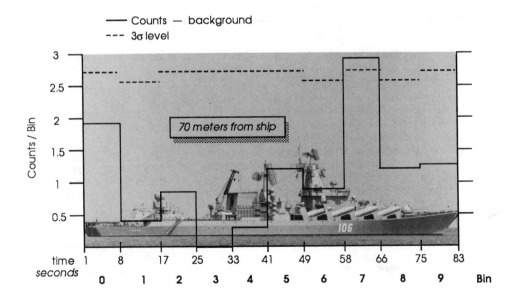

**Figure 14.2:** Plots of tables 14.2 and 14.3: helicopter about 30 meters from ship (above) and 75 meters from ship. The missile is in the forward launcher (see ship outline)

# THE AUTHORS

**David Albright,** a physicist, is a member of the staff of the Federation of American Scientists, 307 Massachusetts Avenue NE, Washington DC 20002

**V.M. Armashov** is a member of the physics division of the I.V. Kurchatov Institute of Atomic Energy, Moscow

**Spartak T. Belyaev** is director of the physics division of the I.V. Kurchatov Institute of Atomic Energy, Moscow

**Thomas Cochran,** a physicist, is a member of the professional staff of the Natural Resources Defense Council, 1350 New York Avenue NW, Washington DC 20005

**Harold Feiveson,** a policy analyst, is a member of the staff of the Center for Energy and Environmental Studies, Princeton University, Princeton, NJ 08544

**Steve Fetter,** a physicist, is an assistant professor in the School of Public Affairs of the University of Maryland College Park, MD 20742

**Valery A. Frolov,** a physicist, is a researcher at the Space Research Institute of the Soviet Academy of Sciences, 84/32 Profsoyuznaya U., 117810 Moscow

**Mikhail Gerasev,** a political scientist, is a senior researcher at the Institute of US and Canadian Studies of the Soviet Academy of Sciences, 2/3 Khlebny per., Moscow

**Lee Grodzins,** is a professor of physics at the Massachusetts Institute of Technology, Cambridge, MA 02139

**Andrei S. Kokoshin,** a political scientist and deputy chairman of the Committee of Soviet Scientists (CSS), is deputy director of the Institute of US and Canadian Studies and a corresponding member of the Soviet Academy of Sciences

**V.I. Lebedev** is a member of the physics division of the I.V. Kurchatov Institute of Atomic Energy, Moscow

**Thomas Longstreth**, a political scientist, is the associate director for Strategic Weapons Policy of the Federation of American Scientists, 307 Massachusetts Avenue NE, Washington DC 20002

**Harvey L. Lynch**, a physicist, is a member of the staff of the Stanford Linear Accelerator Center, Stanford University, Bin 96, PO Box 4349, Stanford, CA 94309

**Marvin Miller**, a physicist, is a professor of nuclear engineering at the Massachusetts Institute of Technology

**Robert Mozley**, a physicist, is emeritus professor at the Stanford Linear Accelerator Center, Stanford University. His home address is 601 Laurel Ave, Menlo Park, CA 94025

**B.A. Obinyakov** is a member of the physics division of the I.V. Kurchatov Institute of Atomic Energy, Moscow

**Sergei K. Oznobishchev**, a political scientist, is a senior researcher in the Institute of US and Canadian Studies, Moscow

**Oleg F. Prilutsky**, a physicist and member of the CSS, is a department head at the Space Research Institute

**Stanislav N. Rodionov**, a physicist and member of the CSS, is a department head at the Space Research Institute, Moscow

**V.A. Ryazantsev** is a member of the physics division of the I.V. Kurchatov Institute of Atomic Energy, Moscow

**Roald Z. Sagdeev**, a physicist and chairman of the CSS, is the head of the Center of Analytic Studies of the Soviet Academy of Sciences' Space

Research Institute, a member of the Academy, and a member of the Congress of Deputies of the USSR

**Richard Scribner**, a physicist, is an associate professor at the School of Foreign Service of Georgetown University, Washington DC 20057-0001

**Theodore B. Taylor**, a physicist, was formerly a nuclear weapons designer. His home address is 2232 Weatherby Road, PO Box 39, West Clarksville, NY 14786

**Valerie Thomas**, a physicist, is a research associate in Princeton University's Center for Energy and Environmental Studies

**Alexei A. Vasiliev**, a political scientist who died in November 1989, was a department head at the Institute of US and Canadian Studies, Moscow

**Frank von Hippel**, a physicist and Professor of Public and International Affairs at Princeton University, is chairman of the research arm of the FAS

**S.A. Voshchinin** is a member of the Soviet Navy

**M.V. Zemlyakov** is a member of the physics division of the I.V. Kurchatov Institute of Atomic Energy, Moscow

**Martin S. Zucker**, a physicist, is a member of the staff of Brookhaven National Laboratory, Upton, NY 11973

# GLOSSARY

*The glossary is divided into two sections: weapons and arms-control related, and physics-related. Cross-references to other definitions are indicated by italics.*

## Weapons and Arms Control

**ABL.** Armored Box Launcher. A launcher for SLCMs located above deck on many US surface ships.

**ABM treaty.** Among other constraints, the Antiballistic Missile Treaty of 1972 limits the US and USSR to 100 single-warhead ground-based interceptor missiles each, bans the testing or deployment of spaced-based ballistic-missile defenses, and requires that high-powered radars that could be used for battle management be located on national boundaries and facing outward.

**ALCM.** Air-launched *Cruise Missile.*

**ASM.** Air-to-surface Missile. Usually refers not to *ALCMs* but to *SRAMs.*

**ASW.** Antisubmarine Warfare.

**B-1.** A US strategic bomber deployed during the 1980s. Carries bombs and short-range attack missiles (*SRAMs*).

**B-2.** A US strategic bomber first flight tested in the late 1980s. Incorporates *stealth* features designed to make it less detectable by radar or infrared sensors.

**B-52.** A US strategic bomber deployed during the late 1950s and early 1960s. Carries a mix of *ALCMs*, *SRAMs*, and bombs.

**ballistic missile.** A nuclear missile that spends most of the time between its launch and impact point(s) in space where its trajectory is determined primarily by inertia and the influence of the earth's gravity.

**barrage attack.** An attack in which many warheads are used to attack the area around a bomber or mobile-missile base in an effort to destroy the scattering aircraft or missile carriers.

**Bear.** US name for the Soviet Tupolev-95 prop-jet powered strategic bomber that has been produced since the late 1950s. The most modern version, the Bear-H, produced during the 1980s and still in production, can carry six air-launched cruise missiles.

**Blackjack.** US name for the Soviet Tupolev-160 strategic bomber first deployed in small numbers in the late 1980s.

**BMD.** Ballistic-missile Defense. A system to defend against ballistic missiles.

**boosted-fission explosion.** A *fission* explosion whose yield is increased in its final phase by a burst of neutrons liberated by the *thermonuclear* reaction, T (tritium) + D (deuterium) → He⁴ (helium-4) + *n* (neutron).

**breakout.** The rapid deployment of weapon systems beyond limits agreed to in an arms-control treaty.

**C³ system.** Command, control, and communication system for a military organization.

**CLS.** Capsule Launch System for SLCMs that is deployed on some US attack submarines.

**counterforce weapon.** A weapon designed to attack the weapon systems of the other side.

**cruise missile.** A small unpiloted subsonic jet aircraft designed to crash into its target carrying a conventional or nuclear *warhead*. Modern cruise missiles achieve high accuracies using radar or optical sensors to recognize ground contours or contrasts.

**declared warheads or facilities**. Warheads or facilities whose existence are revealed to the other side as part of an arms-control agreement.

**deMIRVing**. Reducing the number of warheads on a multiple-warhead (*MIRV*ed) ballistic missile.

**early-warning system**. A system for detecting attacking weapons well before they reach their targets. Ballistic-missile early-warning systems currently include satellites in geosynchronous orbits that can detect the infrared emissions of booster rockets shortly after launch and powerful radars that can detect *re-entry vehicles* once they rise above the radar's horizon. Bomber early-warning systems use both radars that are limited by the horizon and long-range "over the horizon" radars.

**EMT**. Equivalent *Megatons*. The number of 1-megaton *warhead*s that could cover with a given level of peak blast overpressure the same area as the warhead or collection of warheads being discussed.

**enrichment plant**. A facility that produces *enriched uranium*.

**enriched uranium**. *Uranium* in which the concentration of uranium-235 has been increased above the natural level of 0.711 percent.

**externally observable differences**. An observable (usually from an observation satellite) difference between a treaty-limited weapons system (e.g., a strategic bomber equipped to carry *ALCM*s) and a different weapon system not subject to the same limits (e.g., a bomber with the same basic airframe but not equipped to carry ALCMs).

**fingerprint**. Any method for reliably differentiating a system (e.g., a nuclear warhead) of a particular type from other types or from counterfeits.

**finite-deterrence**. Used in this book to describe a posture in which the US and USSR each have enough survivable *strategic nuclear weapons* to hold each other hostage but do not deploy thousands of extra warheads to attack each other's strategic forces.

**first strike**. Ordinarily refers to a hypothetical surprise attack against the other side's strategic nuclear forces.

**functionally related observable differences**. *Externally observable differences* of a treaty-limited weapon system that contribute to the specific capabilities that distinguish it from other weapon systems.

**generated alert**. The posture that a military force would assume if it had sufficient warning of a possible attack. In the case of the strategic nuclear forces: all ballistic-missile submarines that could put out to sea would do so; bombers that were operational would be dispersed to a larger number of bases and prepared for rapid takeoff, and all operational mobile-missile launchers would be dispersed from their bases.

**GLCM**. Ground-launched Cruise Missile. GLCMs with a range of more than 500 kilometers were banned by the *INF treaty* of 1987.

**hardened military system or facility**. A system or facility that has been made resistant against certain nuclear-weapon effects—usually blast. Missile silos are hardened by sinking them into the ground and lining them with reinforced concrete. The hard mobile launcher of the Midgetman missile covers the missile with a strong shield that is designed to be lowered and seal itself to the ground. Often hardness against blast is measured in terms of the number of atmospheres or pounds per square inch blast overpressure that a system can survive.

**heavy bomber.** A strategic bomber.

**highly enriched uranium (HEU).** Uranium enriched to above 20 percent in uranium-235. Although *weapon-grade uranium* contains more than 90 percent uranium-235, HEU can be used to produce a nuclear explosion.

**IAEA.** The International Atomic Energy Agency, headquartered in Vienna, has among its responsibilities the verification that fissile material has not been diverted from civilian to weapons use in countries that have signed the *Non-Proliferation Treaty* or have placed some of their nuclear facilities under IAEA safeguards as a result of another agreement.

**ICBM.** Intercontinental *Ballistic Missile,* by convention taken to be a missile with a range greater than 5,500 kilometers (the approximate shortest distance from the 48 contiguous US states to the USSR).

**INF treaty.** In the Intermediate (and Shorter) Range Nuclear Forces Treaty of 1987 the US and USSR agreed to eliminate all of their ground-launched ballistic and cruise missiles with ranges between 500 and 5,500 kilometers.

**IRBM.** Intermediate-range (1,000–5,500 km) (land-based) Ballistic Missile. Banned by the *INF treaty.*

**kiloton.** A unit of energy used to measure the power of explosions: $10^{12}$ calories or $4.2 \times 10^{12}$ joules of energy. (The approximate amount of energy released by a thousand short tons [907,000 kilograms] of chemical high explosive.)

**launch-on-warning posture.** A posture in which vulnerable systems are poised to be launched before attacking warheads can destroy them. In non-crisis conditions, about 30 percent of US bombers are so poised but they can be recalled if the alarm is found to be false. This is not true of ballistic missiles.

**MARV.** A MAneuvering Re-entry Vehicle. The MARVs that the US has tested maneuver within the atmosphere aerodynamically.

**megaton.** A thousand *kilotons.*

**Midgetman.** The name for a single-warhead mobile *ballistic missile* under development by the US.

**Minuteman.** The first US solid-fuel *ICBM;* developed during the 1960s. The US currently has approximately 450 single-warhead Minuteman II and 500 three-warhead Minuteman III ICBMs deployed in silos.

**MIRVs.** Multiple Independently Targetable Re-entry Vehicles for ballistic missiles. The individual re-entry vehicles are put on separate trajectories in space by a "post-boost vehicle" or "bus." The maximum distance between the targets that can be attacked by re-entry vehicles launched by the same booster rockets (the missile "footprint") depends upon the amount of fuel carried by the post-boost vehicle and the time of flight of the re-entry vehicles on their separate trajectories.

**mobile missiles.** Usually refers to *land*-mobile missiles mounted on either railroad cars or trucks. Mobility can improve missile survivability by making the missile's location uncertain to the other side's targeters. If the missile carrier is not frequently moved in peacetime, however, this survivability depends upon a certain amount of warning of attack. The short time of warning of ballistic-missile attack that would be available from early-warning systems would be most useful for a hardened mobile launcher which could survive all but nearby explosions.

**MX**. The newest US *ICBM*, deployed during the 1980s, which carries 10 warheads. Fifty MX missiles are currently based in silos. Rail basing has been proposed by the Bush Administration and was being debated by the Congress at the time that this book went to press.

**national technical means**. Refers to systems for gathering information useful for verification from space (e.g., imaging or telemetry interception satellites) or from beyond a nation's borders (e.g., radars or seismometers).

**NPT**. Non-Proliferation Treaty of 1970, under which over 100 non-nuclear weapon states have committed themselves to open up their nuclear activities to *IAEA* verification that fissile material is not being diverted to weapons use. In exchange, the US, USSR and UK committed themselves to "pursue negotiations in good faith on effective measures relating to cessation of the nuclear arms race at an early date and to nuclear disarmament, and on a treaty on general and complete disarmament under strict and effective international control." The parties to the NPT will meet in 1995 "to decide whether the Treaty shall continue in force indefinitely, or shall be extended for an additional fixed period or periods."

**NRDC**. Natural Resources Defense Council, a US environmental group that has cooperated with the Soviet Academy of Sciences to demonstrate arms-control verification arrangements.

**nuclear warhead**. The part of a nuclear weapon that contains the nuclear explosive, such as a bomb.

**nuclear weapon**. The entire system including the launcher (e.g., a missile silo, submarine or bomber), the delivery vehicle (e.g., a missile) and the *nuclear warhead*.

**overpressure**. The brief increase in air pressure associated with the shock wave generated by a nuclear explosion in the atmosphere.

**Partial Nuclear Test Ban Treaty**. The 1963 treaty that banned tests of nuclear explosives in the atmosphere, space and in the ocean but not underground.

**penetration aids**. Usually light objects carried by ballistic missiles which are designed to confuse the attacked country's *BMD* radars and other sensors in order to degrade its ability to intercept the ballistic-missile re-entry vehicles.

**Pershing II**. A US intermediate-range ballistic missile which was deployed in Western Europe but has been withdrawn and destroyed under the terms of the *INF treaty*. The Pershing II was the only deployed ballistic missile equipped with a maneuvering re-entry vehicle (*MARV*) to improve its accuracy.

**plutonium**. An artificial element produced as a result of neutron absorption on uranium-238 and subsequent radioactive transformations.

**portal**. An opening in a controlled perimeter where objects passing in or out can be checked to see if they conform to regulations, laws, or treaties. Passengers and carry-on baggage are checked at a portal for concealed weapons before they enter commercial passenger aircraft.

**Poseidon**. The oldest deployed US *SLBM* which typically carries 10 warheads. Poseidon-carrying submarines have 16 launch tubes. As of 1990, twelve had been re-equipped with eight-warhead *Trident I* missiles.

**primary**. Usually the fission explosive that ignites the *thermonuclear secondary* of a nuclear warhead.

**production reactor**. A nuclear reactor in which a significant fraction of the neutrons is used to produce *radionuclides* such as weapon-grade plutonium and tritium.

**re-entry vehicle**. A *ballistic-missile* warhead complete with heat shield for surviving high-speed re-entry through the earth's atmosphere.

**reprocessing plant**. A facility where *spent nuclear fuel* or uranium that has been irradiated in a production reactor is dissolved and the uranium and plutonium are separated by remote chemical processing from fission products and other materials.

**safeguards**. Typically refers to arrangements supervised by the *IAEA* to verify that *fissile material* has not been diverted from peaceful to weapon use.

**SALT**. Strategic Arms Limitation Treaty. SALT I (1972) limited strategic *ballistic-missile* launchers (*silos* and launch tubes in submarines). The unratified SALT II (1979) limited bombers as well and imposed sublimits on numbers of deployed ballistic missiles that could be equipped with multiple re-entry vehicles (MIRVs) and bombers that could be equipped with air-launched cruise missiles (ALCMs).

**seal**. A device that reveals whether a container has been opened up or not. Seals could be used to reveal, for example, whether a declared non-nuclear cruise missile had in fact been opened up and possibly converted to a nuclear cruise missile.

**secondary**. The nuclear explosive that is ignited by the fission *primary* in a *thermonuclear warhead*.

**shielding**. A material placed so as to attenuate the penetrating radiation emitted by a nuclear warhead or *nuclear reactor*.

**silo**. A vertical reinforced concrete structure set into the ground and used to house ICBMs.

**SLBM**. A Submarine-launched *Ballistic Missile*.

**SLCM**. Sea-launched *Cruise Missiles* can be launched from a variety of launchers including, in the case of small modern long-range cruise missiles, torpedo tubes.

**spent reactor fuel**. Nuclear fuel that has been withdrawn from a *nuclear reactor* core and replaced because most of its *fissile material*—usually uranium-235—has been fissioned.

**SRAM**. Short-range Attack Missiles are launched from aircraft. SRAMs are supersonic and generally powered by solid-fuel rockets; they can therefore be targeted ahead of the bomber to destroy air-defense installations.

**SS-18, SS-20, SS-24, SS-25, etc.** US designations for different types of Soviet land-based surface-to-surface ballistic missile. The 10-warhead SS-18 is the largest Soviet ICBM and its deployed numbers (308) are to be halved by the START treaty; the SS-20 was a three-warhead intermediate-range ballistic missile that was eliminated under the INF treaty. The SS-24 is a 10-warhead missile that has been deployed both in silos and on railway cars. The SS-25 is a single-warhead truck-mounted missile.

**SS-N-20, SS-N-23, etc.** US designations for Soviet submarine-based ballistic missiles.

**START**. **ST**rategic **A**rms **R**eduction **T**alks. These talks were in an advanced stage at the time this book went to press. The proposed START treaty would limit *counted* strategic warheads to 6,000. However, because of agreed "counting rules" that count a non-*ALCM* bomber as carrying only a single warhead and undercount warheads on some ALCM bombers, it would allow thousands of additional uncounted strategic warheads to be deployed.

**stealth**. The use of special materials and other techniques to reduce the detectability of radar reflections and/or infrared emissions from an aircraft, cruise missile or re-entry vehicle.

**strategic nuclear weapon**. A US or Soviet nuclear weapon that can be used to attack the other nation's homeland. The US has refused to include in strategic arms limitations negotiations either aircraft carried weapons on bases and aircraft carriers located within range of the USSR or nuclear *SLCMs*.

**tactical nuclear weapon**. Nuclear weapons systems that generally have a shorter range than strategic weapon systems and are designed for attacks against the other side's weapons (e.g., submarines at sea or tanks on the battlefield), forward bases, transport bottlenecks, etc.

**tag**. A unique label that can be attached to a weapon or intrinsic characteristic of that weapon that can be used to identify it. A tag should be very difficult to reproduce and should reveal if it has been moved from its original location.

**tamper**. A layer of material between the chemical explosives and the *fissile material* in a fission explosive that develops inward momentum as a result of the chemical explosion and transfers this momentum to the fissile

material. A tamper can also serve as a *neutron reflector* to reduce the number of neutrons escaping the fissioning core.

**thermonuclear warhead**. A nuclear explosive in which a significant fraction of the energy release comes from fusion reactions.

**Tomahawk**. The name of the US *SLCM*.

**Trident**. The most modern US *ballistic-missile* submarine, has 24 missile-launch tubes.

**Trident I**. An *SLBM* carrying eight warheads that was deployed on 12 *Poseidon* and eight *Trident* submarines during the 1980s.

**Trident II**. An *SLBM* with twice the launch weight as the Trident I which is to be deployed with eight heavy warheads on future *Trident* submarines.

**VLS**. Vertical Launch System embedded in the decks of US ships, stores and launches cruise and conventional missiles.

**weapon-grade plutonium (WgPu)**. Defined by the US government to be plutonium that contains less than about 6 percent of the spontaneously fissioning isotope Pu-240. Taking into account the presence of plutonium-238, plutonium-241 and plutonium-242, the percentage of plutonium-239 is about 93 percent. Nuclear weapons can be made with virtually any concentration of plutonium-240. Therefore almost any isotopic mix of plutonium is weapon useable.

**weapon-grade uranium (WgU)**. Defined by the US government as containing about 94 percent uranium-235. Uranium enriched to above about 20 percent in uranium-235 is weapon useable.

# Physics

The International System of Units (SI) uses a set of standard prefixes to denote the magnitudes of its base units. Here are the prefixes used in the book together with their abbreviations.

| tera | T | $10^{12}$ |
|------|---|-----------|
| giga | G | $10^{9}$ |
| mega | M | $10^{6}$ |
| kilo | k | $10^{3}$ |
| - | | $10^{0}$ |
| milli | m | $10^{-3}$ |
| micro | $\mu$ | $10^{-6}$ |
| nano | n | $10^{-9}$ |
| pico | p | $10^{-12}$ |

Each is separated from its neighbor by a factor of 1,000.

**accelerator**. A device in which charged particles (e.g., electrons) are accelerated by use of oscillating electric and magnetic fields to energies of typically *MeV*s or greater.

**activation**. The transmutation of a nonradioactive into a radioactive isotope by a nuclear reaction such as a neutron absorption.

**active detection**. The bombardment of an object with radiation and measurement of either the radiation that passes through (radiography) or the detection of a different type of radiation emerging from the object as a result of a nuclear reaction between the bombarding particle and nuclei inside the object (e.g., induced fission).

**alpha ($\alpha$) particle (or ray)**. The *nucleus* of an ordinary helium atom containing two *protons* and two *neutrons*. Many heavy nuclei, such as those of uranium and plutonium, are unstable and spontaneously decay into the nuclei of other elements by emitting an alpha particle.

**background radiation**. *Radiation* from natural (e.g., naturally occurring *radionuclides* or cosmic rays) or artificial sources other than the object whose radiation output is to be detected.

**barn**. A (non-SI) unit of area equal to $10^{-24}$ square centimeters used in nuclear physics to measure nuclear reaction *cross sections*.

**beryllium**. A light metal which is often used as a *neutron reflector*.

**beta ($\beta$) particle (or ray)**. A high-energy electron or positron that has been emitted by a *nucleus*. The emission of a negatively charged electron is accompanied by the emission of a *neutrino* and the transformation of a *neutron* in the nucleus into a *proton*. The emission of a positively charged positron is accompanied by the transformation of a proton into a neutron.

**bremsstrahlung**. Literally, "braking radiation" in German. When a fast electron is suddenly stopped or its direction altered, some of its electric field can be transformed into free *photons*. This is the way in which x-rays are ordinarily produced.

**chain reaction**. Usually a *fission* neutron chain reaction in which at least one neutron released by a fissioning nucleus is absorbed and causes another fission.

**collimator.** An aperture in a shield around a *radiation detector* that can be used to determine the direction from which the radiation is coming.

**Compton scattering.** A scattering of a *photon* (e.g., an *x-ray* or *gamma ray*) off an electron—usually with a considerable transfer of energy to the electron.

**critical mass.** A mass of material whose *fissile* content (e.g., of uranium-235) is large enough and concentrated enough so that it can sustain a neutron *chain reaction*.

**cross section.** Represents the equivalent target area of nucleus for a given type of reaction.

**curie (Ci).** A unit quantity of any radioactive nuclide in which $3.7 \times 10^{10}$ disintegrations occur per second.

**decay constant.** The reciprocal of the average lifetime of a radioactive isotope. The average lifetime is approximately 1.44 times the halflife. Therefore, for example, tritium with a halflife of 12.3 years has a decay constant of 0.056 per year, corresponding to a 5.6 percent annual decay rate.

**delayed neutrons or gamma rays.** A neutron or gamma ray resulting from fission that is not produced by the fission process itself but is instead released as a result of later spontaneous de-*excitations* or radioactive transformations of the fission products.

**depleted uranium.** Uranium with less than the natural percentage (0.711 percent) of uranium-235. Depleted uranium is produced as a byproduct of uranium enrichment.

**deuterium.** A naturally occurring heavy isotope of hydrogen (1 in 10,000 hydrogen atoms) that contains a neutron as well as a proton in its nucleus.

**deuteron.** A deuterium nucleus—a proton–neutron pair.

**electron volt (eV).** The kinetic energy of an electron that has been accelerated through a potential drop of one volt. About $1.6 \times 10^{-19}$ joules or $1.6 \times 10^{-12}$ ergs.

**energy level.** An internal energy state of a nucleus. Quantum systems can exist in discrete states with specific amounts of internal energy. This is why the *gamma ray* emitted in a transition between two such states carries away a characteristic amount of energy.

**energy resolution.** The accuracy with which a *gamma-ray* detector measures the gamma-ray energy.

**excited state.** An *energy level* which is higher than the lowest energy state of a nucleus.

**fissile isotope.** An isotope such as uranium-235 or plutonium-239 that, in sufficient quantities (a *critical mass*) can support a fission chain reaction even when the neutrons have been slowed by collisions with the nuclei of a *neutron-moderating* material—as in a nuclear reactor. Some nonfissile artificial isotopes, such as plutonium-240, can nevertheless support a chain reaction with the fast neutrons produced by fission and therefore could be used to produce a nuclear explosion.

**fission.** A nuclear reaction in which a heavy nucleus—usually of uranium or plutonium—splits into two medium-weight nuclei and typically 2–3 neutrons. Spontaneous fission occurs as one of the radioactive decays of these nuclei. Induced fission occurs as a result of an input of a particle (usually a neutron) or energy from a collision or the absorption of a *gamma ray*.

**fission product.** One of the intermediate-weight atoms that results from *fission*.

**fusion.** A reaction in which two light nuclei combine to produce a heavier nucleus. The most important fusion reaction in nuclear explosives is tritium + deuterium → helium-4 + neutron.

**gamma (γ) ray.** A high-energy photon emitted by a nucleus.

**gigawatt (GW).** One billion ($10^9$) *watts*. Large nuclear-power or production reactors generate 2–4 gigawatts of thermal power.

**graphite.** A form of carbon used as a *neutron moderator* in some nuclear reactors. Carbon slows neutrons by collisions without absorbing them.

**gray.** Corresponds to the deposition of 1 joule of energy by *ion*izing radiation (e.g., gamma rays, x-rays, fast electrons, neutrons or protons) per kilogram of a target material.

**halflife.** The length of time in which one half of a given number of the atoms of a specific radioactive isotope will decay radioactively.

**heavy water.** Water ($H_2O$) in which most of the hydrogen is the heavy isotope *deuterium*.

**ion.** A charged atom—usually one that has lost one or more of its electrons or has an extra one or more attached.

**isotope.** An atom of a particular element with a specific number of neutrons, such as uranium-235 or uranium-238. The designating number equals the number of neutrons *plus* the number of protons (92 in the case of uranium).

**kelvin (K).** The SI unit of absolute (not degree of) temperature. Zero K is as cold as it can get. Zero degrees celsius (centigrade) = 273 K.

**keV (kiloelectron volt).** A thousand electron volts.

**kilowatt (kW).** One thousand *watts*.

**megawatt (MW).** One million *watts*.

**megawatt-day.** The amount of energy released in a day by a 1-megawatt flow of power. Approximately the amount of energy released by the fission of 1 gram of uranium or plutonium.

**MeV (megaelectron volt).** One million electron volts.

**millibarn.** $10^{-3}$ barns.

**nanosecond.** $10^{-9}$ seconds.

**nuclear reactor.** A facility in which a *fission chain reaction* is sustained at a controlled rate.

**neutrino.** A massless uncharged particle associated with, inter alia, nuclear reactions.

**neutron (*n*).** A neutral particle that, along with the positively charged and approximately equal massive proton, occurs in the nuclei of all atoms other than ordinary hydrogen. Nuclei that contain the same number of protons but different numbers of neutrons are *isotopes* of the same element.

**neutron fluence.** The cumulative number of neutrons that have crossed a unit area—e.g., $10^{21}$ neutrons/cm$^2$.

**neutron flux.** The number of neutrons crossing a unit area per unit time—e.g., $10^{14}$ neutrons per cm$^2$ per second.

**neutron moderator.** A material in

which neutrons are slowed by collisions with atomic nuclei. Materials containing high concentrations of hydrogen are particularly good neutron moderators since, after a collision, the light hydrogen nucleus can carry away a large fraction of the neutron energy.

**neutron reflector**. Material that surrounds a mass of *fissile material* and reduces its neutron losses, thereby reducing the amount of fissile material required for a *critical mass*. The nuclei in a neutron reflector should have a low probability for absorbing neutrons. Some, like beryllium, can release two neutrons after absorbing one.

**nucleus**. The relatively small core of an atom (nuclear radius $\approx 10^{-13}$ centimeters; atomic radius $\approx 10^{-8}$ centimeters) that is made up of *neutrons* and *protons* and contains about 99.9 percent of the atomic mass.

**pair production**. The conversion of the energy of a high-energy ( > 1.02 *MeV*) photon into the mass and kinetic energy of an electron-*positron* pair.

**passive detection**. The detection of spontaneous radiation from an object.

**photoelectric absorption**. The complete transfer of a *photon*'s energy to an electron. (*Compton scattering* involves only a partial transfer.)

**photofission**. A fission caused by the absorption of an x-ray or gamma ray.

**photon**. A quantum of electromagnetic energy. The amount of energy carried by a photon is inversely proportional to its wavelength. Infrared photons carry less than 1 *electron volt*, visible photons on the order of an electron volt, and x-rays and gamma rays can carry energies ranging up to over 1 *MeV*.

**positron**. The positively charged antiparticle of an electron. Positrons and electrons can be created by *pair production* by a high-energy *photon*. Conversely, a positron and an electron can annihilate into gamma rays.

**prompt neutrons or gamma rays**. Neutrons or gamma rays released during a fission process. (Delayed neutrons and gamma rays are released later by the fission products.)

**proportional counter detector**. A *radiation detector* that detects particles through the ionization that they produce in a gas (see appendix 11.C).

**proton**. A positively charged particle that is present in all atomic nuclei. The nucleus of the lightest atom, ordinary hydrogen, is a single proton. Elements are differentiated by the numbers of protons in their nuclei. The positive protons attract equal numbers of negatively charged electrons into orbits around the nucleus. The number of electrons in an atom determines its chemical properties.

**rad**. One hundredth of a *gray*.

**radiation detector**. An instrument for detecting high-energy particles of the types emitted by *radioactive isotopes*.

**radioactive decay**. The spontaneous transformation of one nucleus into another by emission of a particle or by *fission*.

**radioactive isotope**. An isotope that emits alpha particles, beta rays, gamma rays, or neutrons spontaneously.

**radiography**. Determining the internal structure of an object by its differential absorption of penetrating radiation such as x-rays or neutrons.

**radionuclide**. A radioactive isotope.

**rem**. One hundredth of a *sievert*.

**scintillation detector**. A *radiation detector* that detects particles by converting the ionization energy that they deposit into light.

**sievert**. A measure of the concentration of biological damage done by ionizing radiation. One sievert is the amount of ionizing radiation required to cause the same damage in humans as 1 *gray* of high-voltage x-rays. For neutrons, protons, and alpha rays, which cause denser trails of ionization, a radiation dose of 1 gray does tens of sieverts of biological damage.

**solid-state detector**. A *semiconductor radiation detector* that detects particles by converting the electron excitation energy that they produce into a voltage pulse.

**standard deviation**. The average deviation from the mean of a statistical sample. If, for example, a measurement of the energy of a gamma ray is made repeatedly, the standard deviation is a measure of the spread of the measurements around a central value. For a standard gaussian distribution, where the samples are distributed along a bell-shaped curve, roughly two thirds (67 percent) of the measurements will fall within one standard deviation, 95 percent within two standard deviations, 99.7 percent within three standard deviations, etc. of the central value.

**thermal neutron**. A neutron that has been slowed down by collisions with the nuclei of a moderating material to have an average kinetic energy equal to the average kinetic energy of those nuclei. At 300 K, this energy is about 0.025 *electron volts*.

**tonne**. A unit of mass equal to one thousand kilograms. Not to be confused with the unit used to measure warhead yield. (See *kiloton*.)

**transuranic isotopes**. The artificial *isotopes* that result from neutron absorption by uranium and subsequent radioactive transformations. They have atomic numbers greater than 92—that of uranium.

**tritium**. An artificial heavy isotope of hydrogen that has two neutrons and a proton in its nucleus. It has a 12.3-year *halflife*. It is used in "boosted" fission explosives.

**uranium**. The heaviest naturally occurring element, it contains two principal isotopes: fissile uranium-235 (0.7 percent) and uranium-238 (99.3 percent). Uranium-238 is transformed into plutonium-239 by neutron absorption and subsequent spontaneous radioactive transformations.

**x-ray**. A high-energy photon—typically in the *keV* or *MeV* energy range.

**watt (W)**. One joule per second, a unit of power.

# DETAILED TABLE OF CONTENTS

# LIST OF TABLES

## LIST OF FIGURES